The Universe Natural History Series

THE LIFE OF MAMMALS
VOLUME II

The Universe Natural History Series

Editor: Richard Carrington
Associate Editors:
Dr L. Harrison Matthews
Professor J.Z. Young

The Life of Mammals

Volume Two

L. HARRISON MATTHEWS

UNIVERSE BOOKS

NEW YORK

Published in the United States of America in 1971
by UNIVERSE BOOKS
381 Park Avenue South, New York City, 10016

Library of Congress Catalog Card Number: 71-93951
ISBN-0-87663-140-5

Printed in Great Britain

Contents

List of Plates

Acknowledgements

The author and publishers would like to thank the following for providing photographs for this volume: Dr Herbert van Deusen, plate 1; Australia Museum, plates 2a,b; International Photo/Press Office (photo Kees Molkenboer), plate 3; E. Slater, Division of Wildlife Research, CSIRO, Canberra, plate 4; Bruce Coleman Ltd, plates 5a (and Howard E. Uible), 10 (and Des Bartlett), 11 (and Jen and Des Bartlett), 12 (and Jane Burton), 13 (and Jane Burton), 17 (and C.A.W. Guggisberg), 21 (and Dade W. Thornton), 22 (and Simon Trevor), 23 (and C.A.W. Guggisberg), 28 (and Joe van Wormer), 29 (and Joe van Wormer), 30 (and C.A.W. Guggisberg); Edwin Gould, plates 5b,c,d; Lim Boo Liat, plate 6; Dr Roderick A. Suthers, plate 7; Dr David Pye, plates 8a,b; Dr Bruce Hayward, plate 9; The Zoological Society of London, plates 14, 24; Marineland of the Pacific, Los Angeles, plate 15; Nordisk Pressefoto, Copenhagen, plate 16; Fox Photos Ltd, plate 18; Nigel Bonner, plate 19; Aktien-Verein des Zoologischen Gartens zu Berlin, plates 20a,b; Tierbilder Okapia, Frankfurt, plate 25 and British Columbia Provincial Museum (Dr J. B. Foster), plates 26, 27. The illustrations in the text were drawn by Maurice Wilson.

Preface

VOLUME ONE of *The Life of Mammals* discusses various aspects of mammalian life including, among other matters, evolution, adaptations to various habitats, behaviour, reproduction, migration, hibernation, and the ecological relations between man and the other mammals. Volume Two reviews the orders and families of mammals, with the exception of the Primates which are dealt with by Professor Schultz in a separate volume, and incorporates the latest available information about a large number of the approximately 4,200 living species.

In preparing this volume I have constantly been impressed by the enormous amount of information that has been accumulated about the anatomy, systematics, and distribution of mammals. Nevertheless, these subjects are by no means exhausted; there is much work yet to be done upon them – many new facts, for example, are being discovered in the now unfashionable discipline of comparative anatomy. On the other hand the ecology, behaviour, and physiology of the mammals provide wide fields for study; much has been accomplished, but it is merely a beginning compared with what lies ahead, for barely a few dozen species have been so thoroughly investigated that we can say we have anything approaching a full knowledge of their lives and natural history.

In this survey of the orders and families of the mammals I have mentioned all the genera except a few rare and obscure bats and rodents about which, as living animals, nothing is known. I have restricted anatomical and descriptive facts to the minimum necessary to give the reader a general idea of structure and appearance, although I have mentioned many points that I think are of special interest. I have therefore not attempted to repeat the detailed information so thoroughly presented in works such as those of Owen [362], Flower and Lydekker [148], Weber [501], Grassé [183], Walker [495] or Anderson and Jones [7]; I have instead tried to give some idea of the great diversity and many adaptations of the mammals, and to tell something of what is known about their daily lives.

L.H.M.

Chapter 1

Monotremes and Marsupials

THE ORDERS Monotremata and Marsupialia contain some of the mammals most interesting to the zoologist; they are so different from the familiar eutherian mammals that the surprise and curiosity aroused by them in their early discoverers still lives on. With the exception of the opossums and opossum-rats of the New World they are found only in the Australian region, including New Guinea and some surrounding islands.

The discoverers and early explorers of the coasts of Australia saw very few mammals; of course they did not penetrate far inland, but even so they reported remarkably little. Abel Tasman [97] who in 1642 visited the south coast of Tasmania, which he named Van Dieman's land after the governor of Batavia, 'perceived traces of wild beasts, not unlike the claws of a tiger, or some such creature' – it is possible that he and his men saw the tracks of the Tasmanian devil. Forty-six years later in 1688, William Dampier [98] touched on the north coast of Australia and reported 'we saw no sort of animal, nor any track of beast but one; and that seemed to be the tread of a beast as big as a great mastiff-dog.' He revisited the continent in 1699 and charted parts of Sharks Bay on the west coast where he found 'only a sort of racoons, different from those of the West Indies, chiefly as to their legs; for these have very short forelegs; but go jumping upon them as the others do, (and like them are very good meat)'. A month later he saw some 'beasts like hungry wolves lean like so many skeletons, being nothing but skin and bones.' The 'great mastiff-dog' and wolf-like creatures were evidently dingoes, and the 'sort of racoons' must have been banded hare-wallabies, which reminded him of the barred tail of the racoon and of the racoon's peculiar hunched-back appearance.

Banks and Solander, the naturalists accompanying Captain Cook in 1770 were, however, the first to pay scientific attention to the flora and fauna of Australia. Although they were perhaps more interested in botany – hence Botany Bay, and the plant genus *Banksia* – they collected and examined many mammals, and were the first Europeans to describe a kangaroo under that name, although it was a species of wallaby that they found near what is now Cooktown in Queensland. During 1770 Cook [216] sailed along the whole of the east coast from south to north, and saw kangaroos, possums and dingoes. When the penal settlements were established towards the end of the eighteenth century, and in the beginning of the nineteenth, the fauna began to come under closer scrutiny, and the mammals to receive those peculiar misnomers based on their real or fancied resemblances to the mammals of Europe and elsewhere. John White, the surgeon-general to the first settlement, wrote an account of the voyage out to New South Wales and of the flora and fauna of the country. His book, published in 1790, but dated from 'Sydney Cove, Port Jackson, New South Wales, November 18, 1788' [505], illustrated with beautiful hand-coloured engravings, describes and figures a number of marsupials. Thereafter, as settlement spread out, the marsupials were described and specimens were collected for European museums by many people – doctors, governors of colonies, explorers and other amateur naturalists. Then in the 1830s John Gould and his wife arrived to spend two years collecting birds and mammals, which they described and lavishly illustrated in folio monographs after their return to Europe. For the rest of the century the study of Australia's fauna was left mainly to museum men, at first in Europe, and then later in the growing Australian cities; but until recent years museum men were expected by their employers to stay in their museums, and were seldom encouraged or given opportunities to do field work. The more credit to them, therefore, for the amount of interesting and valuable research they managed to accomplish – such as Krefft's *Mammals of Australia* published in 1871; Johann Ludwig Gerhard Krefft (1830–81) was curator at the Australian Museum, Sydney, for some years. [273] His work has been ably followed in recent years by that of Le Souef and Burrell [297] and of Troughton [478].

Meantime the anatomists and systematists of Europe were fascinated by the specimens that were sent home to them, and certainly made the most of their material. Richard Owen, the English anatomist, recorded the results of his researches on the structure of the monotremes and marsupials in many scientific papers – his enormous knowledge of comparative anatomy made his *The Anatomy of Vertebrates* published in three volumes 1866–8 a standard work still current a hundred years after it was written. [363] The amount of information about Australian mammals that was known and published within forty years of the first settlement was, however, astonishing. The articles – many by Owen – in Todd's *Cyclopaedia of Anatomy and Physiology* [475], published over a number of years starting in 1835, describe the anatomy of monotremes and marsupials in great detail. Another early book full of an amazing amount of facts about the systematics, anatomy, habits and fossil remains of these animals is the first volume of *A natural History of the Mammalia* published in 1846 by G.R.Waterhouse [497] who had been curator to the Zoological Society, and was then an assistant in the British Museum's department of mineralogy and geology.

Long before the end of the nineteenth century the professional zoologists, except for the systematists, had lost interest in research on the monotremes and marsupials, except for a brief revival in 1884 when Caldwell, an Englishman, confirmed what had long been doubted, that the monotremes are oviparous, in a dramatic telegram sent to the British Association for the Advancement of Science meeting at Montreal. It is surprising that the confirmation was so long delayed for oviparity in monotremes had been referred to as early as 1850, in the first volume of the *Proceedings of the Royal Society of Tasmania*. It was wrongly felt in Australia that everything worth knowing about the mammals had been found out and, with the exception of the museum men and a handful of gifted amateur naturalists, Australian zoologists turned their interests to the aspects of research fashionable in Europe. At the end of the century, and for the first quarter of the twentieth, it was the English holders of chairs and other posts in zoology at Australian universities who kept the scientific study of Australian mammals alive. Outstanding among them were J.P.Hill the embryologist who discovered the allantoic placenta in the bandicoot *Perameles*, and F. Wood Jones the human anatomist of Adelaide, and a very accomplished naturalist, who wrote the splendid *Mammals of South Australia* 1923–5. During these years the general public showed great indifference to the native fauna, and the destruction of hundreds of thousands of animals for their fur, for sport, or as vermin, was scarcely noticed. The scientific neglect of these creatures is well exemplified by the *Journal of Mammalogy*: in all the forty annual volumes that were published between 1919 and 1959 those interesting marsupials the Tasmanian devil (*Sarcophilus*) and the Tasmanian wolf (*Thylacinus*) are mentioned only once, and then merely in passing. A recent paper [185] points out that although the spiny anteater has been known to zoologists for over 174 years, only one planned study of its natural history has been carried out – and that has not yet been published at the time of writing.

At last, with the end of the Second World War, Australian zoologists suddenly woke up and realised what an immense and fascinating field for research on the fauna lay at their doorsteps. Numerous zoologists in the universities began work, museum curators were stimulated, and the Division of Wildlife Research of the Commonwealth Scientific and Industrial

Research Organization was established; as a result an ever growing stream of first-class research on all aspects of the biology of the Australian mammals is being reported in the scientific journals. Atonement is surely being made for the wasted years—and public opinion, too, has been aroused so that legislation has been passed to stop the destruction of the fauna and prohibit the export of many of the native animals, alive or dead.

ORDER MONOTREMATA

The monotremes are the only known order of the subclass Prototheria. Although they have highly specialised adaptations to their environments and ways of life they are primitive in retaining many features characteristic of reptiles. The most obvious reptilian character which differentiates them from all other mammals is laying eggs instead of bearing live young, a character correlated with the structure of the reproductive tract which resembles that of reptiles and birds. Some of their reproductive peculiarities have already been discussed in Volume I. Many other details of their anatomy are more reptilian than mammalian, the most striking being the structure of the shoulder girdle which retains separate coracoid and interclavicle bones, and the presence of epipubic bones attached to the pelvic girdle. Numerous other points in the anatomy of the skeleton and the soft parts are similarly primitive – they have been studied by many workers whose results are brought together in such encyclopaedic works as the *Traité de Zoologie* [183] where they are admirably set forth. On the other hand the brain of the monotremes is remarkably unlike that of reptiles, and closely resembles that of the metatheria; the fore-brain, especially of the Spiny anteater, is comparatively large, though it lacks a corpus callosum, the great connecting link between the two sides characteristic of the brain of placentals – as Wood Jones [258] remarks, they are well endowed with cerebral cortex, but their brains are under-wired.

The monotremes have not left their reptilian ancestral characters so far behind as have the other mammals; they do not, however, represent a stage in the evolution of the Metatheria and Eutheria but a parallel line that early diverged. Recent studies [264] of the fossil remains of early reptile-like mammals suggest that the monotreme stock diverged from that which gave rise to the other mammals 'in Triassic times at the reptilian grade of organisation', about 200 million years ago. This work was not done on fossil monotreme material, of which none has been found datipg from before the Pleistocene about two million years ago, but on fossil triconodonts. The remains of these animals which are preserved as fossils are generally their minute teeth, but modern methods of separating small fossils from their matrix have revealed some fragments of bone measuring only a few millimetres overall which under painstaking examination have yielded this information. It also suggests that there is no evidence to support the widely accepted fallacy that there was a violent spurt of evolution among the mammals in the Tertiary. The earliest satisfactory fossil eutherian mammals appear in the Paleocene when nearly all the modern orders, and many extinct ones, were already differentiated – but mammals had had at least a hundred million years to differentiate, and there is no reason to suppose that mammalian evolution quickened its pace at the beginning of the Tertiary some 80 million years ago. The defective fossil record denies us a knowledge of mammalian evolution during the Mesozoic, but the great variety of forms present at the end of that era shows that there must have been no discontinuity.

The living monotremes are confined to Australia with Tasmania, and New Guinea; the fossil forms, which belong to existing genera and are of no great age, have been found only in Australia. There is nothing to show whether they were once more widely distributed as, for instance, were the triconodonts, which are known from England and North America, or were always confined to the Australian region. The monotremes carry the two diagnostic characters of mammals – they have both hair and mammary glands – and in view of the remote separation of the monotreme stock from that of the other mammals it could be suggested that mammals must have evolved twice from reptilian ancestors. On the other hand, the size of the brain in the spiny anteater and the similarity of monotreme and metatherian brains appear to show a closer relationship than that suggestion would imply. In addition, a recent study [40] of the amount of DNA contained in the cells of the monotremes has shown that these animals have a closer affinity with the other mammals than with reptiles; this was confirmed by measuring the total area of the chromosomes. The DNA content in monotremes ranges from 93 to 98 per cent of that in eutherian mammals; the content in the only marsupials that have been examined (*Potorous* and *Didelphis*) was 81 and 94 per cent. In contrast the value in birds is about 50 per cent, in snakes and lizards 60 to 67 per cent, and in crocodilians and chelonians 80 to 89 per cent. In assessing these results, however, it must be remembered that the question of the amount of DNA becomes increasingly complicated through polyploidy of various sorts. Moreover an electron microscope study of the spermatoza of the spiny anteater reveals that although the shape of the sperm superficially resembles that of reptiles much of the ultrastructure is typically mammalian [91]. The discovery of more fossils may one day give us fuller information, and in the meantime there is no doubt much to be learned from a close study of the living animals.

The monotremes are classified in two families, the Tachyglossidae and the Ornithorhynchidae.

FAMILY TACHYGLOSSIDAE. THE SPINY ANTEATERS OR ECHIDNAS.

The spiny anteaters (*Tachyglossus*) are not big animals, the largest species being no more than about thirty-two inches (800 mm) in total length. The compact body is covered with coarse hair among which numerous sharp-pointed spines are set on the back from crown to tail, but not on the belly. In some species the spines conceal the fur, but in others the fur is long and almost conceals the spines. The colour of the fur is dark brown or nearly black and that of the spines yellowish at the base and dark at the tips. The short ear pinna is partly concealed among the fur and spines. The tail is a short stump but the long spines on it make it appear longer than it is; as the log book of Captain Bligh's *Providence* recorded in 1792, 'it has no tail but a rump not unlike that of a penguin' [478]. The legs are short but can raise the body well clear of the ground when the animal is on the move. The hands and feet are provided with large spatulate claws but the digits are not separate, so that the hands and feet are paws with diffuse pads, beyond which the claws project; the claws on the second digits of the hind feet are much longer than the others and are twisted inwards toward the axis of the limb. This claw is the toilet digit which can reach down between the spines so that the animal can groom its fur – or at least scratch its skin. The elongated snout covered with naked skin is shaped somewhat like the beak of a bird, but it does not open, for the mouth is a small slit at the end through which the long narrow tongue is darted in and out. The tongue is covered with sticky saliva so that any insect touched by it adheres and is drawn into the mouth. The jaws are completely toothless but insects are 'chewed' by being broken up between horny ridges at the back of the palate and horny papillae on the base of the tongue. Spiny anteaters live on insects, and it is said that they use their stout claws for breaking open termite nests to feed on the inmates; writers [258, 478] who have studied the animals closely, however, find that the staple diet consists of true ants (*Hymenoptera*) to such an extent that the animal and its flesh are strongly tainted with formic acid. The nostrils are sited at the end of the snout just above the mouth and the eyes at the base of it – the whole anatomy of the snout is highly adapted to searching out small insects under stones, logs, and in crevices of the ground and similar places.

Griffiths and Simpson [185] found that spiny anteaters dig down into the

nests of the meat-ant (*Iridomyrmex*) at the time when the nests are swarming
with virgin queen ants, which contain over 47 per cent of fat. One animal
after feeding 'gave a great display of animation, rolling over, scratching
furiously at his chest and abdomen with the grooming-toes to rid himself of
tormenting ants.' Temperature regulation in the spiny anteater and other
monotremes is poor, and in cold winter weather the animals hibernate – the
attacks on the nests containing fat ants occur at, and shortly after, the time
of emerging from hibernation and provide a good energy-rich breakfast.

One of the most striking features of the spiny anteaters is the stance and
action of the limbs. When the animal stands the front toes are turned in and
the hind ones turned out, presenting an awkward and ungainly appearance
that is belied by the speed with which the animal can move. The posture of
the hind limb, rotated outwards with the femur horizontal, is reptilian, but
Parrington [366] has shown that the horizontal pose of the femur in monotremes is secondary, and not the retention of a primitive condition, as the
form of the pelvis and the mammalian musculature of the limb
demonstrates.

Although the spiny anteaters can dig with their stout fore claws, they live
in hollow logs, cavities among roots and among rocks, but they do not dig
burrows as habitations. The home range appears to be small; a marked
animal recaptured after 350 days was only 800 yards from the point of
release, and another only 80 yards away after 346 days [185]. When at home
they wedge themselves in with their spines much as a sea-urchin wedges
itself into a crack in a rock, so that it is almost impossible to dislodge them.
If alarmed in the open they rapidly bury themselves in the ground – as
Wood Jones said, their powers of clinging to, and sinking into the earth can
hardly be appreciated by those who have not witnessed the process. They
do not burrow head first, but bury themselves by digging and sink straight
down. Where the ground is too hard for digging they curl up somewhat in
the manner of a hedgehog. The remarkable digging powers are not primarily used for obtaining food but for protection, 'given any irregularities of
surface of which it can take advantage, it will manage to wedge itself so
securely, by the aid of its stout feet and its bristles, that it is practically
impossible for an enemy to dislodge or even injure it' [258]. Self burial is
also the spiny anteater's way of keeping cool when cover is not available;
unless it can escape thus its temperature rises with the ambient above 30°C
and it dies.

The spiny anteaters are classified in two genera, *Tachyglossus* and
Zaglossus. *Tachyglossus* contains the short nosed forms in which the snout
is slightly upturned at the tip. They are densely spined, and have five

clawed digits on both fore and hind limbs. They are widely distributed over differing types of country throughout Australia, Tasmania and New Guinea; the Tasmanian form, which has longer hair tending to cover the spines, is regarded as a distinct species. *Zaglossus* contains the bizarre looking long-snouted spiny anteaters of New Guinea, three species of which have been described. The snout is long and downwardly curved, and the spines are much fewer, shorter and blunter than in *Tachyglossus*, so that they are not closely packed but scattered among the fur (Plate 1). Although five digits are present on fore and hind feet sometimes only the middle three bear claws. It is possible that animals of this genus do not eat ants, but feed on earthworms and similar soft invertebrates [486]. The anteaters are relished as food by the natives of their countries, but Troughton [478] found food fried in their fat 'the most bilious experience to which an enquiring mammalogist can be subjected.' The presence of a spur, connected with a poison gland, on the ankle of male spiny anteaters and platypus was mentioned in Volume I; its function remains unknown.

FAMILY ORNITHORHYNCHIDAE. THE PLATYPUS OR DUCK-BILL
Superficially the platypus is totally unlike the spiny anteaters, though the internal anatomy shows similar reptilian features. The platypus, which reaches a maximum length of about two feet, is clothed in thick dense fur; the most conspicuous feature, however, is the peculiar naked snout resembling a duck's bill in shape. It is covered with delicate skin so that it is a sensitive tactile organ, and at the base of both upper and under jaws the skin is reflected as a loose flap 'like the turned back cuff of a sleeve' [258]. The nostrils open towards the end of the bill, but the small eyes lie deep among the fur behind the flap, immediately in front of the ear hole which has no pinna. The sense of touch is the important one in seeking for food, which consists of water insects and their larvae, crustacea, worms and molluscs, because the eyes and ears in their groove in the fur are kept shut underwater. The platypus, unlike the spiny anteaters, possesses teeth, though of rudimentary form. In the young animal three small teeth become calcified, but they are replaced by two horny plates on each side of each jaw at an early age. In the embryo a more complete dentition, lacking upper incisors but with a replacement premolar, is laid down, but it is resorbed and does not erupt.

The platypus lives in rivers and streams, and as befits an aquatic animal the feet are webbed; the webbing on the hind feet reaches the ends of the toes leaving the claws projecting beyond, but that on the fingers is prolonged so that it reaches far beyond the tips of the claws attached to the ends of the digits. The front and hind edges and the semicircular margin of the web appear somewhat thickened, so that in swimming the power stroke does not turn the web inside-out. When relaxed the web falls into folds between the axes of the digits, and in walking on land the palmar surface is applied to the ground with the edge of the web folded under; according to Burrell [60], the Australian naturalist who devoted years to the study of living platypus, the nails of the fore paw take a grip, and the creature pulls itself forwards, going over on its knuckles as it does so. Burrell also says that the two limbs of one side are rotated outwards and forwards together, but his photograph of a platypus running at its top speed of 200 feet a minute (a trifle over two miles an hour), shows the near fore and off hind brought forward together in the usual mammalian fashion. The body is fully raised from the ground only at top speed; at lower speeds the underside and tail drag on the ground so that the tail and part of the belly are often denuded of hair below. The platypus is very buoyant and floats high in the water; when diving the end of each swimming stroke of the hands presses upwards so that the body is forced down. The hind legs seem to be used more for balancing and steering than for propulsion in water. Wood Jones [258] expresses the opinion that the flattened tail is used to smack the water on diving when alarmed, in a similar way to that of the beaver, and Fleay's [147] observations on wild platypuses confirm its use thus in a crash dive. Wood Jones also points out the peculiar fact that in the entirely terrestrial spiny anteater the skin of the fore paw is prolonged between the digits so that when seen from above it has a distinctly webbed appearance; it thus shows in a rudimentary way a resemblance of structure to the fully webbed paw of the platypus.

Although the spiny anteaters are often wrongly referred to as specialised for burrowing instead of for digging, the platypus is rightly regarded as specialised for swimming. It is not, however, generally emphasised that the platypus is equally specialised for burrowing, and it is indeed remarkable that the animal can dig tunnels up to forty feet long in the banks of rivers – the long tunnels are the nesting burrows made and inhabited by the female alone; the camping burrows, made and used by both sexes separately, are much shorter. The powerful claws of the fore paws do practically all the digging, but Burrell found that they are helped by a 'probing or lateral shearing movement of the muzzle', an unexpected use for so sensitive a part of the body. The excavated earth is not thrown out of the burrow but is tamped down on the floor, being compressed and flattened by the tail, which becomes quite bald at the end from this use. This unusual method of burrowing produces the characteristic tunnel-shape of an arch over a flat floor; the tail is also used by the female for carrying in the wet vegetation for constructing her nest – she holds the material between the belly and the underside of the forwardly tucked tail [60]. The wet vegetation and plugging of the nest burrow with earth help to maintain a damp atmosphere believed to be necessary to prevent the egg or eggs from drying up.

The single species of platypus, *Ornithorhynchus anatinus,* is widely distributed in the eastern part of Australia and in Tasmania, inhabiting streams and lakes from tropical rivers to cool mountain streams. It was formerly much persecuted for its rich fur and was greatly reduced in numbers, but under the protection now given it is not uncommon in many places.

ORDER MARSUPIALIA

The marsupials form the only order, living or extinct, in the infraclass Metatheria. They have evolved a great diversity of forms; many of them superficially closely resemble eutherian mammals of various orders. The young are born at a very early stage of development and are carried by the mother attached to the nipples, generally but not always within the protection of an abdominal pouch. The rudimentary placenta is formed of the yolk-sac, except in the genus *Perameles* where a simple allantoic placenta is found. The uterus and vagina are double and open into a cloaca; in the males the penis is forked at the end and lies posterior to the scrotum. Some of the peculiarities of marsupial reproduction have been discussed in Volume I. The pelvis bears epipubic bones as does that of the monotremes and many reptiles; they were formerly termed 'marsupial bones', a misnomer for they do not support the pouch and are present in both sexes. In the skull the palate is commonly incomplete, the brain-case is comparatively small, and the angle of the jaw is usually inflected inwards. In the dentition, which varies greatly between families, there is never more than one deciduous tooth, a milk molar present in all species. In carnivorous species the incisors are numerous, but in herbivorous ones the lower incisors are reduced to two; these characters were formerly used to classify the marsupials into two suborders, the Polyprotodontia and the Diprotodontia, a classification now abandoned. The first digit of the foot is commonly reduced or absent; where present it generally bears no claw and may be opposable to the other digits. In many species the second and third digits of the foot are reduced in size and bound together; they form a comb or toilet digit used in grooming the coat. This feature has also been used as a basis for dividing the marsupials into suborders, the Didactyla and

Syndactyla, but this arrangement too has been abandoned, for classification either by tooth or foot characters alone leads to anomalies. The brain is rather small in comparison with the size of the animals, and like that of the monotremes, lacks a corpus callosum.

Although there are so many widely different marsupials a commonly recurring character is the way the hind end of the body gradually tapers into the tail, in a manner reminiscent of that in lizards and in contrast to the usual abrupt setting-on of the tail as an appendage in the eutherian mammals. Another character that frequently occurs is the cross-banding pattern of the body, often reduced to a loin band, and the dark patch extending from the muzzle through the eye to the ear. A gland in the skin of the chest, the sternal gland, is present in many marsupials; its secretion is probably used for marking territory in some arboreal forms, but in others such as the kangaroos where the secretion sometimes stains the fur in its neighbourhood orange-red, its function is obscure. Compared with the eutherians the marsupials are voiceless mammals; most of them hiss when alarmed, many give low grunts and growls, some make a subdued twittering, but only a few, in particular the greater gliders or flying phalangers have loud voices and produce cries that can be heard at a distance. Most marsupials, too, have a characteristic position for sleeping; they sit on the haunches, drop the head down on to the belly and bring the tail forward between the legs to wrap over all. The sitting position is also that adopted by the females in giving birth to the young, which make their way unaided to the nipples. Although marsupials are able to regulate their temperature more efficiently than the monotremes, many of them are more susceptible to the ambient temperature than eutherian mammals, and become torpid during cold weather, or aestivate during drought.

The living marsupials are classified into eight families containing between them eighty-one genera, and nearly 250 species.

FAMILY DIDELPHIDAE. OPOSSUMS

The family Didelphidae is confined to the Americas, and members of it are found from southern Canada to Patagonia. It is the family least differentiated from the primitive marsupial stock, all four feet having five digits, the first on the hind foot being generally clawless and opposable to the others. The guard hairs of the coat are sparse so that many species are practically without them. The tail is long, generally nearly naked, and prehensile. The snout is pointed and the gape of the jaws wide; the ear is prominent. A pouch is present in some species but absent in many; in others it consists of a fold of skin on each side of the mammary area – the teats may be as many as twenty-five. In size didelphids range from that of a mouse to that of a cat. There are twelve living genera, all mainly insectivorous and carnivorous.

The genus *Didelphis* contains two species, similar in general appearance, *D. marsupialis* ranging from Canada to Argentina, and *D. azarae* confined to South America. The first is 'The' opossum or the Virginian opossum, about which most opossum lore has gathered. It is well known for feigning death when alarmed, a phenomenon which was formerly thought to be a form of cataleptic fit, but is now known not to be accompanied by loss of consciousness, as already discussed in Volume I. The fur, which has longer guard hairs than most species so that it has a rather unkempt appearance, has some commercial value; the flesh, too, is consumed by negroes and others in the southern United States, but few people would care to try that of *D. azarae* because of the animal's horrid stink. Opossums of this genus are not only carnivorous and insectivorous but eat fruit and other vegetable matter. They are good climbers but slow on the ground; they are crepuscular and nocturnal in habit. The single species of genus *Lutreolina*, *L. crassicaudata*, the thick-tailed opossum, is a rather smaller animal with short dense fur of a deep orange-brown colour. It not only climbs well, but also swims; its lithe action is more like that of a stoat than that of other opossums. The ears and limbs are short, and the basal half of the thick tail is densely covered with fur; there is no pouch. The animal lives in widely varied habitats in eastern South America from the Guianas to Patagonia. Another opossum that swims, the yapok, *Chironectes minimus*, the only species of its genus, is the only truly aquatic marsupial. The dense fur is black above with grey patches on the sides confluent with the white below; there is a white spot on the black face above each eye. The tail is long and nearly naked, and the hind feet are fully webbed. The pouch can be shut with a sphincter muscle when swimming, and the young inside it are presumably as tolerant to high concentrations of carbon dioxide as the pouch young of some other marsupials are known to be [16] – perhaps more so. Yapoks inhabit streams and lakes from Mexico to northern Argentina; they are nocturnal, and feed on fish, crustacea and probably anything else they can catch.

The other didelphids are smaller in size; the four-eyed opossums *Philander* and *Metachirus*, both single-species genera, like the yapok, have white spots above the eyes, from which they take their name; the fur is grey or brown. They inhabit forested regions throughout most of South America, and live in burrows or nests built of vegetation among tree branches. They are omnivorous and nocturnal. *Philander* can swim well;

the females have a pouch, but those of *Metachirus* have only lateral folds. The little murine or mouse opossums of the genus *Marmosa*, which contains many species, are found all over South America south of northern Mexico. They have long prehensile tails, and the common marsupial eye-patch or eye-stripe is prominent. The dense soft fur is grey to various shades of brown. There is no pouch; when the young are detached from the nipples they cling to the fur of the mother's back – the old yarn of opossums carrying their young on their backs with the mother's tail arched above them, so that the young hold fast by winding their prehensile tails around it, is a myth. Mouse opossums live among trees and bushes, and some species make nests of plant fibres; they are omnivorous and feed especially on insects and fruit. Another widespread genus with numerous species is *Monodelphis*, the short-tailed opossums, with bare tails and short fur of drab brown or grey colour in some species, brighter and with dark markings in others. They are nocturnal and omnivorous, and make their nests in hollow trees and other crevices. The females have no pouch, and carry their young clinging to their fur after they detach from the nipples. *M. domesticus* commonly lives in and near human habitations in Brazil, to feed on mice and cockroaches. The three species of woolly opossums in the genus *Caluromys*, between them cover South America from Mexico to southern Brazil. They are rather larger than the short-tailed opossums, and have soft fine fur extending along the basal half of the very long prehensile tail. The colour is brown or greyish above, lighter below, with a dark stripe running from nose to the top of the head or farther back, and darker eye-patches. These opossums are quick and agile in movement, and, like *Lutreolina*, remind one more of mustelids than marsupials in their action. They are arboreal, nocturnal and omnivorous; the pouch of the female is vestigial. The single species of *Dromiciops*, *D. australis*, the colocolo, is a small animal with dense soft fur, brown with lighter patches on the sides, and a dark eye-patch; it is found only in Chile where it lives in mountain forests, feeding at night mainly on insects. The females are pouched; and the long tail is fur-covered except for a small part of the underside near the tip.

Few animals of the three other genera of didelphids, *Caluromysiops*, *Glironia*, and *Lestodelphis*, have ever been seen – they are known only from a handful of stuffed specimens in museums. *Caluromysiops*, the black-shouldered opossum from forests in Peru, closely resembles a woolly opossum, with grey fur marked by wide black lines running up the fore limbs and joining over the shoulders, and by two black lines running along the back. Two species of *Glironia*, the bushy-tailed opossums, are known from Peru, and one of *Lestodelphis*, the Patagonian opossum, from the south of Patagonia. All these animals are probably plentiful enough in their proper habitats, which have yet to be discovered.

FAMILY DASYURIDAE

Members of this family are found in Australia, Tasmania, New Guinea and some of its nearby islands. The family contains nineteen genera of animals, of such widely differing size and general appearance that it is surprising to find that their structure shows them to be closely related. It is strange to see classified in one family animals with such varied vernacular names as mouse, rat, cat, wolf – and devil. They are all carnivorous or insectivorous, and in correlation the dentition has small incisors, large canines, and molars with sharp cusps. The five fingers and the four or five toes are all separate, and the tail is long, furred and not prehensile. A pouch is lacking in some species, but where it is present it opens backwards. Many species are known from only a few specimens, either because they have been nearly exterminated by man and his introduced animals in Australia, or because they inhabit remote regions which have been little visited by mammalogists. First there is a large number of different kinds of marsupial 'mice' and 'rats', so called from their size and general appearance; one species at least enters human habitations in New Guinea, and is as destructive as a rodent. Most of the species are terrestrial, but they climb well and some are exclusively arboreal. Although they look so mouse-like the wide gape and sharp teeth including canines at once distinguish them. The genus *Phascogale*, the broad-footed marsupial 'mice', contains a large number of species, including the smallest living marsupial; they have sometimes been placed by classifiers among several different genera. They are active alert creatures, as befits mammals that catch living prey. In the brush-tailed 'mice' the terminal half of the tail is clothed with long black silky hairs that can be erected. In some other species the tail is peculiarly fattened or incrassated, a feature found in other genera and believed to be a way of storing fat for future use during food scarcity.

In contrast with the numerous species of *Phascogale* there is only one species, *D. byrnei*, in the genus *Dasyuroides* – Byrne's pouched 'mouse'. This animal, like some species of the last genus, has the outer half of the tail adorned with long black hairs which, in this species, form a well marked dorsal and ventral crest. Little is known of the habits of this animal, which lives in burrows on the stony tablelands of the desert in central Australia. The crest-tailed pouched 'mouse', *Dasycercus cristicauda*, is also the sole species of its genus. It is about the size of a weasel, bright brown above, white below; the tail is incrassated and bears a crest of black hairs above and below; it lives in the desert of central Australia. It is described as 'a sturdily built, short limbed, compact little animal' by Wood Jones, who kept some specimens in captivity and was obviously delighted with them. He found them absolutely fearless, and one of the most intelligent of the marsupials. He tells of the lightning stroke with which they kill a mouse, and how they then eat it starting at the nose, where the skin is turned back, and continue from head to tail, leaving the skin complete and turned inside out 'as if a skilled taxidermist had been at work.' There are several species of narrow-footed pouched 'mice' of the genus *Sminthopsis*, small greyish brown mouse-like creatures, some species, but not all, having fat tails; the hind foot is proportionately much longer and consequently narrower than in the broad-footed pouched 'mice'. They are nocturnal and insectivorous – Wood Jones calls them slender and delicately built little animals of a general shrew-like aspect. The pouch is more developed than in members of genus *Phascogale*.

The jerboa pouched 'mouse', *Antechinomys*, several species of which live in the deserts of Australia, is something of a zoological practical joke. Superficially it bears a strong resemblance to the eutherian rodent jerboa mice (*Notomys*) which share its habitat in great numbers, and to similarly adapted rodents in other parts of the world. It has very long legs and feet, a long tufted tail, large ears and inflated auditory bullae; as Wood Jones says they show a maximum of characters adapted to a highly developed jumping habit. 'It is a remarkably specialised saltatory animal which spends the hottest hours of the day in burrows in the sandhills.' How interesting his remarks would have been had he ever seen a living example of this scarce animal, for Ride [403] has found that it is invariably quadrupedal, bounding along with hind and fore feet alternately striking the ground, whereas *Notomys* is bipedal and saltatory. Wood Jones gives an excellent drawing from a dead specimen, but shows the animal standing bipedally; when drawing the long conspicuous forearms he must have noticed how different they are from the short forearms of saltatory rodents, in which the hands are held under the chin so that the fore limb is practically invisible – perhaps he did, but missed the point (Plate 2).

Passing from mice and rats we turn to cats, as the animals of genus *Dasyurus* are commonly called in their native land, though they are sometimes referred to as 'dasyures' in books. There are several species, which some zoologists place in separate genera (*Satanellus, Dasyurops, Dasyurinus, Dasyurus*); they are about the size of the domestic cat, some rather smaller, and all have white spots scattered over their generally dark coats. They were formerly common and widespread through Australia, but a mystery epidemic of the early years of this century exterminated them in many places. The eastern native 'cat' *D. quoll*, occurs in a light and dark colour phase, both often in the same litter, as recorded by Fleay [146] who bred the animal in captivity. He noted that

> tame dasyures retain their playful habits in the adult period . . . they delighted in chasing tennis balls about the floor, and in leaping high into the air in pursuit of blowflies during the daytime . . . they loved sunlight and basked in it for hours at a time, despite their nocturnal habits. . . . They also sat bolt upright with ears forward and fore-paws folded down on the chest in the typical attentive attitude of the adult dasyure.

Native cats make nests of dry grass, leaves and so on under a boulder or in a hollow log. In the days of their abundance they often lived beneath settlers' houses, and 'their bumping heads and squabbling cries were familiar sounds at night.' All the native cats can climb well, and the largest species, the 'tiger-cat', is the most arboreal. Their voices are very uncat-like, and consist of guttural coughing noises and hissing sounds. When abundant they were much harassed by farmers and others because of their raids on domestic poultry.

Although the Tasmanian devil, *Sarcophilus*, looks very different from the native cats, the details of its anatomy and its dentition show its close affinity with them. It is a much larger, more heavily built animal, superficially resembling a small bear with a long tail. The only species, *S. harrisi*, is black with a white band across the chest and sometimes another across the rump. It is confined to Tasmania, though subfossil remains of it have been found at widely separated places on the mainland. The devil's general appearance and facial expression, and its habit of opening its large mouth to display an array of sharp teeth, and of hissing when disturbed, have given it the reputation of being untameable and extremely savage. Troughton, however, reports of Tasmanian devils that 'when reared in captivity they make delightful pets, being frolicsome and affectionate, even adults caught wild becoming docile for handling. They are cleanly, being fond of bathing and basking in the sun.' The smaller mammals, birds and lizards form its diet, and the early settlers killed it because it stole their poultry and was accused of sheep-stealing; the first convicts to arrive, however, were said by Harris [206] to have found its flesh 'not unlike veal'. It dens among logs and rocks, does not climb trees, and is nocturnal. Although it has been exterminated near towns it is still plentiful in the wilder parts of the country.

The same cannot be said of the thylacine or Tasmanian 'wolf' *Thylacinus cynocephalus*, sometimes called the Tasmanian 'tiger' or 'hyaena' because the hinder part of the body is striped. It is the largest carnivorous marsupial and because it attacked the sheep of the farmers it was deliberately exterminated, just as the earlier settlers had exterminated the Tasmanian aborigines. As early as 1830 the Van Dieman's Land Company, incorporated to breed sheep for fine wool, offered bounties for killing thylacines. The company received a grant of a quarter of a million acres in northwestern Tasmania in 1825, and established its first settlement, at Circular Head, in October 1826. From 1888 to 1909 the government paid a bounty of £1 for each adult killed and 10/– for each half-grown animal or 'pup' [188]. For some years it was thought to be extinct but it is now known to survive in small numbers in remote regions difficult of access. In the early years of this century it was getting scarce but its nearness to extinction was not appreciated – the writer can remember, when a child about 1910, seeing a living thylacine in a provincial English zoo where it attracted no particular attention; such an exhibit now would attract wide publicity and ballyhoo. Like the devil, the wolf is found only in Tasmania, but fossils show that it once lived throughout the mainland and also extended into New Guinea [484]. It is the sole species of its genus; in general appearance it resembles a large greyish dog, with striped hindquarters, and like the dog it is digitigrade. It is nocturnal, but little is known of its behaviour in the wild.

Figure 1 Numbat (*Myrmecobius*).

The little banded anteater *Myrmecobius*, often called the numbat, one of its native names, is a very different creature, and is described by Calaby [67] the only naturalist who has carefully studied it in the wild, as 'probably the most attractive of all the marsupials' because of its bright coloration and interesting habits. The single species, *M. fasciatus*, about the size of a large rat, is rusty red above, whitish below, and the hind third of the almost black back is crossed by about six white transverse bands; a black stripe runs through the eye from snout to ear, and the tail is long and bushy. The teeth are small and the molars are regarded as degenerate although there are five on each side in the upper jaw and six in the lower, a higher number than in any other heterodont mammal, giving a total of fifty-two. The tongue is long, cylindrical and sticky. The numbat is solitary and lives in scrub woodland, where it shelters and makes its nest in the hollow fallen logs strewing the woodland floor. Unlike many small mammals it is diurnal and feeds on termites which it licks up after scratching open their underground tunnels, but it does not dig into or open termitaria – some hymenopterous ants are also accidentally eaten at the same time. The insects are swallowed without being chewed and the premolar and molar teeth are thus found not to be worn down, although a good deal of sharp sand is incidentally taken into the mouth. There is no pouch, and the four young cling to the nipples by their mouths, and to the hair of the mammary area by their hands. The numbat was once fairly widespread in the south of Australia but its numbers declined with white settlement, and it was thought to be on the verge of extinction until Calaby recently found that it is still 'fairly common over a sizeable area' and is among the more abundant of the small mammals of south-western Australia.

FAMILY NOTORYCTIDAE

The single genus of this family, *Notoryctes*, the marsupial 'mole', is perhaps the most striking example of convergence of form, correlated with similar habits, between marsupials and eutherian mammals, the superficial resemblance to the golden moles of southern Africa being close. The body is cylindrical and stout, with short limbs, like that of eutherian moles, and is clothed in light coloured fur ranging in colour from nearly white to red-brown, the darker colour being perhaps produced by staining from the soil; the hairs are iridescent like those of the golden moles. The eyes are degenerate beneath the skin, there are no ear pinnae, and the nose is protected by a transversely grooved horny plate. The third and fourth digits of the fore foot carry large scoop-like claws under which the small first and second digits are folded, the fifth being rudimentary, and the hind feet have five short digits. The short tail is naked and ringed. The marsupial mole was discovered in 1888, but Wood Jones seems to have been the only zoologist to record observations of its behaviour, from one he held in captivity for a short time. He found the animal full of 'feverish activity' and that it voraciously ate the large quantities of earthworms offered to it – thus resembling the European mole in its restlessness and need for large amounts of food. It also resembles the European mole in a way that was apparently unknown to Wood Jones, for he records that 'in all its activities it carries its extraordinary stump of a tail erect' without comment; the erect carriage of the tail is characteristic of the true mole, and is believed to enable it to act as a tactile organ when burrowing underground. The foraging burrows of the marsupial mole are said to be near the surface, and the belief has been expressed that the animal also makes deeper permanent burrows, but so few observations have been made that such statements record specific incidents, and cannot be taken as applying generally – they have been copied from book to book, becoming progressively modified the farther they get from the original. The anatomy of the animal is well known from the careful dissections of European naturalists – a peculiar feature is that the last five vertebrae in the neck are fused together. *Notoryctes* lives in the sandy semi-desert of central and southern Australia, and has also been found in the far north-west whence the two known specimens have been described, perhaps prematurely, as belonging to a separate species.

FAMILY PERAMELIDAE

The members of this family, the bandicoots, are found throughout the Australian mainland, Tasmania, New Guinea and a number of off-lying islands. They range in size from that of a rat to that of a hare, and live on insects and earthworms, which they scratch out of the ground. The snout is long, the ears medium to large in size, the middle three digits of the fore foot are clawed, the others reduced, and on the hind foot the fourth is the largest, the fifth smaller, the first reduced, and the second and third together form a syndactylous toilet claw. Most of the species have rather coarse bristly outer fur, and some are barred on the hind quarters. The number of lower incisor teeth is not reduced, so that the animals are both polyprotodont and syndactylous. The pouch in female bandicoots opens backwards. They are the only marsupials to have a chorio-allantoic placenta, as discovered seventy years ago by J.P. Hill [228].

The name bandicoot belongs properly to a large rat-like rodent (*Bandicota indica*) of India and Ceylon, where it is destructive to crops and is also a house pest. The native Telegu name Pandi-koku, meaning pig-rat, is derived from the grunting noise it makes, and was corrupted by Europeans into bandicoot. The name seems first to have been transferred, as early as 1831, to the Tasmanian barred bandicoot *Perameles gunni*, which is also found in southeast Victoria. Waterhouse says [497] Gunn reported that the settlers called it bandicoot, and that it was numerous in all parts of Tasmania – it is still common today; the original misnaming must have been made by someone familiar with the true bandicoot in Asia. Zoologists, when using the vernacular names of Australian mammals, are in the habit of writing such misnomers with inverted commas to show that they know that the native 'cat', for example, is not a true cat. But the name bandicoot has become so firmly attached to the peramelids that its wrong application has been forgotten, and is no longer given commas – indeed, Simpson lists the true bandicoot (*Bandicota*) as the 'Bandicoot' rat. The early settlers were not zoologists and knew nothing of the peculiar native names of the new and strange marsupials that they came across; they naturally named them after the animals most closely resembling them with which they were familiar in Europe or elsewhere. The zoologists themselves are not above reproach when inventing scientific Latin names, for *Perameles* means 'pouched badger'.

About twenty species of bandicoots are known, distributed among several genera. Little was known about their habits, behaviour and ecology until recent studies on a few species [223, 459]. The genus *Perameles*, the long-nosed bandicoots of Australia and New Guinea, contains about ten species of which two, *P. gunni* and *P. nasuta*, have been studied in some detail; the first is abundant in Tasmania, the second all along the east coast of Australia. They are crepuscular and nocturnal, and when foraging busily dig up earthworms and insects, nearby animals taking little notice of each other except briefly when a female is in oestrus. In searching for food they dig conical pits a few inches deep into which they push their snouts to obtain their prey. The home ranges of individuals extend to several acres, sometimes to many acres, and overlap; the animals are occasionally aggressive in chasing a stranger away. In confinement when not provided with sufficient space for dispersal they are very aggressive, and the dominant animals kill the subordinate ones which cannot escape; they are thus overcrowded in conditions in which many other mammals of similar size would be quite comfortable. Although they dig for their food they do not burrow, but make nests of grass and other vegetable debris under scrub or other shelter or even in a shallow trench-like hollow scratched in the ground. In gathering nest material the animals scratch a bundle of dry grass backwards with the fore feet and do not use mouth or tail in collecting it. The genus *Isoodon* contains the short-nosed bandicoots, short nosed only in comparison with those of genus *Perameles*, for their snouts are sharply pointed. There are several species widely spread through Australia and Tasmania, most of them rather smaller than those of *Perameles*. In general their habits appear to be similar, but the common, or southern, short-nosed bandicoot, which has been carefully studied in Tasmania, where it is plentiful, appears to be more a creature of scrubland and hedgerows, and less inclined to forage in the open away from cover than the long-nosed species.

Two species of rabbit bandicoots or bilbies, *Thylacomys*, known as pinkies in some parts of their range, formerly inhabited most of the southern half of the Australian continent but are now greatly reduced in numbers and extinct in many places. The bilbies have long silky grey fur, long well-furred tails, naked pink noses, and large ears shaped much like those of a rabbit. Like rabbits, too, they live in burrows which they dig themselves, following a spiral direction for several feet, but they are solitary and do not live in warrens. They are generally nocturnal, but have also sometimes been seen foraging out of their burrows by day. They are insectivorous and carnivorous, eating insects and their underground larvae, and small mammals especially mice. Wood Jones, once again, seems to be the only zoologist who has recorded anything of their habits from careful observation. He says that when walking or running the fore legs move alternately but the hind ones together; the toilet of the coat is performed meticulously, always with the syndactylous claw of the hind foot – an animal that had lost a hind foot in a rabbit trap repeatedly tried to use the stump and never substituted the fore foot. When sleeping the ears are laid back and then folded forward over the eyes; the animal squats on its hind legs and tucks the long snout between the fore legs 'so making itself into a round silky ball'. The rapid disappearance of the bilbies since the beginning of this century is thought to have been caused through crowding out by rabbits, and through killing by man for fur and for sport.

A single rare species forms the genus *Chaeropus*, the pig-footed bandicoot. It is widely but thinly distributed in the centre of Australia, but few living specimens have been seen by white men. It is remarkable for the extreme reduction in the number of digits; in the fore foot only the second and third digits with a minute rudiment of the fourth are present, and being equal in size, form with their large apical pads and long claws a foot superficially resembling that of an artiodactyl. In the hind foot, however, the fourth digit alone forms the functional foot, with single apical pad and claw superficially resembling the foot of a perissodactyl; the fifth digit is a minute rudiment, but the second and third form a small though functional syndactylous toilet digit. These peculiarities naturally lead to the conclusion that the pig-footed bandicoot is specialised for digitigrade movement, and Wood Jones remarks that it may be regarded 'as a specialised cursorial' animal, but Krefft [273] described its gait as like that of a 'broken-down hack in a canter, apparently dragging the hind quarters after it'. When we remember the fallacious conclusion reached about the gait of *Antechinomys* from a consideration of anatomy alone, unsupported by observation on the living animal, it is as well to keep an open mind and await more information. *Chaeropus* is a small mammal with long snout, ears and tail; it appears to be less nocturnal than other bandicoots, and more omnivorous in its diet, which is said to include much vegetable matter – it makes nests of dried grass and so on in shallow hollows, as do *Perameles* and *Isoodon*.

Little is known about the life of the two or three species of New Guinea spiny bandicoots of the genus *Echymipera*, which are widely distributed in New Guinea, and one of which has been found in northern Queensland. They have very long snouts and fur which is stiff and spiny, thus differing from the other New Guinea bandicoots of the genus *Perameles*, which are placed in a separate genus *Peroryctes* by some systematists.

FAMILY CAENOLESTIDAE

The Caenolestidae are a small family inhabiting the Andean region of South America, from western Venezuela to Chiloe Island. They are small creatures and although named rat-opossums, they superficially resemble shrews with long tails. They appear to resemble shrews too in their habits and diet, being terrestrial and feeding on small invertebrates. They were discovered only little over a hundred years ago, and comparatively few specimens have been available for examination by zoologists. Osgood [358] described their anatomy in detail and Simpson [445] has discussed their affinities. Fossil caenolestids are known from the Eocene formations of South America; they show that they form an order as ancient as the didelphids, with which they have no particularly close relationship. Osgood thought that their affinities appear to be closer with the peramelids, and suggested that the North American ancestor of the caenolestids probably extended throughout the holarctic, and may therefore have given rise to the Australian diprotodont marsupials because the first of the three lower incisor teeth is large and procumbent. Simpson, however, points out that they may equally well have been derived from primitive didelphoids. There are three genera in the family; the largest, *Caenolestes*, contains five species, some inhabiting subtropical forest and others more open country at high altitudes on the border of the tree-line. Although specimens are not plentiful in museums the animals are common in their habitats. Tate [468], writing in 1931, said,

It was believed until a few years ago that *Caenolestes* was an exceedingly rare animal. The fact is that in Ecuador, as soon as one becomes acquainted with its habits, it can be caught as readily as any other fairly abundant animal ... *Caenolestes fulig'nosus* is abundant in small bushy ravines near water. Frequently the runways, which are used in common with mice, are found to pass up and down the steepest parts of the gullies.... A specimen taken alive on Mt. Pichincha did not use its tail in any way as a prehensile organ.

These peculiar little marsupials have no pouch.

FAMILY PHALANGERIDAE

In contrast to the last family the Phalangeridae is a large one and contains forty-six species classified into sixteen genera. The family is widely distributed through the Australasian region and, although not found over the whole of Australia, extends through New Guinea and many islands of the region from Celebes to the Solomon Islands. The family contains the Australian possums, the gliders, cuscuses and the koala. In order to distinguish some of these animals from the American opossums of the family Didelphidae, Australian zoologists have adopted the strine abbreviation 'possum' as the English name; the book-name 'phalanger' applied to many species refers to the syndactylous second and third digits of the hind foot, present in all members of the family. The phalangerids range in size from the tiny pygmy possum no bigger than a mouse to animals the size of a large cat. They are all clothed in soft woolly fur, have a well developed pouch opening forward, except in the koala, and are diprotodont with a single large lower incisor on each side, and the second and third, if present, very small. The animals are nocturnal or crepuscular, and most of them are herbivorous. The first toe of the hind feet is opposable and acts like a thumb in climbing; in most phalangerids it has no claw. Many of the smaller phalangerids are pretty, delicate little creatures; in captivity they are docile and tolerate handling so that they make elegant though rather unresponsive pets.

The genus *Phalanger* contains the six or seven species of cuscus, found in New Guinea and the East Indian islands and the north of Queensland. They are robustly built about the size of a cat, and have dense woolly fur, short ears and large prominent eyes with vertical pupils which give them a peculiar staring expression (Plate 3). There is great variety of colouring in this genus, some species are spotted with black on white, others are white, black or various shades of brown or reddish – in some the sexes are strikingly different in colour. In all the tail is prehensile and has only the basal portion covered with hair. Cuscuses are arboreal and have strong claws on their digits except the first toe; they feed mainly on fruits and leaves but are said also to take insects, small mammals and birds and their eggs. They are sluggish animals and rather slow and deliberate in their movements which have been likened to those of the slow loris. A.R.Wallace [496] said they live in trees and devour large quantities of leaves.

C

33

They move about slowly and are difficult to kill, owing to the thickness of their fur and their tenacity of life. A heavy charge of shot will often lodge in the skin and do them no harm. . . . The natives everywhere eat their flesh, and as their motions are so slow, easily catch them by climbing; so that it is wonderful that they have not been exterminated. It may be, however, that their dense woolly fur protects them from birds of prey.

There are only two or three species in the genus *Trichosurus*, the brush-tailed possums, but many subspecies of the common and widely distributed *T. vulpecula* have been recognised; all the species are native only to Australia, but the last has been introduced into New Zealand where it has flourished and become a pest. Brush-tailed possums vary considerably in size but the head and body averages about 18 inches in length; the colour of the soft dense fur also varies widely from nearly black through grey and brown to buff. The long bushy tail is prehensile and naked on the underside towards the end. Although millions of these animals have been killed for their valuable pelts *T. vulpecula* is one of the commonest Australian marsupials, and the only one that has been able to adapt itself to the changed ecological environment produced by man; it even lives in his city parks, and nests in the roofs of suburban houses. Brush tails are nocturnal and arboreal, and nest in holes in trees; in treeless areas they shelter in hollows among rocks or in the burrows made by other animals. The diet is mainly vegetable – shoots and leaves of eucalyptus and farmers' and gardeners' fruit trees. A study of the animals near Canberra [123] showed that the males hold territories from which they drive other males, including their own sons when about a year old – the territory of females is smaller and less well defined, and those of neighbours generally overlap within the territory of a single male. The breeding season falls at different times in different parts of the country, as does the number of young reared, which may be one or two. The mother sits up to give birth, and the young, about half an inch in length, crawls through the fur to the pouch by alternate over-arm movements of the fore limbs, on which the digits have well developed claws. These foetal epitrichial claws are retained much longer than those of most marsupials which shed them soon after entering the pouch [307].

The family Phalangeridae contains three genera of gliding possums, two of them in the subfamily Phalangerinae, but the third in subfamily Phascolarctinae, which includes the koala and the ring-tailed possums. All have a fold of skin along the side between fore and hind limbs forming a patagium which is extended when gliding, similarly to that of the eutherian flying squirrels. The pygmy or feather-tail glider *Acrobates pygmaeus*, the only species of its genus, is found throughout eastern and south-eastern

Australia, where it lives in woods and forests and nests in hollow branches. It is a minute, mouse-like animal, grey brown above and white below. The patagium is narrow and the tail, as long as the head and body together, is slightly flattened and has a fringe of stiff bristly hairs on each side which increase the area of the gliding surface. Although the toes bear claws the greatly expanded apical pads of the digits appear to provide the main support in clinging to surfaces. Feather-tails are nocturnal, and live in small parties in nests of leaves built in hollow branches; their diet is said to include insects and the flowers of eucalypts. Although *Acrobates* is a common animal it is inconspicuous, and its biology still awaits detailed study. The three species of *Petaurus*, the squirrel-, sugar- and yellow-bellied gliders, are larger and have long bushy tails and well developed patagia; their fur is fine and soft with a dark dorsal stripe and facial stripes. They inhabit north-eastern, eastern and south-eastern Australia and one of them extends into New Guinea. They are nocturnal inhabitants of open forest where they make nests in hollow trees. They feed on insects and flowers and the sweet sap of various eucalypts; they are active and agile and have been recorded as gliding a distance of 100 yards. Mention of the third genus *Schoinobates* is made below.

The minute dormouse- and pygmy possums *Cercartetus* and *Eudromicia* superficially resemble the eutherian dormice and, like them, have pre-hensile tails but, unlike them, they are insectivorous. They become torpid in winter and draw upon the fat stored in the incrassated tail and under the skin. Several species are found widely in the forests of Australia and one species in New Guinea. Practically nothing is known of the biology of *Distoechurus pennatus*, the pen-tailed phalanger of New Guinea, a small animal with conspicuous black markings on the white face over the eye and behind the ear. The tail is practically naked except for the basal part, but bears a fringe of stiff hairs along each side similar to that of the feather-tail glider *Acrobates*. There is, however, no patagium and the animal does not glide; consequently one may question whether the explanation that the tail of *Acrobates* functions to increase the gliding surface is adequate. *Wyulda squamicaudata*, the scaly-tailed possum of north-west Australia, is known from only four specimens, and nothing appears to have been recorded of its habits – it is alleged to feed on eucalypt blossoms. Another rare species, *Gymnobelideus leadbeateri*, Leadbeater's possum of Victoria, was known from less than half a dozen specimens in museums, and was thought to be extinct until 1961 when a colony of them was found living in 'mountain ash' country with heavily scrubbed undergrowth. They are small possums with long bushy tails and no patagium [51].

35

The three species of *Dactylopsila* are found in New Guinea, and one extends into northern Queensland; the closely related *Dactylonax* is confined to New Guinea. These are the striped possums. The colour of *Dactylopsila* is pale yellow or white with three dark black longitudinal stripes, giving a skunk-like pattern. Like skunks, too, these animals produce an 'extremely unpleasant and penetrating' stink. They are arboreal and feed on wood-boring insects, tearing the wood open with their strong incisor teeth and fishing the grub out with the elongated and slender fourth finger. The specialised incisors, the long finger and mode of use, show a striking resemblance to corresponding structures and habits in the aye-aye of Madagascar, one of the primates.

Tarsipes spenserae, the honey-possum of south-west Australia, the only species in the genus, is the only marsupial, and one of the very few mammals, specialised for a diet of nectar. It is about the size of a mouse; in colour it is greyish above with three longitudinal black stripes along the back, yellowish white below. It has a long snout and long prehensile tail. *Tarsipes* is arboreal, and the apical pads of the digits are expanded and bear small flat claws, thus resembling those of the primate *Tarsius* from which it derives its name. The prominent eyes are set close together, and the tongue is long, narrow and slightly serrated at its edges near the tip. The teeth are minute and degenerate except the procumbent lower incisors. The animals

Figure 2 Honey Possum (*Tarsipes*).

36

are nocturnal and climb among twigs, inserting the long tongue into flowers and eating the nectar just as do many of the honeyeater birds (Meliphagidae) which have 'brush' tongues reminiscent of the serrations on that of *Tarsipes*. Captive honey possums have also been recorded as eating flies.

The subfamily Phascolarctinae contains three genera. All species are arboreal and feed on leaves; the crescent-shaped ridges on the molar teeth are correlated with this diet. In the fore limb the first and second digits are opposable to the third, fourth and fifth, so that in the koala, for example, the hand appears to have three fingers and two thumbs. *Phascolarctus cinereus*, the koala, the single species of its genus, is familiar to all through the popularity of its model as a children's toy. It inhabits the forests of the eastern side of Australia, feeding exclusively on the leaves and shoots of a few species of eucalyptus. The single young is carried on the back of the mother for about six months after it leaves the pouch, and it feeds upon her faeces for some time before it takes to eating leaves. The species has been much reduced in numbers because it was killed for its valuable fur; it is now protected and increasing, and has been restored to many places it formerly inhabited. The genus *Pseudocheirus* contains some dozen or more species which are known as ring-tailed possums, not from any colour pattern but from their habit of carrying the tip of the long prehensile tail curled into a ring. Various species are found in many parts of Australia and New Guinea; all are nocturnal and most of them are rather slow moving animals which build their nests in trees. One species, however, *P. (Hemibelideus) lemuroides* of northern Queensland is an agile leaper among the branches, and has a rudimentary patagium in the form of small folds along the sides of the body. The greater gliders, *Schoinobates*, of which there are two species, have well developed patagia extending from the elbow to the knee. They are medium-sized animals with long furry tails and have been likened when gliding to 'aerial frying-pans' [478]. One was seen to cover a third of a mile in six glides, from the top of one tree to the foot of another, which it climbed to make the next glide. Greater gliders feed on the leaves of only a few species of eucalypts, but have been accused of damaging apple orchards; they are abundant along the eastern side of Australia.

The subfamily Burramyinae was formed to contain the extinct genus *Burramys* described by Broom [53] in 1896 from fragmentary fossils found in New South Wales. The single species *B. parva* differs from other phalangers in having long straight incisors and large sectorial premolars somewhat like those of the kangaroos. Then in August 1966 *Burramys* was

37

brought back from the dead, for a living specimen was found high up in the mountains of Victoria [13]. Externally it resembles a dormouse possum, but has a long thin tail. This animal, the only living one ever seen, is at the time of writing flourishing in captivity, and the anatomists and systematists are awaiting its demise to examine its internal structure; it is a male, and no doubt further specimens will be found in the habitat whence it came.

FAMILY PHASCOLOMIDAE

The family Phascolomidae contains the few genera and species of wombats, all found only in Australia, including Tasmania. Wombats are medium-large animals which weight up to 60 or 70 pounds, stoutly built, with short legs and the tail a mere rudiment. The hand and foot both bear five digits all, with the exception of the first toe which is small, armed with long stout claws; the second and third digits of the foot are syndactylous but are not reduced in size. The dentition is unlike that of any other marsupial and resembles that of many eutherian rodents; it is regarded as a striking example of convergent evolution correlated with similar functions and habits. A single large incisor in upper and lower jaws is separated by a wide gap or diastema from the molar teeth the crowns of which present a grinding surface similar to that of many rodents. All the teeth are rootless and grow continually as they are worn down; the food consists of grasses, roots and bark.

Wombats are nocturnal and live by day in extensive burrows which they excavate with their powerful limbs – Wood Jones described the wombat as 'an excessively powerful animal, its perfection of muscular development being one of the wonders of comparative anatomy. It is this extraordinary muscular power which enables it to dig its large tunnels into the very hardest sort of ground.' Its burrowing brings it into disfavour with Australian farmers, and great numbers have been destroyed as pests, for the fur has little value.

Wombats are classified in two genera – *Phascolomis*, with naked noses, and *Lasiorhinus*, with hairy noses. Of the two species of *Phascolomis*, the common wombat, *P. ursinus* which is found in Tasmania and the islands of Bass strait, was described in 1802 from a specimen captured by Bass on Cape Barren Island. Although the mainland species *P. mitchelli* is larger, it was assumed to be the same as the Tasmanian animal for many years. In 1838 Owen examined some fossil wombat bones from New South Wales and, finding them different from those of *P. ursinus*, described them as those of an extinct species, which he named *P. mitchelli*. It was not until many years later that it was discovered that the common wombat of the mainland, found in the coastal districts of New South Wales, Victoria and the adjacent corner of South Australia, was in fact the supposedly extinct *P. mitchelli* and not *P. ursinus*. Until the discovery in 1966 of living *Burramys*, the common wombat was the only marsupial mammal described from fossils, and thought to be extinct, that subsequently proved to be still living. The hairy-nosed wombat, *Lasiorhinus latifrons*, is similar in general to the common wombat, but has a hair-clad muzzle, longer ears, silkier fur, and a much broader and more flattened skull. It is found in the south of South Australia and also in Queensland where its limited range overlaps that of the allied *L. gillespiei*.

The early settlers named the wombat the 'badger', and the name has lasted in the outback; they found the flesh to 'resemble pork in its fatness and flavour, but not in colour or texture, being red and coarse'. Troughton remarks that had they been called 'wild pigs' they would have soon been eaten out, but adds 'though fortunately for the vanishing animals the flesh does not resemble pork at all: it is of a musky not very agreeable flavour and very sinewy.'

Wombats have often been kept in captivity and even as house pets; not only is it easy to provide them with suitable diet, but they are naturally docile and tolerant of handling, as Bass found when he caught one in 1798 by placing his hands under its belly and suddenly lifting it from the ground. He carried it lying in his arms like a baby for some time but, when he tied its feet together to prevent it escaping when he put it on the ground, its placidity departed and it 'whizzed with all his might, kicked and scratched most furiously.'

FAMILY MACROPODIDAE

The last family of marsupials in our classification, the *Macropodidae*, contains the mammals which are not only characteristic of Australia but are inextricably part of what is now termed the image of Australia in the popular mind – the kangaroo together with the emu even appears in the armorial bearings of the Commonwealth. They are well named the macropods, for their big feet are one of their important adaptations correlated with their characteristic method of locomotion by hopping. The sight of a mob of big kangaroos sailing effortlessly over the plains at a rate of twenty to thirty miles an hour, springing along as though the ground were an enormous trampoline, is a most beautiful and impressive vision of effortless grace. Young macropods live in the pouch of the mother until they are comparatively large, and use it as a refuge long after they are able to run about and feed independently; the sight of a youngster a third the size of its mother climbing into its bassinet never fails to attract the amused attention not only of the general public but also of hardened zoologists.

The family is divided into two subfamilies, the Potoroinae containing the smaller rat- and jerboa kangaroos of many sorts, and the Macropodinae containing the kangaroos and wallabies of larger size and numerous forms. The general characters found in all members of the family are briefly, the long hind legs in which the fourth toe is largest, the second and third syndactylous, and the first, with one exception, is absent; the short fore limbs with five fingers, and the long thick tails. There are three upper incisors and one lower on each side, separated by a diastema from the grinders. Kangaroos and wallabies of one sort or another – about three dozen species – are found practically everywhere in Australia, New Guinea and some adjacent islands. They are all herbivorous, and the larger kangaroos are mainly grazers rather than browsers. There has long been current a belief that they are able to separate the two lower procumbent incisor teeth, and use them with a scissors action for snipping off herbage. Recent studies [402] have shown that this notion is a fallacy; the two halves of the lower jaw are not united, and the teeth can indeed be separated, but they are not used as scissors. When the animal is grazing and bites at the herbage, the lower incisors separate so that they closely approach the upper ones to grasp the food, which is pulled up, not cut off. When the food is in the mouth the lower incisors are approximated so that the food can be chewed by the grinders with a rotary motion of the jaws, without the lower incisors striking against the upper ones. Much of the food is regurgitated for further mastication by means of a spiral groove in the stomach.

In some parts of Australia the larger kangaroos are regarded as pests by the owners of sheep stations, who have them shot for the skin and meat trade – in 1966 one packing plant in Queensland was processing 2,000 kangaroo carcases a week [184]. The old yarn that a kangaroo eats as much as four or even eight sheep has been disproved as, weight for weight, kangaroos eat no more than sheep. They do, however, compete with sheep for some of the most nutritious herbage, although they do not usually browse on certain trees used by sheep. Thus to some people kangaroos are a pest but to others they are a wildlife resource of economic importance.

The breeding of the macropods has been discussed in Volume I and the way the diapause in foetal development is related to lactation, the corpus luteum, the presence of a joey, and external conditions such as drought, was mentioned. The details of these phenomena vary from species to species but the broad picture is similar in all. The larger kangaroos and wallabies, although more active at night, are not strictly nocturnal; the red and grey kangaroos spend much of the day lying down, but after midday their activity increases and grazing starts about an hour before sundown and continues during the night [71]. Although the larger macropods range about in groups, they show little of the social behaviour seen in eutherian gregarious mammals; the animals of any group are continually changing and exhibit no sign of any permanent group-bond. When conditions are good kangaroos occupy a fairly restricted group range – there does not seem to be any defended territory – but when times are adverse, as in drought, they may migrate long distances. The photograph (Plate 4) of kangaroos drinking at a waterhole shows how each likes to keep its distance from its neighbours – in this example, about two feet – a striking contrast to the crowded contact shown by many other gregarious animals under similar circumstances.

Hediger [222] has pointed out that when alarmed a kangaroo stands upright to get a good view of its surroundings and discover whether any danger threatens; the upright attitude is thus characteristic of defence and aggression. If a man walks into a paddock containing captive kangaroos the animals naturally take his upright stance as the expression of an aggressive mood, and may react accordingly – in fighting a kangaroo stands up and scratches with its hands and then may suddenly sit back on its tail and slash downwards with the hind feet. If the man crouches down in an attitude simulating that of an unalerted kangaroo the animals, in nine cases out of ten, immediately relax, their aggression evaporates, and they take little notice of him; they accept him as just another harmless 'roo.

Within the subfamily Macropodinae the name wallaby is applied to the smaller species, and kangaroo to the larger ones; there are several genera of both kinds. The hare wallabies, *Lagorchestes*, are small animals about the size of a hare and as swift – which lie up in forms by day. Of the three species the one (*L. leporides*) inhabiting south-eastern Australia is probably extinct. The similar banded hare wallaby *Lagostrophus* is restricted to a small area in Western Australia and the islands of Sharks Bay, where it was seen by Dampier in 1699, as mentioned above. The rock wallabies *Petrogale* are larger animals which inhabit mountainous regions, where they are very agile in climbing among rocks and leaping from point to point. There are several species, some with conspicuous markings such as facial stripes, light shoulder patches, or ringed tail. They were widely distributed in Australia and are still common in some places though they have been exterminated in some of their former habitats. The three species of nail-tailed wallabies,

Onychogalea, have similarly been reduced in numbers from their former wide distribution. The end of the tail carries a horny excrescence forming a spur or nail, the function of which is not known.

The four species of pademelon wallabies, *Thylogale,* inhabit dense scrub and thickets and, like the more numerous species of scrub wallabies *Protemnodon,* are distinguished by the details of their incisor teeth; some species have been exterminated by stock and sheep farmers and by fur-traders. In both genera some members bear characteristic markings on the face, back, hips and elsewhere. The habitats of the scrub wallabies range from scrub and open forest to savannahs and swamps. The prettily cross-banded fur of the toolach wallaby *P. greyi,* was much prized in the fur trade – toolach is a corruption of the native name – and in twenty years the pelt hunters and the people who coursed it for sport reduced it from being a common animal in southern Victoria and South Australia to a remnant of less than twenty. In 1923 an unsuccessful attempt to capture them for transfer to a sanctuary where they might be preserved resulted in the final extinction of the species. The red-necked wallaby, *P. rufogrisea,* of eastern Australia and Tasmania, where it is still common, is a hardy species that does well in zoos and parks all round the world. The quokka, *Setonix brachyurus,* is a smaller species with a short tail, that was formerly common near the coast in Western Australia but is now greatly reduced, although it survives in plenty on Rottnest Island off Fremantle and Bald Island off the south coast. Quokkas live in thick grass and undergrowth through which they make well worn runways, and lie up in forms as expertly as a hare, concealing themselves even when the cover seems to be inadequate to hide a mouse.

Several species of forest wallabies, *Dorcopsis,* inhabit New Guinea; superficially they are very similar to *Protemnodon* but little is known of their biology.

New Guinea, too, is the home of one of the most peculiar genera of macropods, *Dendrolagus,* two species of which are also found in northern Queensland. Who would expect to find kangaroos living up trees? The macropods are believed to have evolved from arboreal phalangerine stock, and the tree kangaroos of this genus have returned to their primitive home. They are structurally adapted to it in several ways, although they retain their ability to travel rapidly by hopping when on the ground. They are large stoutly built animals, with thick dense fur and long tails. The hind leg is proportionately shorter than in the wallabies, and the sole bears large naked pads that are supposed to give a grip on tree branches; the fore feet have long curved claws. Tree kangaroos not only climb about agilely among the branches, but take long flying leaps from one to another – the tail is used in balancing, but is not prehensile.

There are three species of large kangaroos, *Macropus*, but the large number of subspecies and geographical races has given rise to much nomenclatural confusion, not least in the vernacular names. They are big animals, often as tall as a man, and between them cover nearly the whole of Australia. *M. robustus* is the wallaroo or euro, or hill kangaroo; *M. giganteus* is the great grey, grey or forester kangaroo; and *M. rufus*, now often placed in a separate genus *Megaleia*, is the red or plains kangaroo, the marloo of Western Australia, the female of which is often called the blue doe or blue flyer. Confusion in identifying the species is increased by the fact that the female red kangaroo is grey, and many wallaroos are reddish; the three are most readily distinguished by the character of the snout – naked in *M. robustus*, hairy in *M. giganteus* and with a limited naked area in *M. rufus*. Some points in their natural history have already been discussed.

The subfamily Potoroinae is distinguished from the last by the small size of the animals, and the presence of canine teeth, which are nearly always absent in the macropodinae. Some eight or ten species of rat-kangaroos or jerboa-kangaroos, including the potoroo, separated into four genera, *Bettongia*, *Aepyprymnus*, *Caloprymnus* and *Potorous*, were formerly common, and between them covered almost the whole of Australia. Now some of them are extinct and most of them are very rare; only the potoroo in Tasmania, and the rufous rat-kangaroo (*Aepyprymnus*) in parts of Queensland, are still common. Their disappearance is due mainly to the spreading of poisoned bait for rabbits, predation by the introduced fox, and to their use in coursing for sport. They are about the size of a rabbit, some rather smaller, with grey or brown fur, off-white on the belly; the hind foot is generally proportionately shorter than in the Macropodinae, and the tail in some species is prehensile in that it is used for carrying nest material. The dentition is peculiar not only in having upper canines but in the size of the large sectorial premolar teeth correlated with their diet. The claws of the fore feet are very large and sharp. Rat-kangaroos inhabit forests and woods, except *Caloprymnus* the desert or plains rat-kangaroo; they build nests of dry vegetation among grass tussocks or under bushes, though one species of *Bettongia* is a burrower and sometimes lives in rabbit warrens. All are nocturnal, and feed on plants and on roots and tubers which they dig up with their strong claws. Finally there is the musk rat-kangaroo *Hypsiprymnodon*, named from the odour emitted by both sexes. It differs from all the other macropods in possessing a first toe, which is clawless and, though not opposable to the others, is held to show its affinity to the phalangers; the tail, too, is naked and scaly, with fur only at its base. Some authorities consider that *Hypsiprymnodon* should be classified in a separate suborder. Musk rat-kangaroos are small animals, scarcely larger than a rat, and habitually move on all fours as well as by hopping; the rain forests of northern Queensland are their only habitat. They are active by day as well as by night, and feed not only on vegetable matter but on insects and invertebrates that they find by turning over the litter of the forest floor.

In concluding this review of the marsupials, so many of which are characteristic of the Australian fauna, it should not be forgotten that Australia has also a fauna of many native species of eutherian mammals, mostly belonging to the orders of rodents and bats, but including marine mammals and the pre-historically introduced dingo.

Chapter 2

Insectivores and Dermopterans

ORDER INSECTIVORA

THE INSECTIVORA are generally regarded as the most primitive order of eutherian mammals – that is, of mammals with chorio-allantoic placentas, a corpus callosum in the brain and no epipubic bones. Although they retain a number of generalised characters such as the comparatively large size of olfactory lobes of the brain, the three-cusped molar teeth in many forms, the almost universal presence of clavicles, and of five clawed digits on the plantigrade or semi-plantigrade fore and hind feet, all are adapted to living in particular habitats, and show corresponding specialised characters correlated with their ways of life. Another perhaps primitive character is the frequent but not universal presence of a cloaca, which provides a common opening for the gut, urethra and genitalia. Many insectivores thus show a high degree of specialisation superimposed on the generalised, so-called primitive background. They are small mammals – none of them is larger than a rabbit and most are much smaller – and in general little is known of their habits and natural history, for they quietly live their obscure lives in obscure places; the common shrew and the mole are present nearly everywhere in the countryside of England, yet how many of the fifty-two million inhabitants of the United Kingdom have ever seen a living one? Few indeed have learnt by direct observation anything of the biology of the teeming populations of these animals that are such close neighbours.

The diversity of the insectivores has always been a difficulty to the systematists, and the order has been something of a classificatory rag-bag for tucking away any small mammal with generalised structure that could not be fitted into one of the other orders. They have consequently been subdivided at various times into suborders in an attempt to make a logical classification. Haeckel in 1866 divided them into the Menotyphla – those having a caecum, the cul-de-sac at the beginning of the large intestine – containing the tree shrews and the elephant shrews, and the Lipotyphla, with no caecum, containing the rest. Gill in 1885 divided them into the Zalambdodonta, with a single outer V-shaped cusp on the upper molars, containing the tenrecs and golden moles, and the Dilambdodonta, with two Vs, containing all the others. Neither classification is wholly satisfactory, though Haeckel's Lipotyphla contains most of the insectivores as now accepted – some systematists favour making two separate orders, Lipotyphla and Menotyphla, to replace the old Insectivora. Of the Menotyphla the tree shrews are now regarded as being more appropriately classified as the most primitive family of the Primates, and they have been transferred to that order by Simpson and those who follow him. The elephant shrews, however, are retained in the insectivore order, though a strong case can be made for transferring them also to the Primates, for the resemblances between them and the tree shrews are great – some think that whichever family receives the first should also take the second. In this book the tree shrews are included in the Primates, dealt with in detail in the volume on that order, and consequently they are no more than mentioned here; but Simpson's classification is followed in retaining the elephant shrews. In many insectivores the zygomatic arch, or cheek-bone, is incomplete so that the skull appears to be fragile because the elongated upper jaw seems to have inadequate support. The gap is due to the absence of the jugal bone, and it is interesting to note that the arch has been rebuilt in some species by extensions of the zygomatic processes of the maxilla and squamosal bones meeting and fusing.

FAMILY SOLENODONTIDAE

This family contains only one living genus, *Solenodon*, with two species, *S. paradoxus* and *S. cubanus*, found only in the West Indian islands of Hispaniola and Cuba, where they are reported to be on the verge of extinction. The solenodons are odd-looking animals, about the size of a guinea pig, with very long pointed snouts well provided with vibrissae, small eyes, large naked ears which project above the fur, and a long, nearly naked tail; the large naked feet have five toes armed with strong claws. The pelage is rather coarse and loose, and ranges in colour from smoky browns through grey to blackish. The animals inhabit rocky wooded or bushy country, and den among rocks or in hollow fallen logs. Allen [4] says that solenodon 'is a rather slow-moving creature, constantly on the move with a shuffling gait, sniffing everywhere with its long nose, and scraping and scratching here and there with its long claws, exploring for food.' In addition to invertebrates and small vertebrates, it is said to eat fruit and other vegetable matter. Cabrera [64] reports that when feeding it can sit up on its haunches propped by its tail; Eisenberg and Gould [131], who fed four captive solenodons on white mice, found that when one of them sat on its hind legs and held the prey in its mouth, 'the forepaws are used (alternately or simultaneously) to tear the exposed body distally. The carcass is thus torn to pieces, and each piece is picked up and eaten in turn'. The young are said to number one to three, but there are only two teats, which lie on the hind edge of the thigh below the tail – they are thus crural, as pointed out by Angulo [9] and not inguinal as generally stated.

Although the Cuban solenodon was known to the Spaniards early in the sixteenth century the genus was unknown to science until the thirties of the nineteenth, when both species were described. Thereafter very few specimens were seen or captured until in the early years of the twentieth century the Hispaniolan species were rediscovered and a few living examples were sent to the United States. In 1935 and 1936 a German collector secured about twenty specimens which he sent to Hamburg where a few arrived alive and formed the subject of memoirs by Dr Erna Mohr [327]. Of the Cuban species fewer still have been captured, and the German collector was unable to find any in 1937, although according to Angulo, one was living in the Havana zoological garden in 1947. It is believed that predation by introduced cats, dogs, rats and mongooses has brought the population of both species to their present very low levels. Eisenberg and Gould [131] recently watched the behaviour of four captive *S. paradoxus*, and found that they showed a rather specialised set of foraging patterns, but that in general the behaviour is stereotyped and limited in variety. They also found that *Solenodon* produces high frequency clicks, probably from the larnyx, which resemble the echolocating pulses of some shrews.

The solenodons, however, are not without defences. The name *Solenodon* which means 'sword tooth', refers to the shape and size of the remarkably large anterior upper and second lower incisor teeth. Moreover, the animals, like a few other species of insectivores, are provided with a venomous saliva, a very rare phenomenon among the mammals. The poisonous saliva is secreted by the sub-maxillary salivary glands and conveyed by their ducts to the base of the enlarged lower incisor teeth which are deeply grooved on their inner sides, so that the venom is injected into the wound produced by a bite. Rabb [386] reported the results of experiments with an extract of the submaxillary salivary glands of *S. paradoxus* 'when three of these animals died at the Chicago Zoological Park within two months after their arrival in 1958 from the Dominican Republic'. He injected the extract of the glands intravenously into white mice and found that a dose of ·38 to ·55 mg. per gram produced urination, irregular rapid breathing, protrusion of the eyes, gasping and convulsions before the mice died within two to six minutes. 'Two animals that had intraperitoneal injections of extract of ·56 and ·66 mg. per gram died in about 12 hours, and one injected at the level of 1·02 mg. per gram died in 13 minutes. Urination, cyanosis, and depression were observed in these animals.' Rabb points out that the toxicity appears to be considerably less than that of the saliva from the shrew *Blarina*, but that this may be due to postmortem inactivation. He quotes the experience of the Cuban naturalist Juan Gundlach who, writing in 1877 of the Cuban species *S. cubanus*, said that he was bitten by a tame specimen, 'which gave me four wounds corresponding to the [large] incisors: those from the two upper incisors healed well, but those from the lower ones inflamed.'

The animals are said to be susceptible to their own venom, for captive specimens have died after scuffles which have resulted in superficial bites but no serious wounds; the inference, however, does not seem to be justified from such slight evidence when one remembers how readily stressed captive animals succumb. So little is known of the life of the animals that it is unprofitable to suggest what part the venomous saliva may play in their lives; it appears to have been little use to them in resisting attack from introduced predators, if indeed such attack is the cause of their near extermination. There can be small doubt that the solenodons are on the way to join another family of insectivores, the Nesophontidae, which inhabited Puerto Rico, Hispaniola and Cuba until recent times, but are now believed to be extinct. Several species of these animals are known, ranging in size from that of a mouse to a rabbit, from subfossil remains unearthed in the caves of the islands. Some species were certainly contemporary with introduced rats and mice, and formed the prey of several species of owl both living and extinct. Remains of at least one species have been found in regurgitated pellets of the barn owl; they are so recent in appearance – the bones have fragments of dried soft tissues still adhering – that it is possible that we may yet have a brief snapshot of a living nesophon before that genus and the solenodons disappear for ever.

FAMILY TENRECIDAE
The animals of this family are found only in Madagascar and the nearby Comoro Islands, where a number of species have evolved striking superficial resemblances to insectivores of other families – hedgehogs, shrews, water shrews, and to some degree, moles. Most of the species, about twenty in all, are nocturnal, and den in burrows dug by themselves; their diet consists mainly of invertebrates; some of them hibernate during the winter months, or become torpid for one or more days at irregular intervals.

Little was known about the life histories and behaviour of most species, until Gould and Eisenberg [179] published the results of their recent studies, which form the foundation for future research on these unusually interesting animals. Before their work most of what was known about the lives of the animals came from the field notes of collectors. Gould [178] has also discovered that several species of tenrecs use a primitive sort of echolocation by clicking with the tongue to produce pulses ranging in frequency between five and seventeen kilocycles a second.

The tenrec, *Tenrec* (formerly *Centetes*) *ecaudatus*, the only species of its genus, is the largest living insectivore, about the size of a rabbit, and the most widely distributed and commonest member of the family. It is rather long-snouted, with short stout limbs, and large flattened claws on all the digits. The pelage is a mixture of hairs and slender spines, buff to brown in colour, and the spines are concentrated on the nape to form a nuchal crest that is erected when the animal is alarmed – tenrecs from the rain forest are more furry and less spiny than those from more arid regions. The young differ from the adults in possessing on the back above the rump a row of quills which are lost on moulting to adult pelage. During the dry winter season tenrecs hibernate in a chamber at the end of a burrow plugged with earth – as in other hibernators the temperature, heart rate and breathing rates are reduced. Tenrecs have been found foraging in packs of up to twenty or even thirty; these are family parties consisting of one or two adults and a brood of juveniles – as many as thirty-two embryos have been found in the uterus of a female (Plate 5(a)).

The two species of streaked tenrecs (*Hemicentetes*) are much smaller and spinier, and have comparatively longer snouts and limbs; they are black, longitudinally streaked with white. The spines on the nape form an erectile nuchal crest, and a group of specialised quills on the middle of the back can be caused to vibrate by the muscle below the skin to produce a stridulating sound which Gould and Eisenberg [179] found was used in communication within a group, especially a foraging family party. The similar patch of quills present only on the juveniles of *Tenrec* are used in the same way, but communication by this means is abandoned when the animals become adult and lose the quills. *Hemicentetes* lives mainly on earthworms, and forages for them both by day and night. The parties of up to twenty or more that have been noted must consist of several families, for Gould and Eisenberg who bred *H. semispinosus* in captivity found no litter exceeding seven young. The genus is restricted to the rain forest and its edges (Plate 5(b,d)).

The three genera of hedgehog tenrecs, *Setifer, Dasogale* and *Echinops*, are covered with spines on the back from brow to tail, and superficially resemble the true hedgehogs, *Erinaceus*, of the Old World. They differ in having stout tails covered with spines, and in being only about half the size of the European hedgehog. *Dasogale* is known from only one specimen, but the other genera are locally abundant in various parts of the island. Both are nocturnal and insectivorous, and *Echinops* climbs slowly and deliberately among the branches of trees and shrubs; *Echinops* hibernates in the winter but *Setifer*, according to Rand [391] does not. Like the true hedgehogs, these animals roll into a ball when alarmed and pull the spiny dorsal skin over the whole of the exposed surface by the highly developed skin-muscle. Interesting as the anatomical convergences between these tenrecs and the hedgehogs may be, a convergence in behaviour, observed by Gould and Eisenberg in *Echinops*, is even more remarkable. When the animals are stimulated by encountering an object with a strong scent they salivate copiously and annoint themselves with foam from the mouth. *Echinops* differs from the hedgehogs by applying the foam with the fore limb to the sides of the body and face, whereas the hedgehogs apply it direct to the sides with the tongue. The function of this peculiar action is not known; its cause may be similar to that in the hedgehogs.

A number of species and subgenera of shrew tenrecs, *Microgale*, have been described, but little is known of them or their lives. They resemble shrews with long tails and with ears showing above the fur, which is soft and without spines. They are believed to be terrestrial, and some species are active both by day and by night; their diet and way of life are said to resemble those of shrews. The tip of the long tail is said to be prehensile, but Gould and Eisenberg found no evidence for this in ten captive, *M. (Nesogale) dobsoni*. In this species the tail becomes incrassated (fat-tailed) in the dry season.

The single species of web-footed tenrec, *Limnogale mergulus*, about the size of the water-shrew, lives in burrows in the banks of rivers and streams even high in the mountains. The fore feet are partly, and the hind feet fully, webbed; the animal swims strongly in swift streams by their aid and that of the long tail, which is flattened sideways in its distal half. The diet consists of small frogs, crayfish and aquatic insects but not, as was formerly thought, of water plants. Although *Limnogale* appears to be not uncommon in certain mountain torrents, only about twenty specimens were preserved in museums since it was discovered in 1896 until Malzy [309], late in 1963 and early 1964, captured fourteen specimens and kept them alive for up to eighteen days. He found they were exceedingly quick in their movements in the water, and that they not only swam but walked underwater clinging to the bottom. The fur quickly became waterlogged, and on returning to the bank with prey the animals made a meticulous toilet to dry themselves. Gould and Eisenberg suggest that *Limnogale* does not hibernate because they found droppings, showing that the animals were active, when the temperature was low enough for the ground to be covered with frost.

Three species of rice tenrecs, *Oryzorictes*, are rather molelike; the snout is pointed, the dark fur is dense, the eyes and ears are small, the limbs short and stout, the fingers of the fore limbs are armed with broad stout claws, except that of the small first digit, and the tail is less than half the length of the body. These animals appear to be mainly subterranean and nocturnal in habit, but little is known of their biology. Their burrows are said sometimes to damage the banks of rice fields, in which they tunnel in search of insects and other invertebrates.

The single species of the remaining genus, *Geogale aurita*, inhabits the west part of Madagascar, but is little known. It is small in size and generally resembles a shrew with large ears, brown above and yellowish white below. Nothing is recorded of its habits or life history, but Gould and Eisenberg found one specimen torpid in midwinter, the first record of hibernation in the subfamily Oryzorictinae (Plate 5(c)).

FAMILY POTAMOGALIDAE
This family, classified as a subfamily of the Tenrecidae by some systematists contains only two genera and three species of the west African otter shrews. *Potamogale*, the giant water or otter shrew, resembles a tiny otter, though it is large for an insectivore as the head and body are up to about fifteen inches long and the tail almost as much again; the two species of

Figure 3 Small Otter Shrew (*Micropotamogale*)

Micropotamogale are only about half the size. *Potamogale* has a fairly wide distribution in west equatorial Africa and has been known since 1860, but the two species of *Micropotamogale* have been found only in much smaller areas – they were not discovered until 1951 and 1953.

In both genera the head and body are long and narrow, the limbs short, the ears small, and the eyes minute; the region of the muzzle bearing the long tactile vibrissae takes the form of a large pad, as does that of several aquatic mammals of other orders, discussed in Volume I (pp. 158, 166). The colour of the dense fur is dark brown to black above, whitish, often with a yellow tinge, on throat and belly; the long stiff white vibrissae are conspicuous in contrast with the dark face. The long tail of *Potamogale* is flattened from side to side and covered with short hairs: that of *Micropotamogale* is more oval with slight keels above and below; in both it is the main organ of propulsion in swimming, together with sinuous movements of the whole body. In *Potamogale* and *M. lamottei* neither hands nor feet are webbed, but the feet have a raised ridge of skin, which appears to be a modified hypothenar pad, along the outer edge; it is suggested that the enlarged area thus given to the plantar surfaces assists the grip of the sole on slippery rocks. In *M. ruwenzorii* on the other hand, both the fore and the hind feet are fully webbed, a striking difference in animals otherwise so similar. The most interesting feature of the hind feet of all three species is, however, that the second and third digits are syndactylous to the bases of the distal phalanges. They thus closely resemble the feet of the syndactylous marsupials except that all the other digits, including the first, are also present; there can be little doubt that these toes are used by the otter shrews in the same way as those of marsupials for grooming the coat, although direct observation on the point does not appear to have been recorded. The first upper, and second lower, incisor teeth are large and caniniform, as in many other insectivores, and all the teeth behind them bear sharp cusps; the diet of all three species consists, as far as is known, exclusively of fresh-water crabs.

Potamogale inhabits mountain streams, forest pools and slow-flowing rivers of the lowlands; both species of *Micropotamogale* have been found only in mountain torrents. All are said to live in burrows in the banks with the entrance under water, and all are extremely swift and agile swimmers. Those pregnant females that have been examined contained two foetuses, as might be expected from the presence of only a single pair of inguinal nipples.

Little is recorded of the life of these animals, but Durrell [125] writes of a captive *Potamogale* in west Africa that he offered it live and dead fish, frogs, a water snake, freshwater shrimps, and then crabs.

As each was presented he would approach it, sniffing through a maze of quivering whiskers, and then retreat, sneezing. Only when offered crabs did he pounce forward, turn the crab over on its back with his nose, and then bite through the body. He settled down to eat it, holding down the wriggling legs with his paws, while he tore out the flesh with his strong teeth. Having disposed of the body and the two claws, he left the rest.

This animal ate fifteen to twenty crabs out of the forty or so offered every night. Sanderson [415] also found that only river crabs were eaten, although fish were abundant in the streams, and adds, 'Live and dead fish were positively refused by animals kept in captivity . . . in fact, they seemed thoroughly frightened of a wriggling fish.' Durrell further remarks that on land the otter shrew seems to hunt more by hearing and scent, for the eyes appear almost useless: 'On hearing a crab the animal will prick up its ears and move forward, sniffing prodigiously, its whiskers stuck out ahead of its nose, quivering like the antennae of an insect.' When foraging underwater the vibrissae must be the sole means of discovering prey.

It is probable that the habits of the two *Micropotamogale* species are in general similar. The anatomy of *M. lamottei* has been described by Guth, Heim de Balsac & Lamotte [191] from the five specimens, all young or sub-adult, known up to 1959, and that of *M. ruwenzorii* by de Witte & Frechkop [522] and its distribution discussed by Rahm [389] from the eleven specimens known up to 1960. *M. lamottei* is found in the Guinea forests on the eastern border of Liberia, and *M. ruwenzorii* not only on Mt Ruwenzori, north of Lake Edward, but also on the mountains to the west of the lake, and those west of Lake Kivu.

FAMILY CHRYSOCHLORIDAE

The golden moles are found throughout the southern half of Africa to about five degrees north of the equator, and not elsewhere. Although the Cape golden mole was given the specific name *asiatica* in 1766 by Linnaeus, who was erroneously informed that it came from Siberia, by the law of priority this remains the valid name of the South African *Chrysochloris asiatica*. Five genera, with about eleven species, are now recognised, although several further genera and many more supposed species have been described.

The chrysochlorids present a superficial resemblance to the true moles (Talpidae) in both appearance and burrowing habits. The fur is thick and has a dense soft underfur, but it has a lay from front to back and is not velvety as in the talpids; it ranges in colour from buffish through shades of brown to dark grey, and has an iridescent bronzy sheen of green, violet, yellow or red. The eyes are rudimentary and are completely covered by the skin; the ears have no pinnae, and are small holes hidden by the fur; and the tail does not appear externally, although there are some caudal vertebrae. The fore limb is powerful and bears four digits, the outer ones small but the central two large and armed with huge pointed claws; the number of phalanges is reduced so that the claws appear to be attached directly to the hand. The hind feet are small and have five toes. The claws are used in burrowing, but not in the manner of the true moles although, as in them, the pectoral girdle lies far forward so that the shoulder joint is at the level of the first or second neck vertebrae. The action of the fore limb as shown by Campbell [68], is a scratching or running motion, and a sideways scraping; the thrust is not transmitted to the sternum by the clavicle, as in the talpids, for the inner end of the bone does not articulate with the sternum but is attached by a short fibrous strip, which can transmit tension but not compression. Campbell, too, was unable to confirm the statement first made by Dobson [115] that the anterior part of the thorax is hollowed to accommodate the pectoral girdle.

The hind part of the skull is wide owing to the size of the auditory bullae, and although a zygomatic bar is present the skull is evidently derived from a form in which it was absent. The bar consists of a backward extension of the zygomatic process of the maxilla reaching to the squamosal, with no jugal component. The canine teeth are small, but the first upper and second lower incisors are large, the former are set side by side like those of rodents. According to Shortridge [442] a remarkable feature is that 'the teeth do not cut the gum until the young animal is almost full grown', and that the young are suckled until that age. All species are said to become inactive during the colder part of the year, but there appears to be no information recorded showing that they become torpid and hibernate – it is possible that they do. In seeking their food which consists of worms and other soil invertebrates most of the species burrow close to the surface, so close that the roof of the tunnel is pushed upwards, making a ridge on the surface. In forming these burrows the snout, protected by a rather thick leathery skin on the end, is alleged to be used, as well as the claws of the fore limb. The surface burrows radiate 'in all directions' from a central point where there is a deeper tunnel. The soil is not thrown out of the surface burrows, but some, probably all, species throw up mounds of soil when they are burrowing at a greater depth. Little is known about the breeding of golden moles, but as there are only four nipples on the female the number of young is unlikely to exceed that number – some pregnant specimens have been examined containing only two foetuses. The young are cradled in a nest of dried grass in a chamber well below the surface.

The single species of *Eremitalpa*, the smallest of the golden moles, and one of the two species of *Cryptochloris*, both inhabit the sand dune country of the Namib desert on the coast of South-west Africa, and prey largely upon the burrowing legless lizards that live below the surface of the sands. The moles are said to burrow more deeply during the heat of the day. The giant golden moles of the genus *Chrysospalax* – one of the two species, *C. trevelyani*, is as large as a rat – are reported to feed on the surface more habitually than the others. Roberts [406] reports that members of the smaller species, *C. villosus*, emerge on to the surface only after rains, when they forage about rooting up the surface 'like miniature pigs'; at other times, presumably, they feed below the surface. The larger species is said to live mainly on giant earthworms, which it seeks on the surface in forested areas. Both species throw up mounds of earth and live in nest chambers within them; they are, in addition, believed to make use of the burrows of other animals such as mole rats. It is strange that the biology of the golden moles has been little investigated until quite recent years. The family has been closely studied by the systematists who have even proposed that it should form a distinct order of mammals, but although so many species and races, both living and fossil, have been described, the living animals have generally been regarded as difficult to find in spite of their abundance in many places. Thus van der Horst [482] tells how in 1897–8 Dr Broom, who became the outstanding expert on the taxonomy of the family, collected some owl pellets in Namaqualand, 'and found, in this way, the skull of a new species of golden mole' which afterwards became the type of a new genus, *Eremitalpa*. He adds, 'It was only many years later that this species proved to be very common in the neighbourhood of Port Nolloth, in the extreme north-west of Cape Province' whence he obtained many specimens to investigate the embryology.

FAMILY ERINACEIDAE

Three genera – some zoologists think five – of hedgehogs, and four or five genera of moon rats and allied species, make up this family; all have a zygomatic arch formed by the jugal bone.

The hedgehogs inhabit all Europe and Asia north to latitude 50° to 60°N, east to Borneo but not Japan, and the whole of Africa. All are covered with spines on the back and sides from forehead to rump and with soft hair on the face and underside. When alarmed their defence is to curl up nose to tail, and contract the well developed skin muscle like the draw-string of a purse, thus forming a ball completely covered with erected spines. Their biology is known mostly from studies on the common and widespread hedgehog of Europe, *Erinaceus europaeus*.

Hedgehogs can adapt to the most varied habitats and diets, and to the environmental changes produced by man's agricultural and other activities; this, with the protection from predators afforded by the spines, is doubtless one of the main reasons for their ubiquitous distribution. They are solitary except in the breeding season; and are crespuscular and nocturnal – their activity shows peaks early and late in the night. Hedgehogs will eat almost anything, though the diet consists mainly of invertebrates found by rooting among litter under trees, in hedge bottoms, among tall herbage, and in the open. In addition, Herter [226] reports that vegetable matter such as seeds, fallen berries and fruit are eaten. Insects, slugs, snails, earthworms, the eggs and young of small ground nesting birds, occasionally mice and other small mammals are all taken, but Dimelow [112] found that *E. europaeus* prefers millipedes of various species to all else, perhaps because of their peculiar, and to us distasteful, cyanide-like odour. Hedgehogs are often accused of eating the eggs of game-birds, but Herter found they are unable to break those of domestic hens, pigeons or pheasants – indeed, the mouth, although armed with large caniniform first upper incisors, appears to be too small to encompass such comparatively big smooth objects. Loukashkin [302] found that *E. dauuricus*, the Daurian hedgehog, on the semi-desert shores in the 'naked and dismal landscape' around Dalai Nor lake in the north-west corner of Manchuria, subsisted mainly on small rodents 'living there in countless colonies.'

Some species of hedgehogs dig burrows, others generally make use of natural hollows and crevices; some species make nests in such places for regular use, others only for nursing the young or for insulation during hibernation. In gathering nest material the animal collects a pile of dried grasses or leaves and carries it in its mouth – although some debris may be accidentally picked up by the spines, nothing is deliberately carried on them. The ancient story that hedgehogs carry off fruit by impaling it on their spines is a widespread myth, believed not only by European country people but also, according to Liu [300], by peasants in northern China.

Hedgehogs are nearly naked at birth, but the spines, though soft, already protrude from the skin, into which they are drawn by an oedematous swelling of the tissues which subsides soon after birth. The prenuptial display by the male is prolonged, and consists of repeated butting of the female with the spines over the brow erected by the *fronto-cuticularis* muscle; the female retaliates, and the sparring continues, sometimes for hours, before the female becomes receptive; the display has often been mistaken for fighting between males.

The peculiar habit of anointing the spines with saliva, which species of all genera of hedgehogs share with the similarly spiny tenrecs of Madagascar, is elicited by contact with many strong smelling substances, whether likely to be met with in the wild, such as carrion, a toad, or even a strange hedgehog, or artificial, such as tobacco smoke or scented soap. The animal smells, and often chews, the object and then dribbles copiously, working the saliva into a froth with the tongue and lips, and then spreading it on the spines of the shoulder and back. The meaning of this performance remained a puzzle until Dimelow [111] watched captive hedgehogs, and found that it is part of their normal washing routine; on being removed from its sleeping box and taken outdoors at early dusk a hedgehog soon started washing itself: 'First the hedgehog whipped up his saliva into a foaming froth inside his mouth and then he licked the foam on to the soft fur on the underside or the spines of the back.' The presentation of a strong smell thus stimulates the animal into making its toilet at other times of day; the function of the foam however, is still unexplained – the manufacturers of modern non-foaming detergents have to add artificial foaming agents to make their products acceptable to ignorant housewives, who do not understand that such detergents are not soaps, but hedgehogs do not need to deceive themselves. It could be that the foam tracks down the spines to wet the skin where the animal's tongue cannot reach. It may be that salivation in the presence of food leads to feeding, whereas salivation produced by a stimulating odour in the absence of food leads to washing.

Hedgehogs have been much used in studies on the physiology of hibernation, as discussed in Volume I; species inhabiting regions with cold winters hibernate, whereas those living in arid zones usually aestivate – even during the seasons of activity there is a difference of one or two degrees between the temperature of the body when sleeping and when foraging. Hedgehogs are also used as laboratory animals in the study of foot-and-mouth disease, to which they are susceptible and of which they may be vectors. Hedgehogs are peculiarly resistant to poisons, not only natural ones such as bee, wasp and snake venoms, but artificial ones such as arsenic, cyanide and others. They are much more resistant than other mammals, but not immune; Herter points out that a hedgehog presents the erected frontal spines when attacking an adder so that the snake is unable to penetrate to the skin when striking with its fangs, but if a lucky strike reaches the unprotected skin the animal may sicken and die in the course of a few days.

The genus *Erinaceus*, with about six species and many subspecies is widely spread throughout Europe, Asia and Africa; *Hemiechinus*, the long-eared hedgehogs, has two species which inhabit arid regions from Egypt to Mongolia; *Paraechinus*, containing the three desert hedgehogs, extends from north Africa to India and Afghanistan.

Herter, already referred to, gives a splendid general account of hedgehog biology.

In contrast the moon rat *Echinosorex gymnurus*, and the few other genera segregated with it into a separate subfamily, are entirely without spines. The moon rat is almost as big as the American opossum *Didelphis*, and bears a striking superficial resemblance to it, especially when it opens its large mouth showing a long gape in threat if it is alarmed; it also has similar long, loose and rather harsh pelage with a dense underfur and, as its name implies, a naked tail. The head is whitish with dark bands through eye and ear, and a dark bar across the forehead; the rest of the body is nearly black but the end half of the tail is whitish. The animal is found in Thailand, Malaya, Sumatra and Borneo. Lim Boo Liat [298] found that in Malaya it is widely distributed in lowland forest, and ranges into mangrove swamps, rubber estates and other fringe areas. Moon rats lie up during the day in rock crevices, hollow logs and under roots of trees; they are good climbers. They feed on earthworms and arthropods, but also take fish, crabs and land molluscs – Lim used bananas as bait for trapping them on the banks of streams in the forest, and kept some in captivity for seven years on a diet mainly of fresh fish. He records that they are aggressive animals, and that 'they would dash forward with open mouth if attacked or teased. They have a noticeable odour all the time, but become especially offensive when they are excited or attacked'. The odour is produced by the anal glands, and has been likened to rotten onions or stale sweat.

Hylomys suillus, the sole species of its genus, which has the book name of 'lesser gymnure', is a small shrew-like creature with soft fur, a long snout and very short tail. It has a similar but rather wider distribution in south-east Asia, where it forages by night over the forest floor feeding mainly on insects and earthworms. Like the moon rat, it has a strong unpleasant odour. Three further genera of small shrew-like mammals are included in the subfamily, *Neohylomys* from Hainan, *Podogymnura* from the Philippines and *Neotetracus* from Viet Nam and southern China, each with a single species. Although the last two at least are common in their habitats, little appears to have been recorded about the biology of any of them.

FAMILY SORICIDAE

The shrews are a numerous tribe, and form by far the largest family of the insectivores, some 300 species being classified in twenty to twenty-four genera. They are all small animals, the largest no bigger than a rat, and include the smallest living mammals. They inhabit the whole of Europe, Asia, Africa, North and Central America, and extend into the north-west corner of South America, but they are absent from the remainder of South America and Australasia.

A zygoma is lacking in the skull of shrews, and the dentition characteristically shows a large procumbent first incisor in both jaws, the upper one with two hooked cusps projecting downwards; a number of small unicuspid teeth of uncertain homology lie between them and the last premolar which is followed by three molars – the milk teeth are shed or resorbed before birth. In most shrews, but not all, there is a cloaca; the eyes are small and the fur short and velvety. A long snout well supplied with vibrissae is characteristic, and some species have subcutaneous stink glands on the flanks. The pelvic girdle forms an incomplete ring, because the pubic bones do not meet ventrally so that there is no symphysis, a feature found also in moles and many bats. The jaw-joint of the shrews is peculiar; the condyle of the mandible is large and bears two facets forming hinges with the skull. Fearnhead *et al.* [143] find that this condition is correlated with the absence of a zygoma, together with the great length of the skull behind the jaw joint; of the muscles that close the jaw the *temporalis* is large and pulls mainly backwards in contracting, and the *masseter* muscle is reduced. These authors also point out that although the movement of the lower jaw 'is predominantly in a vertical plane, there is also the possibility of considerable fore-and-aft movement.' One cannot help thinking that perhaps the fore-and-aft movement allows the lower jaw to move forward and the incisor teeth to have a forceps-like action when picking up small prey, which is then chopped up by the sharp-cusped posterior teeth – the jaw joint allows of little or no rotary chewing movement.

Shrews are typically terrestrial mammals foraging in and under the litter in woods and the mat of vegetation below herbage, but some of them are aquatic, one species even having webbed feet. All of them dig burrows and make underground nests of plant material as dwellings and for cradling their young, but none of them hibernate. Although mainly tactile animals they use their voices for communication, probably mostly as a warning to others of their species, because they are habitually solitary, and aggressive to strangers – the high-pitched squeaks, screams and twitterings may be heard by day and by night. Possibly, like some small rodents, they also produce, and hear, ultra-sounds in communication. Gould [177] has recently found that shrews of the genera *Sorex* and *Blarina* emit pulses of ultra-sound which serve as a crude means of echolocation; the sounds appear to be made by the larynx, and the frequency of the pulses ranges from 25 to 60 kcs. They fulfil their designation as insectivores, but also eat any small animal, vertebrate or invertebrate that they can overpower. Crowcroft [90] found that the European common and pygmy shrews (*Sorex araneus* and *S. minutus*) prefer certain species of woodlouse to all other prey, and dislike millipedes – an exactly opposite preference to that of the hedgehog. Shrews also eat some vegetable matter such as seeds and fruits, and this addition to the predominantly animal diet appears to be essential for at least some species. Moore [332] has shown that some North American shrews of the genera *Sorex* and *Blarina* sometimes take a preponderance of vegetable matter, and are particularly fond of the seeds of the Douglas fir in Oregon, where consequently they may be responsible for the relative scarcity of this species among the dominant conifers of other kinds.

Food is found by the tactile vibrissae and is examined by the nose; the eyes seem to be almost functionless. Crowcroft further discovered that shrews 'refect', though the refection is not like that found in the rabbit and some rodents, where faeces are taken from the anus. The shrew refects only when the intestine is free of faecal pellets; the animal doubles itself up and, with some abdominal straining, everts the rectum to as much as a centimeter, whereupon it nibbles and licks it 'for some minutes, after which it is drawn back into the body by muscular contractions . . . refection in shrews may be connected with the conservation of vitamins, proteins and trace elements'.

Shrews are very active animals and consume large amounts of food; a study by Pearson [369] on several species of American shrew showed that their metabolic rate is higher than that of small rodents of equal weight. These results, however, have not been confirmed in the European species which Hawkins and his co-workers [217, 218] examined. They found that the amount of food eaten daily is large, and often exceeds the weight of the animal, but it is juicy animal food and thus its dry-weight and calorific value differ little from the lesser quantities of more concentrated food, such as seeds, eaten by some small rodents. Crowcroft [88] found that *Sorex araneus* and *S. minutus* have alternating periods of activity and rest, lasting one or two hours throughout the day and night, with peaks before dawn and after sunset. Digestion is rapid; the gut may be emptied in three hours, and as the 'small body contains available reserves for only an hour or two more, feeding at frequent intervals is essential.' These facts, then, give an explanation for the notorious voraciousness of shrews, and their liability to die of starvation if left without food for a few hours. Common shrews are often found lying dead in the open during the autumn. These are all old adults that have bred in the previous summer; their teeth are badly worn, they are not moulting into a warm winter coat, and they appear to be generally worn out. The immediate cause of their death seems to be starvation, and their bodies remain uneaten because the secretion of the lateral skin glands makes them unpalatable to many carnivorous animals – they are eaten by owls which, like most birds, have a poor sense of smell, but owls in the wild take only living prey. The life span of most small mammals is short; that of shrews is little more than a year and a half, if they are not killed sooner. The young, born during the summer, become sexually mature the next spring, breed – one to three, usually two litters are produced – and die in the following autumn. A strange thing is that young shrews grow up during their first summer, and then grow down during the following winter; they grow up again in the next spring. Pucek [381] showed that in many species of *Sorex* the height of the brain-case is reduced as much as 17 per cent during the winter, and correlates the changes with a shrinkage in the size and volume of the brain. Another peculiarity of young shrews, recorded of some species of *Crocidura* and *Suncus*, is 'caravanning'; when old enough to leave the nest they form a line, each holding the rump of the one in front by its teeth, and the foremost similarly fastened to the mother, behind which the whole caravan trundles along.

The bite of some shrews is venomous so that the prey is killed or paralysed before being eaten. Pearson [368] showed that the submaxillary salivary glands of a single specimen of the American *Blarina brevicauda* produce enough poison to kill by intravenous injection 'nearly 200 white mice or several cats and rabbits.' Histologically similar glands have been found in numerous other species of shrew, but venom of similar potency has not been demonstrated in their secretion. Nevertheless in the European water shrew, *Neomys*, Cranbrook [86] found that the poison enabled it to minimise the struggles of its larger invertebrate prey, and there is reason to think that some toxic action on invertebrates may occur in other species.

The number of chromosomes in the cells of mammals is usually a constant number characteristic of the species, but some variations are known, particularly in the shrew *Sorex araneus*. Ford, Hamerton and Sharman [149] found that in this species the chromosome number varies from 21–27 in males and 20–25 in females, and that there is a multiple sex-chromosome mechanism, the males being XY_1Y_2 and the females XX. Similar chromosome variation is known in some invertebrates, but the shrew provides the first known example in mammals; its biological significance is still unknown – it makes possible the occurrence of 27 different chromosomal patterns, of which 19 have actually been found.

Finally, before glancing at the different genera of this highly interesting family of insectivores, it should be said that the old stories of shrews dying of shock at a sudden noise or clap of thunder, or from fright, are nonsense; shrews soon die of starvation if they are denied access to food for a few hours, but in other respects they are hardy, active and robust little creatures.

Shrews of the subfamily Soricinae are distinguished by the coloured enamel of their teeth; it ranges from yellow through red to brown and dark purple. *Sorex* is a genus of many species distributed throughout Europe, Asia and North America, and includes the American water shrew, *S. palustris*, which has a fringe of hair on the hind feet thought to increase the area of the foot for swimming but also, according to Conaway [79], used 'almost as a comb' for drying the fur on leaving the water. *S. minutus* is one of the smallest mammals but the American pygmy shrew, *Microsorex hoyi*, the only species of its genus, is probably the smallest of all. *Soriculus* is an Asiatic genus of rather large shrews with very long tails, and *Neomys* contains the two species of Old World water shrews – although they swim well, the water repellant property of the fur is soon lost, and Crowcroft [89] relates how a bedraggled water shrew 'ran to the nearest burrow mouth and disappeared. Almost at once it reappeared at another entrance as completely dry and glossy as it was before entering the water . . . the tight fitting tunnels squeezed out the water and the soil absorbed it like a sponge.' He points out the similarity of this behaviour to that of the Australian platypus. The American greater and lesser short-tailed shrews, *Blarina* and *Cryptotis*, are sympatric in parts of their ranges, as are the European *Sorex araneus* and *S. minutus*, whose relationships with each other are described in Volume I. The numerous species of *Cryptotis* are small shrews, and those that live in South America are the only insectivores found there. The single species of *Blarinella* is an inhabitant of China and northern Burma, and the two *Notiosorex* species live in the southern United States and northern Mexico. *N. gigas*, a rare Mexican species, is comparatively large and has been placed in a separate genus *Megasorex* by some zoologists because, unlike the other soricines, the enamel of the teeth is not pigmented.

The white-toothed shrews form the family Crocidurinae; *Crocidura* is a genus with many species distributed over Europe, Asia and Africa – some of its species are not entirely solitary, as several individuals share a winter nest. The only species of *Praesorex*, *P. goliath* is the giant of the shrews; it lives in the Cameroons, but only a few specimens have ever been seen. About twenty species of musk shrew, genus *Suncus*, are known; they inhabit Africa, Asia and southern Europe – the Italian *S. etruscus* rivals *Microsorex* for the place of the smallest mammal in the world. The musky smell of these shrews comes from the gland under the skin of the flanks; their tails appear to be much thicker than those of other shrews because of the comparatively dense covering of hair. One species, *S. murinus*, is a domestic pest like the house mouse in many parts of Africa and Asia but, in spite of the damage it does to stored food and its unpleasant smell, it is said to be useful in destroying insects. Two species of Ceylon shrews with long claws on the fore feet have been assigned to separate genera, *Feroculus* and *Solisorex*, but little is known of their habits; it has been suggested that the long claws show that they are burrowers, but so are all shrews, and even the habitually subterranean mole shrew *Anourosorex squamipes* of China, Vietnam and parts of Burma, which has very dense fur like that of a mole, minute eyes, small ears and very short tail, has proportionately shorter claws. The peculiarly coloured piebald shrew, *Diplomesodon pulchellum*, of the Kirghiz steppes, is dark grey with a white patch on the back and white underside and feet; although it feeds on insects it is said to be particularly partial to the lizards of its sandy habitat as an article of diet. Two solely African genera, *Mysorex* and *Surdisorex*, contain each a couple of species; the former is confined to South Africa and the latter to the mountains of Kenya; in both genera there is often an extra, small, tooth between the normal second and third teeth of the lower jaw. Several

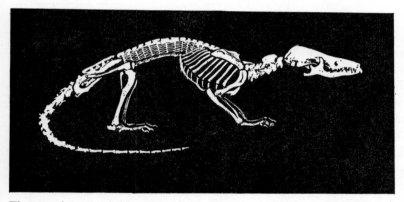

Figure 4 Armoured shrew (*Scuticorex*). Skeleton showing the peculiar bracing of the vertebrae.

species of Asiatic water shrew, genus *Chimarrogale*, inhabit the mountain streams of eastern Asia. Their hind feet are fringed with stiff hairs on both edges, so that the area presented to the water when swimming is increased, but it is only in the Tibetan water shrew, *Nectogale elegans*, that the hind feet are webbed as well as having fringes of stiff hairs along the edges. In addition, the long tail of the last species bears fringes of similar stiff hairs so that its terminal part appears flattened from side to side, and the pads of the feet are said to be adhesive when the animal runs over wet stones.

Finally two species of armoured shrews from tropical Africa, genus *Scutisorex*, though white-toothed, form a third subfamily, the

Scutisoricinae. They are medium-sized shrews with rather long hair, but offer no particularly striking peculiarities to a casual inspection. Their internal anatomy, however, is extraordinary; the details of the vertebral column are like nothing found in any other mammal. The dorsal, and particularly the numerous lumbar vertebrae of which there are eleven in contrast with the five to six of other soricids, are large and provided with numerous excrescences forming a complicated system of articular apophyses, so that the vertebral column forms a strong braced girder. Whether this unique arrangement is correlated merely with the unusually large number of lumbar vertebrae, or with some peculiarity in the animals' way of life, remains unknown. It does not, however, form a rigid structure, for the animals can bend the back both up, down and sideways with some freedom; yet it does provide resistance to great pressure from above, as related by Allen [5] who says that a man weighing 160 pounds can stand barefooted, balancing himself on one leg, upon the back of a *Scutisorex* which emerges from the ordeal 'none the worse for this mad experience'.

FAMILY TALPIDAE

The moles are a small family of less than two dozen species, although classified into fifteen genera. They inhabit most of Europe and Asia south of 63°N, but do not extend into Arabia, Persia or India; they also inhabit much of North America but are absent from some of the mid west. They are small mammals, the largest (*Desmana*) about the size of a rat; they have dense velvety fur and have a true zygomatic arch in the skull. Most species are subterranean, and nearly all those that are surface dwellers are burrowers, even the aquatic *Desmana*. Those most highly modified for subterranean life are the Old World moles of the genus *Talpa*, which includes the common European *T. europea* and its subspecies found over much of Eurasia, the Chinese genus *Scapanulus*, and the North American genera *Parascalops*, *Scapanus*, *Scalopus* and *Condylura*. In all of these the fore limb and hand are specialised; in the more extreme types such as *T. europea*, the shoulder girdle is shifted forwards alongside the neck, and the front joint of the sternum is prolonged forwards and keeled, thus giving a large area of attachment for the pectoral muscles. The humerus is short and wide, quite unlike that of most mammals, and by rotation on its long axis it produces a backward, scratching action of the forearm and hand in tunnelling. All the books tell us that the size of the hand is increased by the presence of a sickle-shaped radial sesamoid bone, but few of them mention that there is an equivalent tibial sesamoid bone in the hind foot so that there is a small second great toe; Yalden [526] notices this bone, but it was known to the anatomists of the late eighteenth century, referred to by Bell [29] in the 1830s. It increases the size of the foot, and may be useful in gripping the sides of the burrow when tunnelling but it has not been shown to be of any special importance to the animal; it is almost as though the genetic code had called for 'an extra toe' but had forgotten to specify 'in the fore limb only'. Most species of mole throw the excavated earth from their tunnels on to the surface as mole-heaps or 'heaves', pushing it up from below with the palm of the hand; some species, however, do not burrow deeply but make sub-surface runs, raising the earth above into a ridge without making mole heaps. The nest, made of dry leaves and grasses, may be below, at, or even above ground level, and is covered with an extra large heap, the so-called fortress; on low-lying land the burrows are liable to flooding, but when necessary moles can swim strongly. They appear to leave their burrows and forage on the surface more often than is generally appreciated.

The food of most species consists mainly of earthworms, and stores of worms, semi-immobilised by biting the head end, are sometimes accumulated; although the salivary glands are large they have not been shown to produce any toxic substance that might paralyse the victims. The diet of worms is supplemented with insect larvae and other invertebrates, which are of equal or greater importance in some species. Moore [331] found that the moles of the genus *Scapanus* on the Pacific coast of North America eat large amounts of vegetable matter, even when worms are plentifully available, and are a nuisance to the growers of bulbs and garden crops. Moles, like shrews, need a comparatively large amount of food daily owing to the high water-content and low dry-weight value of the animals eaten.

The breeding season of moles is generally short, and intense sexual activity is concentrated into a brief period. During much of the year the sex glands and reproductive tracts of both sexes are very small and show no activity, but as the breeding season approaches they undergo enormous proliferation. They regress again as soon as the season has passed. In the female of *T. europea*, and no doubt of other species, the sexual and urinary openings are separate, the latter being at the tip of a perforate peniform clitoris. The sexual passage is open only during the breeding season, after which it closes and remains sealed with scar-tissue until it spontaneously opens again in the following year. As in the shrews, and some other mammals, the right and left pubic bones do not meet or join in the mid line to form a symphysis. The ovary of the European mole and of the Pyrenean desman [376] have a strange cycle of activity; the part that produces egg cells is much reduced except at the breeding season, and at other times the rest of the gland predominates, and shows a remarkable

E

resemblance in some of its details to parts of the testis in the male. The meaning of these ovarian changes, which produce a condition somewhat resembling the abnormal intersexes sometimes found in other mammals, is not understood, but they are probably closely connected with the action or production of hormones.

The eyes of moles are small, and in some species they are covered with skin. The retina, however, contains all the different elements normal in the eyes of mammals but, because it is so small, in far fewer numbers. The eyesight seems poor by human standards, such as the ability of quickly recognising patterns, but a light receptor may be important to an animal living much in the dark to enable it to know when it comes out or closely approaches the surface. The sense of smell seems to be poor so that the animals rely mostly on the senses of hearing and particularly of touch in foraging and in exploring their surroundings. The elongated snout contains great numbers of minute neuro-cutaneous tactile receptors – Eimer's organs – in addition to vibrissae on the nose and wrist. The tactile papillae are concentrated on twenty-two pink fleshy tentacles that surround the nostrils in the star-nosed mole *Condylura cristata* of eastern North America. This species lives in marshy ground near streams and lakes; although it tunnels and throws up heaps it obtains as much as four fifths of its food from the water, in which it dives and swims expertly. Hamilton [196] found that leeches and other water-worms preponderated over earthworms in the diet, and that aquatic insects also formed an important part of it. The tail of *Condylura* is much longer than the short stump typical of the other moles, but according to Hamilton it plays only a minor part in swimming which is done by rapid alternate strokes of the fore limbs.

Much interesting information about the biology of the European mole is given by Godfrey and Crowcroft [171] and in the symposium edited by Quilliam [384].

The desmans are talpids which are highly aquatic; the fur is dense, the tail fairly long and flattened from side to side, the fore feet are partly, and the hind feet fully, webbed and both are fringed with stiff hairs. The eyes are small, and the snout is elongated, flattened, highly mobile, and well provided with long vibrissae. The large Russian desman, *Desmana moschata* which gets its specific name from the strong, musky scent secreted by a gland at the base of the tail, inhabits eastern Europe and western Asia. It digs burrows in the banks of the streams and ponds from which it obtains its food consisting of various aquatic invertebrates – it is trapped for its commercially valuable fur. The smaller desman of the Pyrenees and parts of Portugal, *Galemys pyrenaicus*, is confined to swift running mountain streams; it is one of the few talpids that does not burrow, but lives in natural hollows among stones and tree roots on the banks. Peyre [376] found that the food consists of aquatic insect larvae and small crustacea, and that, like the other moles, it needs a large amount of food daily but that, unlike moles, it is active mostly at night and little by day.

The shrew moles, of several genera, are comparatively long-tailed; their fore feet are not broadened into fossorial palms, though the claws are well developed, especially in the rare 'long-tailed mole' *Scaptonyx fuscicaudus*, of southern China and north Burma. Although they are shrew-like in general appearance the character of the skull, with a zygoma, and of the dentition in which the incisors are not procumbent, at once shows they are talpids and not soricids. The Asiatic shrew mole *Uropsilus soricipes* has a similar range to *Scaptonyx*, but the front claws are not broadened, hence its specific name. The American shrew mole, *Neurotrichus gibbsi*, of the humid Pacific coast of North America is a small species which makes surface runs as well as burrows; it appears, however, to feed mainly underground, for Reed [396] found that a captive specimen ran back and forth unconcernedly over the bunch of writhing worms that he had put into the cage. 'Only when the worms were packed in damp soil in the bottom of a shallow container, where the shrew mole could dig them out, would it eat them.' Two allied species, of the genus *Urotrichus*, are found on the opposite side of the Pacific in Japan; they make shallow burrows but also forage on the surface, and even climb into bushes in search of their invertebrate food.

FAMILY MACROSCELIDIDAE

One glance at a living member of this family at once convinces the observer that the doubts, based on anatomical studies, expressed by systematists about the taxonomic position of the family are fully justified. The large eyes, the long legs and the general stance and comportment are quite unlike the appearance and action of the other insectivores. The development of the zygoma and orbit, among many other characters of the skull, the long pubic symphysis, and the presence of a caecum in the intestine, point to an affinity with the tree shrews of the East, which are now classified as lower Primates – as has been mentioned, some zoologists think that this family should join them in that order. Indeed, van der Horst [481] implied a close relationship with the Primates when he described the abnormal implantation of an embryo which resembled that normal for man, in a paper entitled '*Elephantulus* and homunculus' which, however, was re-labelled by a grim-faced publications committee as 'A human-like embryo of *Elephantulus*'.

The elephant shrews which superficially look like long-nosed rodents, are confined to Algeria, Morocco and Africa south of 15°N, whence something over two dozen species are classified into four or five genera – a large number of species and subspecies of doubtful validity has been named, causing a confusion that needs thorough taxonomic revision.

The elephant shrews are mostly mouse-sized mammals, though a few species are larger; all have a long mobile snout with the nostrils at the end, and long legs, the hind ones longer than the fore. The characteristically large eye is often made even more prominent by a ring of light-coloured hair surrounding it. The coloration of most species is various shades of brown above, whitish below, but *Rhynchocyon* is darker both on back and belly. The ears of all species are prominent, and all species have a gland secreting a musky scent on the under side of the tail.

Macroscelidids are extremely alert and agile creatures, which dash swiftly for cover when alarmed. The very rapid bounding or jumping action has misled many observers into thinking that the animals hop on the hind legs, like jerboas or miniature kangaroos, and has given them their alternative name of 'jumping shrews'. This is a misnomer for recent careful observations [56] on wild specimens have shown that the jump is a four-legged one, not a two-footed hop, and the animals always land on all fours. Throughout their range different species occupy the most varied habitats from sandy plains with sparse vegetation, rocky hills, scrubland and thickets, to tropical forests. All the species follow well-worn runways in moving about their territory; some species make nests in burrows of their own digging, whereas others nest in surface depressions under rocks or logs, or in the shelter of thick vegetation. Most species are diurnal and crepuscular, but some are also active on moonlit nights. The main diet of all species is ants, but other insects and perhaps other invertebrates, are taken as well. Many species like basking in the sun; Shortridge [442] quotes FitzSimons 'I have often noticed what was apparently a small round stone on top of a boulder which would turn out to be a jumping shrew with body bunched up, immovable, but very wide awake; for, on realising that it had been observed it would bound off so rapidly that only a momentary glimpse of it would be seen.'

Although elephant shrews may be numerous, most species are not social animals; each lives by itself. The number of young in a litter is never more than two, and frequently only one. They are well developed at birth, fully furred and with the eyes open, and are able to accompany the mother when only a day or two old.

Of the two species of *Macroscelides*, *M. proboscideus* is found on the karoo and sandy thorn-bush plains of the Cape and South-west Africa, and

M. (Nasilio) brachyrhynchus, with a shorter snout, is widespread north-wards to Kenya and the Congo. Various species of *Elephantulus* are found over the whole area of distribution of the family and include the north African species, *E. rozeti*. The giant elephant shrews, *Petrodromus* and *Rhynchocyon* are forest dwellers and differ in having only four toes on the hind feet; several species of *Petrodromus* are found throughout tropical and subtropical Africa and four or five species of *Rhynchocyon* in central and East Africa. The last is the most bizarre of all the family; it is the largest, and has long legs and tail, and a very long pointed snout. Brown [56] is one of the few zoologists who have recorded observations on it in the wild. Speaking of *R. cirnei*, which, although it runs with considerable speed, was never seen to jump, he says that a pair of animals appeared stealthily out of the undergrowth when a drive to catch specimens was under way, and 'They stood, strangely bizarre on their disproportionate stick-like legs, with every sense obviously alert. They had certainly sensed my presence and were ready to react to it when a further chorus from the beaters sent them rushing into the net.' He found that the coastal species *R. petersi* and *R. chrysopygus* move slowly over the litter of the forest floor:

The head is held up and the long trunk is deflected down to sniff the litter. In this way the animal is very much on the alert and the slightest noise or untoward movement will send it off. Undisturbed, it will quarter the ground and at suitable places the litter is turned over with the nose and scuffled up by the fore-feet. This scuffling of the litter is surprisingly loud and quite distinctive.

Much valuable work has been done on the embryology and placentation of *Elephantulus* by van der Horst and his collaborators, and on various aspects of the anatomy of the macroscelidids by numerous zoologists; more detailed studies similar to those of Clevedon Brown on the behaviour of the animals are much needed.

ORDER DERMOPTERA

This small order contains the single family and genus of the mis-called 'flying lemurs'; as they are not lemurs and do not fly the growing tendency of zoologists to use 'colugo', one of the many native names for the animals, is welcome. The colugos have no close affinity with any other mammals – early Eocene fossils from North America show that they have been differentiated for many millions of years. They are regarded as an offshoot of primitive insectivore stock, but so are most of the eutherian orders; and they have no nearer relationship with the lower primates or the bats, with both of which they have been associated by naturalists in the past.

The skull of the Dermoptera is rather broad and flattened, but its most striking feature is the character of the teeth. The lower incisors are procumbent and pectinate; in particular the first and second bear five to a dozen or more incisions reaching to their base, which divide the teeth so that they are like combs. The canines resemble premolars – Wharton [504] calls them 'shark-like premolars', and says that in some specimens they project from the mouth during life. The teeth are further peculiar in that the posterior incisors and the canines are double-rooted, a feature found also in the hedgehogs and moles.

The outstanding character of the colugos is, however, the patagium by means of which they glide from tree to tree. It is more extensive than in any other gliding mammal; it extends from the side of the neck near the ear to the tips of the fingers which are webbed to the bases of the claws, thence to the similarly webbed toes, and then to the tip of the fairly long tail. The patagium consists of two layers of skin and is furred on both surfaces, but is so delicate that Wharton could 'almost read through it.' A single young is born annually and is carried lying across the underside of the mother until it is as much as a third the size of its parent; the nipples lie in the axilla.

FAMILY CYNOCEPHALIDAE

The family contains the single genus *Cynocephalus*. For long the accepted generic name was *Galeopithecus* and the family name followed it, but unfortunately the law of priority enforced the change, which was reluctantly accepted by zoologists. Although the name implies that the head is like that of a dog, Wharton says that the lips are muscular and calf-like and that 'One is struck by the similarity of a [flying] lemur's head to that of a young ruminant.' Two species of *Cynocephalus* are recognised, *C. volans* from the Philippines and *C. variegatus* from much of Malaysia and Indonesia. They are similar in form but *C. variegatus* has irregular white spots on the upper surface of the soft dense fur which Wallace [496] likens to that of the chinchilla, and which blends with the colour of a lichen covered branch. The fur of the males seems always to be darker and browner than that of females, which is lighter and greyer (Plate 6).

Colugos are animals of the forest, and are strictly arboreal for they have to take off from a height when gliding to another tree by using the patagium. Wallace saw one glide a distance which he paced out at seventy yards, in the course of which the animal lost thirty-five or forty feet of height, or less than one in five. Unlike other gliding mammals such as flying squirrels and flying possums, which tuck their patagia away so that they are inconspicuous when not in use, the colugo cannot do so, and looks as though wrapped in a soft woolly blanket when it is not gliding – Wharton, however, states that when the animal is climbing about a long 'string-like muscle about 2 mm. in diameter connects the shoulder to the edge of the web and draws the membrane down under the fore legs so it will not be caught on branches'. Colugos are so completely arboreal that they are clumsy on the ground, and Wharton has seen them 'flop across the floor and start up a post supporting a house as if it were a tree'. Wallace found the animals sluggish by day, 'going up a tree by short runs of a few feet, and then stopping a moment as if the action was difficult'. Wharton disturbed one in its den from which 'it dropped very steeply for perhaps twelve feet, then swooped across to land on the mamacao tree.... It hitched rapidly upwards, keeping the tree between itself and its observers, and as soon as it had gained sufficient altitude it flew to another tree'.

The diet of the animals is entirely vegetable; Wallace reported of *C. variegatus* that it consists of leaves, but doubtless other things are also eaten, for Wharton found that *C. volans* eats leaves, flowers, fruit and seeds, and is particularly fond of the young seed pods of kapok, *Ceiba pentandra*. Correlated with this diet, the premolar and molar teeth are suitable for shearing or chopping up tender food but not for grinding hard substances, the stomach is large, and the intestine long and voluminous – as Wharton says, 'The entire make-up of the animal is adjusted to a vegetable diet of considerable bulk.' When walking about among branches the colugo always hangs upside-down like a sloth, and uses the fore foot to pull leaves within reach of its mouth; 'He then shears off leaves or portions of fruit, or pulls the leaf off and chews it in like a rabbit'.

Colugos are nocturnal and spend most of the daylight hours asleep. The Malaysian species either hangs suspended from a branch or clings head upwards to a tree trunk, which it clasps with arms and feet armed with needle-sharp claws, as reported by Wallace and by Harrison [213]; on the other hand the Philippine species, according to Wharton, always sleeps in a den in a hollow tree trunk, clinging head upwards to the inner surface.

Many writers on natural history state that the peculiar lower incisors are used by the colugo as combs for grooming the fur, but there seems to be no authoritative original observation on this matter. Wallace, who was familiar with the animals in Sumatra, Borneo and Singapore makes no mention of it; and Wharton who handled a large number in the Philippines says, 'They are clean. When wet or dirty they cannot rest until they have groomed themselves thoroughly', but speaks only of them licking their fur. When defaecating or urinating the colugo turns the tail, which is usually tucked

71

forward over the belly, upwards over the back while hanging head upwards so that the fur is not soiled.

Wharton says, 'Lemurs are known to be of a low order of intelligence' and Wallace, 'The brain is very small, and the animal possesses such remarkable tenacity of life, that it is exceedingly difficult to kill it by any ordinary means'. Nevertheless, colugos were gliding about among the trees millions of years before man appeared on earth, and are common animals over vast areas of the East – they may yet have the last laugh.

Chapter 3

Bats

THE ORDER Chiroptera is the second largest among the mammals; it contains nearly nine hundred recognised species of bats, distributed throughout the world with the exception of the coldest regions. The bats are the only mammals capable of true powered flight, and they have been flitting about the sunset skies for at least fifty million years, for a fossil bat is known from the early Eocene. They almost certainly originated from a primitive insectivore-like stock, but their fragile bones are so unsuited to preservation as fossils that the record is almost non-existent. The beautifully preserved fossil bat, *Icaronycteris index* [255] from the Green River formation of Wyoming, U.S.A. is an exception, and is one of those lucky accidents that lead us to hope that further material may one day be found. Its main point of interest is that although assignable to the microchiropteran bats, it has a claw not only on the thumb but also on the second digit. Other fossil bats are referable to living families. The general structure of bats and their wings, their extraordinary powers of echolocation and the peculiarities of their reproductive processes, have been discussed in Volume I and do not need repetition here, where some of the characters of the various families are considered.

Although the form of all bats is so similar and characteristic that a bat can be recognised at a glance by anyone, the details of their anatomy and the variety of their habits is as diverse as would be expected among such a vast number of species. In size they range from animals with the body size of a rabbit and a wing span of over four feet to minute creatures smaller than a mouse with a wing span of less than six inches. In many families of bats a bizarre elaboration of the tissues of the face puts some of them, by human standards of beauty, among the most ugly – even hideous – of the mammals in facial expression. In some families of microchiroptera the skin of the snout surrounding the nose is proliferated into a peculiar nose-leaf which stands free of the face and is often thrown into complicated folds and convolutions. In others, although there is no nose-leaf, the sebaceous glands of the muzzle are very large and form the pararhynal glands, which throw the surface of the snout into conspicuous lumps and pads of characteristic shape for each species.

In most female bats, and in the males of some species, the pubic symphysis is incomplete, so that the front arch of the pelvis is represented only by a ligamentous connection between the pubic bones. This condition, one would think, makes for easy parturition, but in those instances where parturition has been watched it is accompanied by vigorous straining and muscular contractions much as in other mammals in which the symphysis is present. Weber [501] suggests that this condition is correlated with the power of flight and the attachments of the patagia; this explanation leaves out of account the presence of a symphysis in the males of some species, and furthermore overlooks the similar absence of a symphysis in the moles and other insectivores. On the other hand it may be correlated in some way with the concentration of the locomotive effort in the fore limbs of both, unrelated as they are, although the presence of a symphysis in some male bats and in the females of the family Rhinolophidae still remains unexplained.

The bats fall naturally into two suborders, the Megachiroptera or fruit bats, the larger of which are also known as flying-foxes, and the Microchiroptera or insectivorous bats, many of which are not insectivorous. The largest bats are found among the Megachiroptera, but some of the smaller members of the suborder are exceeded in size by some of the larger Microchiroptera.

SUBORDER MEGACHIROPTERA

The Megachiroptera are distinguished not only by their generally larger size but by the presence in most species of a claw on the second digit, the forefinger, as well as one on the first or thumb, and by the tail which is rudimentary, very short or absent, except in one species. The eyes of the fruit bats are comparatively large, a character doubtless correlated with the fact that these bats do not use echolation with the exception of one species, though they are said to locate their food by the sense of smell. Many species are gregarious and live in large communal roosts, known as 'camps' in Australia, generally fully exposed to view in trees, but some species roost in caves or even the roofs of buildings – a roost of fruit bats hanging from the branches of a tree gives the appearance of a large crop of pendulous fruit like the 'pudding tree' on a willow-pattern plate. Most of the fruit bats feed on fruit, flowers and nectar, some species being restricted to the last. The large fruit-eaters can be very destructive in plantations which they may visit every night in hordes from roosts as much as fifty miles away. In feeding they chew the fruit and then spit out the fibrous matter and swallow only the juice and the softest pulp. The palate is crossed by a number of ridges against which fruit tissues are squeezed by the tongue. Fruit bats have loud voices and utter a variety of grating cries very different from the high-pitched squeaks of the Microchiroptera. The fruit bats, too, show sexual dimorphism; in many species the males are larger and more intensely coloured than the females, and have bigger canine teeth – in some species they have large pharyngeal sacs or pouches extending under the skin down on to the chest. Some of the Australian species of fruit bat are migratory and make long journeys to different quarters according to the seasons.

FAMILY PTEROPODIDAE

There is but one family of fruit bats, the Pteropodidae, divided into four subfamilies containing some 34 genera and about 150 species. They are confined to the tropics and subtropics of the Old World, and extend from the west of Africa through southern Asia to the east of Australia. The genus *Pteropus* contains most of the species including the largest 'flying foxes'; they take their name from the long muzzle characteristic of the family. The colour is black or brown to grey, with a lighter or yellowish mantle on the shoulders of most species. In some there are conspicuous markings of dark bands on the face and round the eyes.

Pteropus giganteus, the flying fox of India, Burma and Ceylon is the largest species, reaching a wing span of 50 inches and a weight of $2\frac{1}{2}$ pounds; it forms an article of diet for some natives. Wallace [496] ate flying fox, probably *P. vampyrus*, with the inhabitants of the island of Batchian of the Moluccas, between Celebes and New Guinea, and said, 'These ugly creatures are considered a great delicacy, and are much sought after. . . . They require to be carefully prepared, as the skin and fur has a rank and powerful foxy odour; but they are generally cooked with abundance of spices and condiments, and are really very good eating, something like hare.'

Although the climate of Colombo in Ceylon is equable, with little variation in temperature and no great seasonal variation in precipitation, Marshall [310] found that *P. giganteus* has a very distinct breeding season, with the young being born in late May and early June; the fertility of the males also fluctuates seasonally. Marshall found that although there are no seasons in the European sense, the breeding season is correlated with the slight decrease in day-length in December. Pregnancy lasts about six months, and for about three months 'the rapidly growing young one is carried over long distances, clinging to the fur of the ventral surface of the mother.' An African species, *Eidolon helvum*, which in Uganda lives in roosts that sometimes contain 250,000 bats, has been found [342] to have a delayed implantation period of about three months; this is a point of particular interest because, as discussed in Volume I, delayed implantation is characteristic of the microchiropteran bats of the temperate regions that hibernate. The ecological importance of its occurrence in a tropical fruit bat is not at present understood.

The most detailed investigations on the biology of fruit bats, however, are those of Ratcliffe [392, 393], who was engaged on research directed to finding ways of controlling the vast mobs of fruit bats destructive to Australian growers' crops. Four species are common along the east coast of Australia where they have their camps in rain forest or mangrove swamps; camps vary in size from a few hundreds of bats to many thousands – one species, *P. scapulatus*, forms the largest, which are said to cover up to 50 acres and to contain as many as a quarter of a million animals. A camp half a mile wide and about four miles long has been recorded. Ratcliffe found that although the bats eat enormous quantities of fruit they are primarily dependent upon the blossoms of the various species of *Eucalyptus* as food; they chew the flowers and spit out the fibrous matter, swallowing only the nectar and juice together with some pollen and stamens. As showing how the bats feed only on juice he mentions a fruit bat that 'succeeded in feeding on peaches, despite the fact that a mosquito net had been spread over the tree, by sucking the fruit through the mesh.' Although the bats feed on plant juice they drink regularly, generally lapping up water as they glide over the surface of a river or pond – some are said also to drink sea-water.

The migrations of the Australian fruit bats seem to be correlated with available food; the animals follow the blossoming of the various species of eucalypts, but in addition there are regular north to south migrations in spring and summer, more pronounced in some species than others. The amount of migration varies from year to year so that the fruit-growers of New South Wales have 'good' or 'bad' fruit bat years. In 1926 there was an exceptionally heavy migration of *P. scapulatus* into the southern orchards, and the regular habit of extensive southerly migration was established, together with a change in habit from being mainly a flower-eater to raiding the soft fruits of the growers, thus becoming a serious economic problem. It is thought that the cause of the first heavy migration was a drought in the northern regions, but the changed habits of the bats have persisted into seasons of normal rainfall.

Although the genus *Pteropus* is so widely spread and occurs in Madagascar, it is absent from Africa where species of the genera *Rousettus* and *Eidolon* are its ecological equivalents; as mentioned in Volume I, *R. aegyptiacus* is the only fruit bat known to use echolocation, by clicking its tongue. Three species of short-nosed or dog-faced fruit bats of genus *Cynopterus*, rather small in size, are found from India to the Philippines, and one of them is said to spread the wild date palm by carrying off the fruit and dropping the seeds at a distance [318]. The dispersal of seeds from other fruit has been recorded for several genera of fruit bats. The epauleted fruit bats, of the genera *Epomophorus*, *Micropterus* and *Epomops*, are African and are distinguished by having large glandular pouches in the skin of the shoulder surrounded by tufts of long and usually light-coloured hair – hence their name. The males also possess pharyngeal air sacs under the skin of the neck and chest. These bats, like most of the smaller fruit bats, are not so gregarious in roosting as the larger flying foxes – many of

Figure 5 Hammer-head Fruit-bat (*Hypsignathus*), showing the pendulous lips and lappets.

the small fruit bats are solitary and roost in caves, hollow trees, the roofs of buildings or in the shelter of thick foliage. The lips of epauletted bats, particularly those of the genus *Epomophorus*, are fleshy, pendulous and sometimes folded, but are far exceeded in extravagance by those of the hammer-headed fruit bat *Hypsignathus monstrosus*, the only species of its genus, which lives in the swampy forests of equatorial Africa from Uganda to the west coast. The head of this bat is bizarre in the extreme and is described [495] as having a thick, hammer-shaped muzzle with enormous pendulous lips, ruffles round the nose, a warty snout and a hairless, split chin. This is one of the noisiest of the fruit bats, the males having a pair of inflatable pharyngeal air sacs, and an enormous larynx that fills most of the chest and pushes the heart and lungs back and sideways; the voice is described as a continuous croaking and quacking and is thought to attract the females. Van Deusen [485] has recently recorded a remarkable discovery about this species made by H.A. Beatty, a field collector for American museums. Beatty, who nightly heard the ringing 'kalank' uttered by the males around his hut in Gabon, found that when he threw out the skinned bodies of his specimens in the evening the bats flew down, hovered over them, picked them up, and took them away. One actually roosted in his hut, and took the bird-bodies to its roosting place where it ate them. Beatty also found these bats attacking his domestic fowls. Van Deusen points out that in tropical rain forest many birds roost at night exposed on bare branches and not concealed in shelter as is the habit of most small birds in temperate regions; he suggests that *Hypsignathus* normally catches and feeds on birds roosting in this manner.

The bare-backed fruit bats, about a dozen species of the genus *Dobsonia*, found from Celebes to the Philippines and north Australia, are peculiar among fruit bats in possessing no claw on the second digit; the lower part of the back is apparently bare of fur because the naked patagia join the body along the line of the backbone, but under the wing the body is fur-covered as usual in other genera. There are some eighteen further genera of fruit bats, many of them containing only a single species, and many known by only a few specimens, so that little can be said about their biology. The Pygmy fruit bat, *Aethalops*, of Malaysia may however be mentioned as its head and body length does not exceed 3 inches, and it is thus the smallest of the fruit bats with the exception of *Syconycteris*, one of the macroglossine subfamily mentioned below. Finally there is the single species of the genus *Nanonycteris*, *N. veldkampi* of west Africa, whose native name has been translated as 'little flying cow' – but any animal, even the tiniest that can be eaten is 'beef' to the natives of those parts. This species, like some others of the smaller fruit bats, feeds on nectar which it extracts from flowers with its comparatively long tongue; it does not eat or chew the flowers, and it serves as the pollinater for the flowers of the tree *Parkia*.

The more specialised of the nectar feeding fruit bats form the subfamily Macroglossinae, the long-tongued fruit bats; about half a dozen genera are distributed from India to Polynesia, and one genus *Megaloglossus* is found in west Africa. They have long narrow snouts and very long slender tongues provided with brush-like papillae with which they extract the nectar from the flowers they visit. Some species are solitary, and those that are gregarious do not roost together in large numbers. Although they feed mainly from flowers, at least one species of *Macroglossus* is destructive to soft fruit. The three species of *Syconycteris* of Australia and New Guinea are the smallest fruit bats, one species scarcely exceeding $2\frac{1}{2}$ inches in length of head and body. The single species of *Notopteris, N. macdonaldi,* of the western Polynesian islands is the only fruit bat with a long tail, which is free and not included in the patagium.

Two genera of tube-nosed fruit bats, *Nyctimena* and *Paranyctimena,* found from Celebes to the Solomon Islands, New Guinea and Queensland, form the subfamily Nyctimeninae. They differ from the other fruit bats in having short snouts, with a rounded profile, and the nostrils prolonged as scroll-like tubes for as much as half an inch. The fur is pale in colour, and the wings and ears are covered with yellow spots. These bats feed on the flesh of fruits, and like the pteropine fruit bats chew the pulp to extract the juice. The function of the extraordinary tubular nostrils is unknown, but the suggestion has been made that they may serve some purpose in echolocation, though there is no proof that echolocation is used by these bats. The single species of *Harpyionycteris* from the Philippines is placed in a separate subfamily on the characters of its dentition, but it is a rare species known from only a few specimens, and nothing can be told of its habits.

SUBORDER MICROCHIROPTERA

The suborder Microchiroptera contains the majority of the bats, and is divided into sixteen families of living species which show great diversity of form and habits. Although they are often known as the insectivorous bats in distinction from the fruit bats, insects do not form the diet of a considerable number of species. Echolocation is highly developed among some members of the suborder, in which it is probably universal, but one must remember that comparatively few out of the vast assemblage of species have been tested for the use of echolocation. Many kinds of bats live in caves and sometimes roost in enormous numbers – caves with bat populations of three million have been recorded – and their droppings then form a thick layer of guano on the floor. The droppings are eaten by the larvae of dermestid beetles, which liberate large amounts of ammonia in their metabolic processes so that the atmosphere of the caves becomes intolerable, and sometimes even dangerous, to human beings. Constantine [80] found that the concentration of ammonia was so high in some caves in Texas that the fur of the living bats was actually bleached, and Mitchell [326] found the concentration in similar caves in Mexico to reach as high as 1,800 parts per million, over three times the concentration that human beings can withstand safely for half an hour. The tolerance of high concentrations of ammonia is known in several species [462] and is probably widespread among cave roosting bats. Oxygen consumption and metabolic rate are reduced in bats exposed to ammonia, but their means of avoiding poisoning are not recorded.

FAMILY RHINOPOMATIDAE

The bats of the small family Rhinopomatidae, which contains only one genus, *Rhinopoma*, of three species, are considered to be the most primitive of the microchiroptera. They are distinguished by the long tail which nearly equals the length of the head and body and gives them the name of mouse-tailed bats. The inter-femoral part of the patagium is narrow, and the long tail issues from its free edge. The eyes are large, as are the ears which are connected at their bases. The presence of two phalanges in the second digit of the wing, in addition to the metacarpal present in the leading edge of the wings of most microchiroptera, is also a primitive character. The nose carries a rudimentary 'leaf', an expansion of skin highly elaborated in some families of bats described below. These bats roost in caves and buildings, clinging to wall or ceiling with all four limbs – feet and thumbs. Bats of this family are found from Rio de Oro through Africa north and south of the Sahara to Arabia, and through India to Sumatra. *Rhinopoma microphyllum* lives in large colonies among the ancient monuments of Egypt and is said [495] to have continuously 'inhabited certain pyramids in Egypt for three thousand years or more.' This species, like all of its genus, is insectivorous, and sometimes accumulates large deposits of subcutaneous fat especially at the base of the tail; it has been suggested that the fat forms a food reserve if the animals become torpid during times of drought. The skin of the face is naked, as is that of the lower back and abdomen so that the bat seems to be without trousers.

FAMILY EMBALLONURIDAE

The nearly related family of Emballonuridae or sheath-tailed bats is much larger and contains a dozen genera and about fifty species. The arrangement of the tail is peculiar; it is always short and its basal part is included in the interfemoral patagium, but it then emerges on the dorsal surface of it so that the end part is free. The skin covering the tail, however, is rather loosely attached, so that it has considerable freedom to slide up and down the tail vertebrae. The method of folding the wings is peculiar because when at rest the first phalanx of the third finger bends forwards at the joint with the metacarpal so that the end of the wing is reflexed. The emballonurids are also known as the sac-winged bats because some species have a small pouch in the antebrachial patagium between the shoulder and the elbow. A pouch is present in the wings of the central and South American genera *Saccopteryx* and the closely related *Cormura* and *Peropteryx*, and in *Balantiopteryx*; about a dozen species of the tomb bats belonging to the genus *Taphozous*, found throughout Africa and southern Asia to the Philippines and Australia, also have wing sacs. The inside of the sac is glandular and produces reddish-coloured sebaceous matter with a strong smell. The secretion may have a sexual significance because the sacs are larger in the males than in the females of many species, and are present only in the males of others [455]. Most species of *Taphozous* have in addition a glandular sac opening under the neck, and *Diclidurus* has a glandular pocket in the interfemoral patagium near the point of emergence of the tail. The colouring of most species in the family is generally brown, black or greyish, but there are some striking exceptions. The peculiar little sharp-nosed bat *Rhynchonycteris naso* of Central and northern South America, in which the snout projects like a small proboscis, bears two curved white lines on the lower part of the back; and the sac-winged bats of genus *Saccopteryx* bear two conspicuous white or pale coloured zig-zag lines on the back from shoulder to rump. Some species of *Taphozous* have pale spots on the body, but the most unusual of all are the three species of ghost bats, genus *Diclidurus*, of tropical America, which are entirely white, including the wings. No one has been able to suggest a plausible explanation of what biological advantage the unusual colour may be to these species, or even determine whether there is any. A related genus, *Drepanycteris*, known from only a single specimen from Amazonia, although not white is very pale in colour.

All the emballonurid bats are insectivorous; they roost in caves, rock crevices, buildings and hollow trees, usually in small groups though some species are solitary, and a few species such as some of the genus *Taphozous* form fairly large colonies. In tropical America *Rhynchonycteris* and *Saccopteryx* sometimes roost in the open on tree trunks, under bridges, or under palm and other tree fronds; they do not bunch together but space themselves with regular intervals between individuals.

FAMILY NOCTILIONIDAE

The family Noctilionidae is a small one containing only a single genus, *Noctilio*, with two species, but it is of outstanding interest because of the fish-eating habits of one species, *N. leporinus*. Very few other bats, noted under their appropriate families below, have a similar diet. The Noctilionidae are found in tropical America from Cuba and Mexico to northern Argentina. *Noctilio* has a pointed snout with the tip of the nose projecting over the nostrils, and fleshy lips that form folds and wrinkles round the mouth; the upper lip is thrown into vertical folds under the nostrils, and the under lip forms semicircular folds under the chin – these feature give the bats their vernacular names of bull-dog and hare-lipped bats. The fullness of the lips produces a pouch in each cheek. The ears are long and pointed, and have a small tragus. The fur is rather short and on the back it runs as a central strip leaving the sides naked. The colour of the males is generally bright rufous and of the females dull brown; both usually have a pale stripe along the centre of the hair-strip on the back. The wings are narrow, and extend only to the knee, not to the ankle as is usual in bats. The interfemoral patagium, however, does extend to the ankle and is supported there by an unusually large and strong calcar on each side. The feet in *N. leporinus* are exceptionally large in proportion to the size of the bat and are armed with strong claws. The end of the tail, like that of the emballonurids, lies free on the dorsal surface of the patagium. *N. labialis*, the smaller of the two species feeds solely on insects which it catches when flying low over water. The larger species, *N. leporinus*, about five inches long from snout to base of tail, is the fisherman bat (Plate 7).

The fish-eating habit of *N. leporinus* has been suspected for nearly a hundred years but was not finally confirmed until much later. Thereafter there was much difference of opinion about how the bats found and caught fish, until the fascinating researches of Bloedel [45], who studied the animals in Panama in 1953, settled the argument with photographic and other evidence. His observations were made both on captive bats and on those fishing in the wild, and show a most interesting correlation between the fishing habit and the peculiarities of structure mentioned above. *Noctilio* is a trawler, or rather a foul-hooker, for it does not locate and grab individual fish but sweeps at random in places where small fish are likely to be. It flies low, and rather slowly, over the water, at about 18 to 25 feet per second, dips the large hind feet with widely extended toes about an inch into the water and makes a sweep through it from 1 to 10 feet long. The narrowness of the wings, extending only to the knee, allows the bat to dip its feet and legs without the wings touching the surface or dragging in the water – and here also is where the use of the long bony calcars is seen. Just as it dips, the bat folds the calcars up parallel with the tibias, and tucks up the interfemoral patagium, which is furled so completely that the legs appear to be entirely free from each other. Bloedel says,

The bats make no use of their interfemoral membrane during their fishing dips – the feet are held straight down, the membrane folded and tucked up between the legs . . . the bat tilts the body slightly more vertically as it approaches the water, then lowers the feet, raises the membrane and makes a dip from one to six feet long. . . . As the feet emerge at the end of a dip, the membrane is again immediately lowered and used for flight manoeuvering, apparently its main function.

When a fish is caught it is at once transferred to the mouth, and even in the dark Bloedel knew when a dip had been successful by the crunching noise of chewing; he adds that both species of *Noctilio* 'chewed their food twice – the first time quickly, depositing the large pieces in their large expansible cheek pouches, then chewing these pieces more finely before swallowing.' As long ago as 1847 P.H. Gosse [176] watched a captive *Noctilio* in Jamaica eating a cockroach – he did not know that the bat was also a fish-eater. He said, 'The eating was attended with a loud and very harsh crunching of the teeth', and noted that the meal was put into the cheek pouches and then 'by a contortion of the jaw, aided by the motion of the muscles of the pouch' some of it was returned to the mouth for second chewing 'till all was swallowed and the pouches appeared empty and contracted out of sight'. He added that eating was 'a rapid succession of choppings with the long canines, through which the tongue was thrust about so nimbly that it appeared a wonder it was not impaled perpetually'.

Although *Noctilio* uses echolocation, Bloedel came to the conclusion that the bats probably do not use it for finding fish below the surface, though it might be possible at very close ranges. He found, however, that any sound of splashing water at once aroused the bats and arrested their attention. The fish caught are small, only an inch or so in length, and Bloedel found that they congregate in shoals close to the surface by night; he calculated that random sweeping would be amply sufficient to obtain the thirty to forty fish eaten by each bat every night.

On the other hand *Noctilio* is not always exclusively a fish-eater, for it also catches and eats insects; Goodwin [173] watched some of these bats apparently fishing over a small pool in St. Thomas in the Virgin Islands, and although they made loud splashes as they swooped over the water, the stomachs of those he killed were filled with insects exclusively. He also noted that the bats had a strong musky smell which he found 'not objectionable' although a specimen from Porto Rico had a very offensive smell. The strong unpleasant odour has been noted by others; Kingsley [116] refers to being 'nauseated by their detestable scent' and Gudger [187] says Goodwin told him that although the most dreadful and overpowering odour '. . . was not distinctively fishy, it was thought that fish had contributed to it'. Benedict [32] saw numbers of *Noctilio* fishing in broad daylight over the sea where pelicans were feeding, and other observations show that the bats fish by night in the sea as well as in fresh waters. Although a museum specimen of the fisherman bat might appear to casual inspection as just another kind of bat, Bloedel's investigations show that the structure and habits of the species are highly specialised and efficient for its unexpected way of life. Finally, Goodwin and Greenhall [175] record that, 'A large fish about three feet long was seen to leap out of the water as a *Noctilio* flew over the surface, and it was apparently trying to catch the bat'. So there are bat-eating fish as well as fish-eating bats.

In the Nycteridae we meet the first of a number of families of bats which bear the peculiar facial adornment of nose leaves. This structure, as we shall see, takes many forms, some of them very complex. The nose leaf is an extension of the skin over the face surrounding the nostrils; it generally extends outwards from its attachment round the nose to form a flap or flaps overlapping the fur-covered parts of the snout and face surrounding the attachment. It is thus a double fold of skin enclosing a thin layer of tissue, and might be likened to a symmetrical growth of lichen set on the snout. In some species it is known to act as a director for focusing into a beam the ultrasonic transmission used in echolocation; it is believed to have this function in most species, though in some, as will be mentioned below, it may have other sensory functions.

FAMILY NYCTERIDAE

The Nycteridae or hollow-faced bats are small to medium sized bats found over most of Africa and Arabia, and in Malaya, Sumatra and Borneo, but not in the regions between; it contains but one genus *Nycteris* with about a dozen species. Between the eyes and the nostrils there is a deep hollow on the front of the skull which in life is filled by the cutaneous outgrowths of the nose leaf, a complex structure with overlapping double blades and a wrinkled boss. The tail is long and the last joint is T-shaped, the head of the T lying in and supporting the edge of the interfemoral patagium; the nycterids are the only mammals to have this peculiar modification of the tail end. The ears are large but the tragus small. These bats feed on insects and spiders, and *Nycteris thebaica* is said to live mainly on scorpions [495]; they generally roost in small groups in hollow trees, caves, in the shelter of overhanging banks and such places, but *N. arge* is solitary, and *N. luteola* has been found in East Africa roosting in burrows occupied by porcupines [305]. The last species, and *N. hispida* from the same regions, are among the few bats that are known to be polyoestrous – after giving birth to a young one in December they experience a post-partum oestrus, and become pregnant with a second young while still suckling the first [317].

The nycterids are regarded as a subfamily of the Megadermatidae by some systematists.

FAMILY MEGADERMATIDAE

The Megadermatidae or false vampire bats are found throughout tropical Africa, southern Asia, Malaysia and tropical Australia. They are medium to large sized bats, grey or brown in colour, with large ears joined at the bases – in *Lavia frons* of Africa the ears and wings are reddish yellow. The family is a small one and contains only four genera; *Megaderma* with two species in Asia, *Macroderma* with one species in Australia, and *Cardioderma* and *Lavia* both with single species, in Africa. They are called false vampires to distinguish them from the true vampires of the New World, for although they are carnivorous they do not feed exclusively on blood. The eyes are large; perhaps this character is correlated with the habit of flying by day as well as by night found in *Lavia*. The nose leaf is a long blade reaching upwards between the eyes towards the base of the ears from its origin on the nose. The tragus of the ear is long and pointed, and is bifid, carrying a shorter point on its inner side. The bats are solitary or roost in colonies of moderate size, though the cave roosting species of *Megaderma* in India and Ceylon form large ones. There is no tail, but a conspicuous blood vessel occupies the centre line of the interfemoral patagium; in the females there is a pair of false nipples on the pubis, one at each side of the vulva, which are grasped by the mouth of the young when clinging to the mother. They are not connected with mammary glands, although they contain ducts resembling milk ducts; the functional nipples and the mammary glands lie on the chest, as in all bats. The males have rudimentary false nipples in a corresponding position.

Lavia frons is, as far as known, exclusively insectivorous, but the species of the other genera are less restricted in their diet, and eat many animals besides insects. *Megaderma lyra* of India may be a fisherman like *Noctilio* of the New World, for the scales and bones of fishes have been found in their stomachs, although no one seems to have recorded seeing these bats actually catching fish – their feet are large and might be used for fishing in a similar manner to those of *Noctilio*. Their usual prey, in addition to insects, consists of small vertebrates such as mice, birds, lizards, frogs and smaller species of bats. They have been called cannibals because they eat other bats, but this is not correct; they would be cannibals if they ate their own species, but they do not – hawks feeding on other birds are not cannibals, and even the human cannibals of the Cameroons at the present day are cannibals only when they eat human flesh, not when they eat gorilla. In some places, or at certain times, the diet has been found to be mainly frogs, but that may be because frogs were plentiful when the observation was made. Prakash [380] found that *Megaderma lyra* always seized its victim, whether mouse, bird or bat, by the head and ate from head to tail, discarding the wings and feet – when feeding on a bat often the head only was eaten. He also found that these bats were not interested in dead food, a fact supporting the suggestion that the species catches fish; on the other hand some observers have seen them take a dead mouse when in captivity. Prakash also offered a house rat (*R. rattus*) to a cage full of false vampires, but this was too large a prey; although the bats attacked it at once and fought furiously, they did not eat it, and the next morning he found the uneaten rat dead with head wounds and one bat dead with a deep wound in the breast. *Macroderma* gigas of tropical Australia is the largest of the microchiroptera, the length of the head and body together being about six inches. It, also, feeds on other bats as well as other small vertebrates, and insects. It is a cave-roosting species and it formerly inhabited southern Australia, where various cave deposits of bat guano containing the mummified remains of the species have been commercially exploited. *Cardioderma cor* of tropical Africa is said to have carnivorous habits, but the present writer could find nothing but the remains of insects, probably coleoptera and orthoptera but not lepidoptera, in the stomachs of those he examined in what was then Tanganyika.

FAMILY RHINOLOPHIDAE

The Rhinolophidae are the horseshoe bats of the Old World and are found from western Europe through temperate and tropical Asia to Japan, the Philippines, New Guinea and Australia. There are but two genera, *Rhinolophus* with about seventy species, and *Rhinomegalophus* with one. The vernacular name is derived from the shape of the complex nose leaf, the basal part of which is raised as a rim round the nostrils, and is fancifully likened to the shape of a horseshoe with the gap upwards. Above the gap in the horseshoe a free pointed blade, the lancet, extends upwards between the small eyes towards the base of the ears. Its sides are sculptured into pocket-like hollows, and at the middle of the base there is a narrow flattened structure, the sella, set vertically, that more or less divides the depression of the shoe into right and left halves. These bats transmit their echolocating ultra-sound pulses through the nostrils with the mouth shut, and the nose leaf focuses the beam. Their method of echolocation thus differs from that of the vespertilionid bats, as well as in the type of pulse used, as explained in Volume I. The ears are large and pointed, and have no tragus. The tail is comparatively rather short, and when the bat is at rest it is turned up over the back, not tucked under the belly as in the vespertilionid bats. When rhinolophid bats roost, usually in caves, and the roofs or cellars of buildings, they hang freely suspended from the ceiling and do not squeeze into crevices and cracks, so that they are easily seen. They hang by the claws of the feet and wrap the wings round the body very precisely so that the fur is entirely covered, and they resemble the hanging fruits or pods of some vegetable. The smallness of the foothold needed is remarkable, for they can grip the least irregularly of surface; furthermore, they can fly straight to the perch, turn a somersault in the air, take grip and fold the wings in a fraction of a second in complete darkness.

Like the megadermatid bats the horseshoe bats have false teats on the pubic region, which are gripped by the young one when carried by the mother – as in the former the mammary glands and functional teats are on the chest. Another reproductive peculiarity is that in the females the left ovary is small and apparently inactive, and the right one alone produces ova – and gestation takes place always in the right cornu of the uterus – at least in the few species which have been studied in detail. Several species, *R. ferrum-equinum* and *R. hipposideros*, the greater and lesser horseshoe bats, and *R. euryale* the Mediterranean horseshoe bat, are common in many parts of Europe where, owing to the ease of finding them in their roosts, they have been studied closely by zoologists for many years. Banding – placing numbered split rings on the forearm – has given much information on the habits of these bats. This technique has shown that hibernation is not continuous through the winter, and that bats wake at frequent intervals, and not only move about in the hibernating caves, but emerge and fly about even on nights when the temperature is low. Within the caves they are very selective in choosing their roosts, and evidently need a micro-climate with narrow limits of temperature and humidity. Although they do not migrate they do shift their quarters from time to time, and may travel as much as fifty miles on such excursions. Caves are favoured for hibernation, but are usually deserted in the summer when the bats form colonies in the roofs of buildings, and perhaps hollow trees, which they must presumably have used before roofs were available. Hibernation in bats, however, is really no more than the extension of normal sleep [59]; during normal summer sleep the temperature of bats often drops to that of the air, and if the ambient temperature is low the temperature of bats falls with it, but rises again when they wake up in the evening, as discussed in Volume I. In the summer colonies the sexes are segregated – the colonies usually consist of pregnant and nursing females with only a few males nearby; there is little exact information about where the males go in the summer. Banding has also revealed the surprising longevity of these bats in the wild; the tiny and fragile lesser horseshoe bat has been shown to be able to live at least fifteen years – the life of other small mammals of similar size such as mice and shrews is generally less than two years in the wild and not more than three or four in the protection of captivity.

Horseshoe bats are entirely insectivorous, as far as is known, and their nourishment is derived from the soft tissues of their prey; the hard exo-skeletons of insects are crunched up and the fragments appear in the bats' droppings together with the scales from the wings of moths. Nevertheless it has recently been discovered [257] that the gastric juice of the greater horseshoe bat contains large quantities of chitinase, an enzyme which breaks down chitin, the characteristic component of insect exoskeletons and one of the most durable and indestructible of natural productions. It is not known whether the bats derive any nutriment from the digestion of chitin, but it is probable that the break down of the substance helps in making the other tissues available to digestion – chitinase certainly does not produce solution of all the chitin ingested. The numerous and widely distributed species of the genus *Rhinolophus* differ in their size, colour and the details of the structure and shape of the nose-leaf, but they are closely related phylogenetically and form a well defined group. *Rhinomegalophus paradoxolophus*, the only species of its genus, is known from a single specimen from a cave in Tonkin, Indo-China; it was first described in 1951. It differs from the bats of the genus *Rhinolophus* in the character of the nose-leaf and in its enormous ears; these features combine to give it a most grotesque appearance.

FAMILY HIPPOSIDERIDAE

All the other leaf-nosed bats of the Old World belong to the closely related Hipposideridae, a family containing nine genera and about seventy species, over forty of which, however, belong to the genus *Hipposideros*. They are found in tropical and subtropical Africa and southern Asia eastwards to the Philippines and Australia. They differ from the rhinolophids in the form of the nose leaf and in details of skeleton and dentition, but resemble them in lacking a tragus in the ear. They are mostly small to medium sized bats, but *H. gigas* of west Africa is one of the larger of the microchiropterans. In the genus *Hipposideros* the nose-leaf consists of a number of more or less concentric flaps round the nose and others forming a broad leaf above the nose and between the eyes; some species have an evertible sac secreting a waxy substance behind the leaf. In *Asellia* and *Cloeotis*, the trident-nosed bats, the upper blade forms a three-pronged fork, and that of *Asellicus* is three lobed and has further elaborations, but in the other genera such as *Anthops, Coelops, Rhinonicteris* and *Triaenops* the complication of the details of the nose-leaf, with overlapping leaflets, accessory blades, hollows, knobs and other sculpturings, is so complicated as to defy verbal description – it can be represented only by graphic illustration. These complex nose-leaves, occupying much of the face, give the animals a most bizarre appearance when examined closely. Many species throughout the family are comparatively brightly coloured, usually with shades of reddish brown to yellow, and some are blotched with patches of lighter colour on a dark ground. Some species, too, are dimorphic, some specimens being red, others grey or brown. The tail in bats of this family is proportionately longer than that in the rhinolophids, and as in that family it is carried cocked up over the back when the bats are at rest. All the hipposiderids are insectivorous; they roost in caves, trees, buildings and such places, some species congregating in very large numbers.

FAMILY PHYLLOSTOMATIDAE

The only other family of leaf-nosed bats is a large one, the Phyllostomatidae containing all the American species, found from the southwestern United States to northern Argentina and most of the West Indian islands; it contains nearly a hundred and fifty species classified into fifty genera. This numerous tribe is conveniently divided into seven subfamilies. The nose-leaf throughout the family is generally much less complex than in the preceeding families, and consists of a plate on the nose above which rises a pointed blade which during life is held erect and free from the face when the bat is active; in a few species the nose-leaf is reduced or absent. A tragus is always present in the ear.

The three genera of the subfamily Chilonycterinae are without nose leaves, but one, *Mormoops*, has foliaceous outgrowths on the chin, and fleshy projections on the lower lip; the nose is strongly turned up and bears ridges and grooves. *Chilonycteris* has a leaf-like expansion of the lower lip and moustache-like tufts of hair in the upper, and *Pteronotus* is peculiar in having the back from shoulder to tail quite naked. In this genus the wings are attached almost at the centre line of the back, which thus appears naked although the back of the body under the wings is furred. All these genera are insectivorous.

The subfamily Phyllostomatinae contains eleven genera and about thirty species; all have nose-leaves with well developed vertical blades or 'spears'. The bat-god in the pantheon of the ancient Zapotecs in southern Mexico was evidently based upon a phyllostamatine bat, perhaps *Lonchorhina*; he is represented bearing a nose-leaf with a very prominent erect spear, but his dentition is not chiropteran, for his aggressive open mouth with lolling tongue shows human incisors. The three species of big-eared bats, *Macrotus*, have long ears shaped like those of a rabbit, and the sword-nosed bat, *Lonchorhina*, has a very long narrow, erect nose leaf. The long-legged bat, *Macrophyllum*, widely distributed over tropical America, is distinguished by its very long tail and legs, whereas in the spear-nosed bats *Mimon* and *Phyllostomus* and some others the tail is very short, and the interfemoral patagium is extended mainly by the calcars. Some species of the subfamily are insectivorous, others also eat fruit, and some are carnivorous. Huey [245] investigated a colony of some 500 *Macrotus californicus* roosting in the semi-darkened stope of an abandoned mine shaft in California, where a good deal of light penetrated. He found abundant remains of large insects – moths, grasshoppers, beetles and others, together with willow leaves, and showed that the bats are exclusively insectivorous; 'It is possible that some of their food, such as the grasshoppers and harvest flies, which are diurnal insects, were taken from their resting places on the willow trees, where they were seized, leaf and all, by the bat and then carried to the roost to be devoured'. The little big-eared bats of the genus *Micronycteris* eat fruit; Goodwin and Greenhall [175] found that *M. megalotis* in Trinidad is fond of small ripe guavas and 'plucks the fruit while hovering in the air and carries it to a nearby tree to eat'. The spear-nosed bats, *Phyllostomus*, and the American false vampire, *Vampyrum*, are omnivorous; they sometimes eat fruit and insects, but to a large extent they are, like *Megaderma lyra* of the Old World, carnivorous, eating small birds and mammals including other species of bats.

Goodwin and Greenhall [175] found that although *P. hastatus* is carnivorous, in Trinidad it also feeds eagerly on the seeds of the sapucaia nut – it is gregarious and leaves its roosts in flocks of up to a hundred to visit the trees; 'when the seeds are ripe great numbers of these bats fly together into a grove of nut trees about dusk making loud vocal sounds, tear off the seeds to eat the fleshy attachment, and fly off again'. They found the species 'bold and even aggressive, and on several occasions a large individual has deliberately left a cluster of roost mates to walk down a cave wall head first and bite the gloved hand of the junior author'. In captivity it has 'no hesitation in killing and eating most other species of bats put into its cage, although it appears to be uneasy in the presence of *Desmodus*'. *Desmodus* is the much smaller true vampire, discussed below. These authors also found that *T. cirrhosus* of the nearby genus *Trachops*, which has warty excrescences on the chin and lips, feeds on geckos and perhaps other lizards.

Dunn [122] kept specimens of *P. hastatus* in captivity and offered fresh ripe banana daily as well as small animals and found that some of the fruit was eaten on all except one of 168 nights. 'When there was no mouse, bird or bat on which to make a meal more of the fruit was eaten. On the night the three mice were devoured the banana was not touched'. During this period the bat killed and ate twenty-five mice, thirteen bats and three birds. The bat seizes the victim's head and kills it by crushing the skull, and while eating it holds it between the wrists and thumbs. The bat eats practically the whole of the prey, leaving only the larger wing and tail feathers of birds, and sometimes the teeth and tail of small mammals. *Vampyrum spectrum* catches, holds and eats its food in a similar manner – it was believed to feed on blood when Linnaeus gave it its misleading name. It is a large species, with a wing-span of about thirty inches, large ears and no tail. Although it has eaten fruit in captivity it appears to be solely carnivorous in the wild, feeding on birds and small mammals. Some captive specimens kept by Goodwin and Greenhall ate four mice each every day; these authors say that this species 'drinks water and soon becomes tame and gentle'. They also tell that it appears among the apparitions and ghosts of Trinidad folk lore as 'the *douens* that flit about among the branches of the silk-cotton trees. They are believed to be little children who have died before they were baptized. They have no sex and no faces, their feet turn backward, and they are either naked or wear long white robes and large hats'.

Although some bats in this family roost in large colonies, many are generally found only in small clusters in caves, culverts, roofs or hollow trees. Some of them, too, seem to be very tolerant of light in their roosts – Huey's big-eared bats were in only semi-darkness, and although Dunn's *Phyllostomus* were in a cave the present writer has caught a bat of this genus hanging on a tree branch fully exposed in the deep shade of a Brazilian forest.

The subfamily Glossophaginae, with about thirty species distributed among a dozen genera, contains the nectar feeding phyllostomatids. These bats are primarily feeders on nectar and the juice and pulp of ripe soft fruit, but some species at least, such as *Glossophaga soricina*, also eat insects. Most of them are recorded as roosting in caves, sometimes in considerable numbers, and a few species have been found in the roofs of buildings, but the habits of those in several genera are imperfectly known as very few specimens have been collected. All the species have pointed snouts and in some, such as the long-nosed bat, *Choeronycteris* of central America, the snout is extremely long and narrow. The nose-leaf at the end of the snout is small and the spear is carried erect or even pointing forwards; it is possible that it may have a tactile function, useful when feeding from flowers. The tongue is narrow but very long and extensible; its surface bears conspicuous long bristle-like papillae which in some species are concentrated towards the tip as a sort of brush. All the bats of the subfamily are small in size; the ears are small, and the interfemoral patagium is very narrow as in some genera there is no tail, and in others only a very short rudimentary one. These little bats hover before the flowers they visit and probe them for nectar with the tongue which can be protruded for an extraordinary distance – in *Anoura geoffroyi* it actually exceeds the length of the head and body together – somewhat like the proboscis of a butterfly or moth. They are said also to feed on pollen, though the ingestion of pollen may be accidental in taking nectar, and are alleged to pollenate agave, calabash and other flowers that remain open at night – *Glossophaga* and *Musonycteris* also visit the flowers of the banana. Wille [507] has shown that the musculature of the throat is highly modified in the glossophagine bats for protruding and withdrawing the long tongue. The masticatory muscles, on the other hand, are small and weak because 'mastication and grasping play only a minor role in eating'.

Figure 6 Nectar bat (*Leptonycteris*) one of the glossophagine bats.

The subfamily Carolliinae contains only half a dozen species, four of them in the genus *Carollia*, widespread and plentiful throughout much of Central and South America. They are medium sized, stoutly built bats with a conspicuous nose leaf and a warty patch on the lower lip at the chin. They roost in sometimes large colonies in caves, hollow trees and other sheltered places from which they issue at dusk to feed on fruit. Goodwin and Greenhall found that in Trinidad *C. perspicillata* fed on a wide variety of fruits. If the fruit is large it is eaten hanging on the tree but if small the bat carries it to a temporary roost to eat it. 'The choice of these feeding stations varies. One favoured spot appears to be mosquito nets, and the bats drop debris down the sides of people's beds'. The bats are wasteful feeders, but are important in the natural dissemination of a great many fruits and seeds.

The Sturnirinae are another small subfamily of about half a dozen species, all except one in the genus *Sturnira*. They are medium-sized bats with well developed nose-leaves and warty lower lip. There is no tail, and the interfemoral patagium is extremely reduced so that it forms a narrow fringe along the inner sides of the legs; both it and the legs are well covered with fur. The males have light coloured tufts of fur covering a glandular area on each shoulder, from which comes a strong sweetish musky odour – they thus resemble the megachiropteran epauletted fruit bats, and are similarly named American epauletted or yellow-shouldered bats. They resemble their Old World counterparts, too, in feeding on fruit and fruit juices.

In contrast the subfamily Stenoderminae is a large one, containing some forty species classified into eighteen genera – these are the fruit bats of the New World. They are small to large-sized, stoutly-built bats with no tail and a narrow interfemoral patagium. The snout is short so that the muzzle is broad, and there is a nose leaf with a spear in most genera; most genera, too, are marked with light coloured longitudinal lines on each side of the face, running back above the ears; many also have a light stripe along the middle of the back. In several genera, such as *Uroderma*, *Vampyrodes* and *Chiroderma*, there is a further light stripe on the face below the eye – and in *Centurio* there is a light spot on the shoulder. The most unusual colour, however, is found in the two species of white bats in the genus *Ectophylla*; the fur of *E. alba* is white with a ring of dark grey round the eyes, the wings are pigmented but the ear-margins, chin, nose-leaf and the skin covering the wing bones, are bright yellow. The second species is more pigmented, being dull brownish white. These species have rarely been captured and there are few specimens in the great museums of the world – the same remark applies also to several other genera of the subfamily. Some of the species form large colonies, but many roost in only small groups, and some appear to be solitary. Many roost in caves, hollow trees or buildings, but in general they tend to be less partial to dark places than many bats, and some species are commonly found by day hanging on the branches of shrubs, under large leaves, or the eaves of houses. Two genera, *Uroderma* and *Artibeus*, the tent bats, actually improve on the natural shelter. The first makes a series of cuts across the pleated surface of the fan-shaped leaves of certain palms so that half of the leaf bends at an angle and makes a protected shelter; at least one species of the second genus is reported to have a similar habit.

The food of nearly all species consists of soft fruit; the bats swallow only the softest pulp and, like the fruit bats of the Old World, chew the rest to extract the juice and spit out the squeezed remains – this habit has been observed in one species of *Artibeus* and is probably widespread; indeed, the throat of *Centurio* is so narrow that nothing but liquids or semi-liquid pulp could be swallowed. Some species carry small fruits away to eat them at feeding roosts, and play a large part in disseminating their seeds; where *Artibeus* is plentiful it is a great pest to fruit-growers in some parts of the West Indies. Goodwin and Greenhall report that this bat often stains its fur with fruit juices by actually eating its way inside large mangoes – they found 'four yellowish green females, each with a yellowish green young' with the fur dyed to the roots all over the body. They also say that sometimes the bats dropped mangoes on the galvanised iron roofs of houses and 'the incessant bombardment on the roofs of dwellings has made sleep for human beings impossible, making the cutting down of mango trees in the vicinity a necessity'. On the other hand the four species of *Brachyphylla*, a genus which has no erect spear on the nose-leaf, which is thus much smaller than in other genera, are believed to be insectivorous judging from the insect remains found in the guano below a roost of about 2,500 *B. cavernarum* found in an abandoned building. This genus is confined to various islands of the West Indies.

The wrinkle-faced bat, *Centurio senex*, now divided into two geographical subspecies, is the most extraordinary of all the bats in this subfamily. The muzzle is very short, with underhung lower jaw, but the skin of the face, which is naked, is thrown into a number of folds and ridges, making an intricate pattern which gives the bats a most bizarre and grotesque expression according to human standards, hideous or comical as you prefer. The ears are widely spaced and erect, and the heavily furred skin under the chin hangs in several semi-circular folds. The wings, too, are peculiar for the patagium between the second and third finger is translucent, and that between the fourth and fifth is crossed by a number of dark and light transverse bands, as is a smaller area internal to the fifth – hence their alternative vernacular name of lattice-winged bats. The skin inside the lips is covered with papillae which are believed to act as strainers when the bat is imbibing fruit juice, for the diameter of the gullet is less than one-and-a-half millimetres so that nothing but liquid or soft mush could pass through it. The facial wrinkles and pattern are more strongly developed in the males than the females; the latter have only rudimentary folds of skin under the chin. These bats roost singly or in groups of two or three hanging from the branches of large trees, and when sleeping the males put their heads under the bedclothes; they sleep with the skin of the chin pulled up over the face where, as reported by Goodwin and Greenhall:

it covered the top of the head and extended over the flaps of the ears which lie flat across the top of the head. A little bump on the crown of the head acts as a sort of 'doorstop' and the wrinkled chin skin is stretched taut at this point. In some specimens there are two translucent areas in the middle of the stretched facial mask, devoid of hair, which cover the bat's eyes, and presumably enable it to see light, and perhaps objects, even when the face is covered. It was possible to pull this skin mask away from the bat's face and to see light and moving objects such as a waved finger or pencil through these translucent windows. When aroused, the bat unshrouds itself, and the skin mask covering the head and face slips back in wrinkled folds to the normal position under the chin, with the two windows appearing as bumps. . . . The large lappets of loose skin on the chin of the males, not part of the mask, probably contain scent glands, as there is a distinct, musky, skunk-like odour about the head. *Centurio* flies with a some-what wobbly, jerky motion . . . and . . . resembles a large butterfly (Plate 8).

These bats have been found in widely separated places in tropical America and their distribution probably extends [364] from Mexico to Venezuela, Trinidad and Tobago. An allied genus, *Ametrida*, has a similar short muzzle, but no facial lobes and a small but normal nose-leaf; specimens have seldom been seen and nothing is known of its habits.

Two genera with a total of five species form the small subfamily Phyllonycterinae; they are found only in some of the West Indies and the Bahamas. They are long-snouted bats and in general resemble the Glossophaginae, for the tongue is long and narrow and bears backwardly directed bristly papillae, but the nose leaf is rudimentary and is little more than a bare area round the nostrils. The tail is comparatively well developed and projects beyond the margin of the narrow interfemoral patagium. One species of *Phyllonycteris* and both those of *Erophylla* roost in caves, some-times in colonies of many hundreds, and fly out at night to feed on fruits and perhaps also the nectar of flowers. *P. poeyi* is reported to be the prey of more than one kind of snake in Cuba; the Cuban boa seizes the bats when they fly through the narrow opening to their cave roosts. The other species of this genus are either very scarce or extinct.

FAMILY DESMODONTIDAE

Turning to the family Desmodontidae we come to the true vampires in which the diet consists exclusively of the blood of vertebrates. They are found only in tropical and subtropical America, hence it is peculiar that the legend of the nocturnal, blood-sucking, vampire transformed from a re-animated corpse, arose in the Old World long before the discovery of the New – though it is possible that the assumption of a bat-like shape did not occur until afterwards. Although the family is small, containing only three genera, each with a single species, it is of great importance to man because these bats, owing to their blood-feeding habit, can transmit paralytic rabies to their prey, including man and his domestic animals. They also transmit a trypanosome disease fatal to horses. *Desmodus rotundus* is a medium sized bat, dark brownish above, lighter beneath, with dermal outgrowths round the nose forming a rudimentary noseleaf, short pointed ears, no tail and a very long thumb with a rounded pad at its base. In all three genera, *Desmodus*, *Diaemus* and *Diphylla*, the interfemoral patagium is short and the calcar is reduced or absent. The central upper incisors, which are very small teeth in most bats, are very large and triangular in shape, with razor-sharp cutting edges. The canines also are large but are set well behind the incisors and appear not to be used in making the incision for feeding; the lower incisors are widely spaced so that there is a gap through which the long tongue can be protruded. Vampires often roost in large colonies in caves, hollow trees and many man-made structures; when disturbed they generally do not fly but nimbly scuttle into the shelter of crevices.

When feeding on a sleeping mammal or bird at night the bat alights on its victim and runs over the surface to find a place where the skin is sparsely covered; it runs quickly and agilely on the feet and wrists with the body held well above the surface, using a gait unlike the usual rather shuffling one of most bats – they have been likened to large black spiders running over their prey. They run similarly on the ground and have also been seen to 'hop like toads'. Ditmars and Greenhall [113], describing a captive vampire, said it walked across the floor of the cage 'with wings so compactly held that they looked like the slender fore-limbs of a four-footed animal. Her rear limbs were directed downward. In this way her body was reared a full two inches from the floor . . . anyone not knowing what she was would have been unlikely to suspect her of being a bat'. When the area for attack is selected a quick slash with the sharp incisors makes an elliptical shallow wound; the running about and biting are frequently not felt by the victim who is not even awakened. Before biting the bat opens the mouth rather slowly, 'as if to gauge precisely the sweep of the incisor teeth'; Ditmars and Greenhall found that when the bat is selecting the site for biting and occasionally lifts its head showing 'the leer that disclosed its keen teeth' it produced a sinister and impressive effect.

As soon as the blood starts to flow the tongue is pushed out so that the tip touches the welling blood, and its edges are turned down to produce a concave under surface which with a deep groove on the middle of the lower lip forms a tube through which the blood is sucked into the stomach; the flow is stimulated from time to time by agitation with the tip of the tongue, and clotting is delayed by an anti-coagulant in the bat's saliva. Goodwin and Greenhall say that when feeding on an animal the tongue is usually stationary, 'and the upper surface remains completely free of blood. The bat's chest and throat movements clearly indicate that the blood is flowing to the digestive tract. From time to time the tongue may be seen to make in-and-out lapping motions which are thought to produce a partial vacuum in the buccal cavity and further assist the flow.' On the other hand Ditmars and Greenhall found that a captive vampire lapped defibrinated blood from a dish with the relatively long tongue which 'moved at the rate of about four darts a second. At the instant of protrusion it was pinkish, but once in action it functioned so perfectly that a pulsating ribbon of blood spanned the gap between the surface of the fluid and the creature's lips'.

The vampire often gorges to such an extent that its body becomes almost spherical; it may then drop to the ground, whereupon it has to climb up some object to launch itself into flight. The bat then retires to a temporary shelter low down where it is lightened by the quick digestion of part of its meal. The permanent roosts are heavily fouled by the tarry ammoniacal droppings resulting from blood-meals. A vampire can take up to two ounces of blood in a single night's feeding; Goodwin and Greenhall relate that one captive vampire will take about five and three-quarter gallons of blood in a year, 'During 1959, approximately 1,800 vampire bats were collected in Trinidad. Therefore, the assumption is made that these bats, while free, could under favourable conditions be responsible for a maximum of 10,350 gallons of blood during the year!'

Rabies appears to be endemic in vampire bats, and at intervals it becomes epidemic, when many other kinds of animals and birds, including even the phyllostomatid fruit-eating bats, become fatally infected through being bitten, not only by vampires but by other infected animals. One of the most serious epidemics occurred in Trinidad between 1925 and 1935 and caused the death 'of eighty-nine human beings . . . and livestock losses numbering into thousands'. Naturally the authorities had to take action, and control measures were started which included the destruction of vampires by shooting, netting, smoking and poisoning, and the vaccination

G

of livestock. Vampires have the habit of returning to the same animal night after night, singling the same victim out of even a large herd, and can thus be poisoned with a drop of sugar-water containing strychnine placed on the wound of the night before. Increased aggressiveness by vampires, and strange behaviour by other bats, such as the biting of people by fruit-eating bats, herald the approach of an epidemic. It has now been found that many species of bat, and of other small mammals, may be symptomless carriers of rabies between the outbreaks of epidemics.

The white-winged vampire, *Diaemus youngi*, and the hairy-legged vampire, *Diphylla ecaudata*, are in general similar to the common vampire, but have shorter thumbs; their feeding habits also are similar. *Diaemus* is brownish in colour, but the leading and trailing edges of the wings are white, as is the patagium at the wing-tip. This species appears to prefer feeding on the blood of birds; Goodwin and Greenhall report that in captivity it refused cattle blood but fed readily on that of fowls. They also say, 'When disturbed this bat opens its mouth and can, at will, bring to the commissures of the mouth two cup-shaped glands ... when further disturbed the stoma of the glands can be directed forward, and with the accompaniment of a noisy 'psst,' emit a powerful and peculiar odour which is offensive and nauseating to many people'. – it has been likened to a mixture of sulphur and mushrooms. *Diphylla* also appears to prefer the blood of birds. Like some other bats, vampires appear to be long-lived animals, for Trapido [476] has recorded one that lived in captivity for twelve years and nine months – its age must have been greater than this because it was an adult of unknown age when captured.

FAMILY NATALIDAE

We now come to four small families only one of which contains more than one genus. The first is the Natalidae, small insectivorous bats of tropical and subtropical America – about half a dozen species form the single genus *Natalus*, the funnel-eared bats. The ears are not long but are wide and somewhat funnel-shaped, there is no nose-leaf, the body is slim and the wings, legs and tail are long and slender, the tail reaching the edge of the wide interfemoral patagium – the tail and legs are longer than the head and body. These bats are unique in the possession by the adult males of a thin glandular structure under the skin of the forehead between the eyes, lying rather loosely between the skin and underlying muscles. It was named the natalid organ by Dalquest [96] who points out that it is not connected with the pararhinal glands which are greatly developed; 'the anterior portion of the muzzle is so greatly swollen by the glandular masses that the nostrils are forced far anteriorly, on to the upper lip'. The natalid organ is not conspicuous externally, and dissection shows that it varies in size from species to species; it is completely independent from the skin. Dalquest found the organ consists of a mass of intertwined tubules lined with columnar cells which he described as 'cells that closely resemble sensory cells, but also show evidence of glandular function'. What part the organ may play in the life of the natalids remains unknown. The lower lip is fleshy and thick, but has no glandular structure.

The colour of the fur of several species ranges widely from dark to light, and in *N. mexicanus* the differences are so great that they form two colour phases, a greyish brown dark phase, and a bright orange or reddish brown light phase. Novik [351] recorded the echolocation ultrasounds of this species and found that they are of very high frequency, with a harmonic as high as 170 kc; he suggests that this may be correlated with feeding on very small insects. Natalid bats roost in caves, generally in the darkest parts far from the entrance, but some species are known also to roost in hollow trees; they often form small colonies of up to about a hundred animals. Goodwin [173] found *N. mexicanus* very alert in its roosts so that 'it was necessary to shoot to procure specimens', but *N. tumidirostris* in Trinidad [175] was 'a swift flier but delicate, as the concussion of a gun fired in a cave drops specimens even though they have been untouched by pellets.' – perhaps this sensitivity may be correlated with the funnel-shaped ears and the perception of extremely high frequencies.

FAMILY FURIPTERIDAE

Two genera, *Furipterus* and *Amorphochilus*, each with a single species form the family Furipteridae, the thumbless or smoky bats. They are small insectivorous bats of tropical Central and South America; they have been found roosting in caves and cellars, but little is known of their habits. The thumb is present but rudimentary, and enclosed in the patagium except for the tiny vestigial claw. The ears are short but very wide and funnel-shaped so that the eyes appear to lie within their bases. The tail, unlike that in the natalids, is short and does not reach the edge of the interfemoral patagium.

FAMILY THYROPTERIDAE

Two species of *Thyroptera* fill the family Thyropteridae or sucker-footed bats. They are small, slender insectivorous bats of the American tropics, with funnel-shaped ears, and the tail extending slightly beyond the edge of the patagium. Their most striking characters, however, are the sucker-discs on the wrists and feet; the discs are hollow and set on short stalks, those on the wing at the base of the thumb, which bears a claw, and those of the legs on the soles of the feet, in which the third and fourth toes are fused together. The wing discs are larger than those of the feet. These bats roost singly or in groups of up to about half a dozen head upwards in the large curled, faded leaves of heliconias, bananas and other plants, clinging to the smooth polished surface by means of their suckers. Dunn [121] came across several *T. albiventer* while searching for tree frogs in Panama; they were in the curled leaves 'in the form of a cornucopia' of an unidentified green plant about three feet from the ground. He tested the sucking power of the discs after one of the bats had been killed and found that 'the disc of one thumb had sufficient suction to support the entire weight'. The suction must therefore be maintained automatically by the concave shape of the suckers without muscular effort by the sleeping bat; on waking, muscle action would release the partial vacuum to detach the animal. It is strange that such peculiar wing and foot sucking discs should be found in another species at the other side of the world.

FAMILY MYZOPODIDAE

The rare and little known species *Myzopoda aurita*, the only species in the family Myzopodidae, is known solely from Madagascar. It is a medium sized bat with long, narrow ears, thumb with a vestigial claw, and the tail projecting beyond the edge of the patagium. The tragus is fused along its anterior edge with the ear conch, and the ear passage is partly closed by a conspicuous mushroom-shaped process – a disc flush with the surface and attached to a narrow stalk – which, as Miller [323] remarks, is unlike anything known in other bats. The sucker discs lie at the bases of the thumbs and on the soles of the feet but, unlike those of *Thyroptera*, they are sessile and have no stems. Their use is presumably similar, but the habits of the species are not known. The structure of the skeleton shows a number of points in common with that of *Thyroptera* and, indeed, with those of closely related furipterids and natalids; the presence of sucker discs may be an example of convergent evolution, or the result of evolution from common stock, as believed by Thomas [473] – if the latter, it sets a nice problem in distribution. Three genera of vespertilionid bats also have sucker pads or discs on their feet, but these suckers are very much less perfectly elaborated.

FAMILY VESPERTILIONIDAE

The Vespertilionidae are the largest family of bats; they comprise about thirty-five genera and nearly three hundred species. They are world-wide in distribution, from the limit of tree growth in the north to the southernmost tips of the continents, with the exception of the Antarctic. They are small to moderately large in size, and the majority are insectivorous – most of the common bats of Europe and North America belong to this family. The bats of this family have a well developed tragus in the ear, and a nose leaf is absent in nearly all; the pararhinal glands are prominent and often forms pads on the side of the muzzle. The tail is generally long and the interfemoral patagium wide; when a vespertilionid is feeding on large insects the tail may be curved forward so that the interfemoral patagium forms a pouch in which the insect is manipulated by the mouth, after it has been swept into the pouch by a stroke from the wing tip. The colour of the fur is generally dull, ranging from nearly black to pale brown, the underside often lighter, but there are some unusually brightly coloured species. Roosting places are very varied, from caves and hollow trees and other natural cavities to man-made shelters both above and below ground – a few roost in the open hanging from the branches of trees. Some species are colonial, either seasonally or permanently, and in some colonial species the sexes segregate when the females are giving birth and nursing young. A few species undertake regular seasonal migrations over great distances.

A family so large is conveniently divided into six subfamilies, although the first, the Vespertilioninae, contains by far the greatest number of genera, and its first genus, *Myotis*, nearly seventy species, a greater number than that in any other.

The little brown bat of North America, *M. lucifugus*, the mouse-eared bat *M. myotis*, the whiskered bat *M. mystacinus* and the Natterer's bat *M. nattereri* of Europe and Asia are widely distributed and well known members of this genus. Most of the species are dull grey or brownish in colour but one, Hodgson's bat *M. formosus*, from southern China, Japan and northern India is bright orange. Many species roost in caves or the roofs of buildings, and some are colonial. They are all insectivorous – *M. daubentoni*, the water bat of the Old World, is noted for its habit of skimming low over ponds and rivers to feed on the insects flying near the surface. Species native to the cooler parts of the range of the genus hibernate in winter. The single species of the allied genus *Pizonyx*, *P. vivesi*, is of particular interest because, like *Noctilio*, it is a fisherman. Superficially it resembles *Myotis*, but differs in the relatively enormous feet with the long toes and claws strongly flattened from side to side. Although one or two Asiatic species of *Myotis* with large feet and claws have been suspected of feeding on fish occasionally, *Pizonyx* is the only bat besides *Noctilio* that is known habitually to feed on fish. The species is limited in range, for it is found only on the islands of the Gulf of California and the adjacent coasts of the mainland, and in one area on the west coast of Lower California; it was first discovered in 1900. It roosts in crevices and the interstices between the stones of rock falls and slides on the hillsides, and especially in the gulleys, in company with two species of small petrels; bats and petrels share the same homes during the breeding seasons of the birds. Although restricted in distribution the bats are common enough in their habitat for Burt [61] who landed on one of the islands before dawn found that they were 'all back in their retreats and could be heard squeaking over the entire hillside. This chorus continued until daylight when it gradually died down'. Reeder and Norris [398] found that *Pizonyx* feeds on shrimps – the stomachs of two were filled with small crustacea – as well as on small atherinid fishes and anchovies. Unlike *Noctilio*, *Pizonyx* has a long tail and interfemoral patagium, but it does have long calcars. Norris was able to watch *Pizonyx* fishing at a short distance from a light hung over a ship's side, and concluded that the bats use the patagium as a scoop in catching their prey. Bloedel [45] doubts that this is the method used, and thinks that Norris misinterpreted what he saw owing to the difficulty of exact observation at distances of more than three or four feet. Although Bloedel could not get *Pizonyx* to fish in captivity like *Noctilio*, he thinks that *Pizonyx* uses its feet for fishing, and drags them through the water in making fishing sweeps. He suggests that the tail is curled forward so that the patagium is folded against the belly during the dip – this might well be aided by the long calcars – and that the patagial pouch may be used for manipulating they prey after it is caught. He found that the food is about equal quantities of shrimps and fish, eaten separately, not mixed.

A character of *Pizonyx* that Miller [323] described as 'unique in the family' is the presence of what was thought to be a 'large glandular mass near the middle of the forearm'. A study published by Quay and Reeder in 1954 [382], however, showed that the mass is not glandular nor constant in position. These authors found that nodules may be present in almost any part of the patagium of the wings and tail, but that they most commonly occur in three positions on the wing, behind the bones of the upper and fore arm. The nodules are the sites of chronic haemorrhage, inflamation, and the formation of new blood cells – they usually occur at the intersections of bundles of skeletal muscle and elastic fibres. Bleeding into the nodules takes place through the breakdown of small blood vessels, and is followed by inflammation and by the formation of new red and white blood cells. The cause of this extraordinary condition may be a mild deficiency of vitamin C, possibly combined with back pressure in the veins and engorgement of the vessels during excessive heating. It is difficult to regard this phenomenon as anything other than a pathological condition, yet it is apparently always found in every individual. The suggestion has been made that it may be correlated with a diet of fish and crustacea, but if that is correct why is no similar condition found in *Noctilio*? These observations cannot be properly understood until the results of experimental work are available.

The single species of the genus *Lasionycteris*, the silver-haired bat of Canada and the United States, is one of the species that migrates. It has frequently been found in Bermuda, and has also been recorded as migrating in flocks of about a hundred far from the east coast of the United States, in company with the red bat, *Lasiurus borealis*.

Pipistrellus is another large genus, containing about fifty species, distributed practically all over the world, except in South America. Pipistrelles come out in search of insects early in the evening, and often appear erratically in full daylight. The genus includes some of the smallest and also the largest of the vespertilioids. Their roosting sites are very varied and include all sorts of natural shelters in caves and trees, as well as roof spaces and crevices in buildings; in the colder parts of their range pipistrelles hibernate in winter. Three of four species of *Glischropus*, from south east Asia to Australia and Tasmania, differ from the pipistrelles in the character of the feet and thumbs – there are callosities or pads on the sole of the foot and the base of the thumb which suggest a primitive development of suckers, as found in the thyropterids. Similar pads on thumb and foot are present in the three species of *Tylonycteris*, from south east Asia and the islands to the Philippines; the foot pads are functional suckers. These bats are very small and have curiously flattened skulls, a character which is said to facilitate their creeping into their roosts in the hollows of bamboo stems, where they stick to the inner surface with the suckers. *Eudiscopus denticulatus* is another genus with well developed foot discs. Only a few specimens are known, from Laos, and its roosting habits have not been observed.

The genus *Nyctalus*, inhabiting Eurasia from the Azores to Japan, contains about half a dozen species of noctules. They are stoutly built robust bats with rather narrow pointed wings and rather short dense fur. Like the pipistrelles they are early on the wing, and often come out before sunset. They feed on insects, especially beetles, crickets and the larger moths. A banded specimen of *N. noctula* has been recovered over four hundred and fifty miles from its release point, though it is doubtful whether there is a regular migration; the smaller *N. leisleri* also is known to cover long distances. Noctules live in caves, buildings, and in summer generally prefer hollow trees, where the females segregate into maternity colonies. It was in the European noctule that the histology of the pararhinal holocrene glands was first investigated, by Harrison and Davies in 1949 [211]. The glands are modified and greatly enlarged sebaceous glands which discharge their secretion on to the face through ducts opening into hair follicles. The glands give the muzzle a padded appearance – in some other genera they are much larger than in the noctule, and cause grotesque swellings on the face. Pararhinal glands have been rather overlooked by zoologists, although the glandular swellings of the nose have been known as long as bats have been studied; they are present in many, and probably in all, of the microchiroptera. The shortest-winged of the vespertilionids is the single species of *Mimetillus*, *M. moloneyi*, from west and central Africa to Rhodesia; it has been suggested that the short wings may be correlated with feeding by creeping about in search of insects, but this is no more than conjecture.

Eptesicus is another large genus, with about thirty species; it is worldwide in distribution. The most familiar members of it are the big brown bat *E. fuscus* of North America, and the Serotine *E. serotinus* of Europe and Asia to Korea, and west Africa. There is a wide range of body size in the genus; the colour is generally some shade of dark brownish above and lighter below. These bats have rather broad wings, and fly with a slow fluttering flight near the ground, where they feed on insects of many sorts but appear to take few moths – in captivity they will often attack and eat other bats, but there appears to be no record of such carnivorous habits in the wild, where they frequently share roosts with other species. There is one record of *E. fuscus* being preyed upon by a snake living in the roof space of a farmhouse occupied by the bats [443]. The big brown bat, common in the United States, has been the subject of much study by American zoologists; with its aid they have learnt much of bat biology – food, habits, hibernation, echolocation and reproduction. Two species from the Sudan and from Cape Town are sometimes placed in a separate genus *Rhinopterus*. *E. (R). floweri*, the Sudan species, is peculiar in having the upper ends of the forearms, legs and tail, the inner half of the wings and the interfemoral patagium thickly studded with wart-like papillae – in *E. (R).* *notius* of the Cape the papillae are rather fewer. Four species of *Hesperoptenus*, from India through Malaya to Borneo, are similar to *Eptesicus* in appearance and habits, and differ only in details of the dentition and skull. The big-eared brown bats of the genus *Histiotis* differ from

Eptesicus mainly in the very large ears which are as long as the head and united by a ridge across the forehead. Four or five species are known, all from South America. The single species of *Laephotis* from East Africa and the Congo is very similar, but the ears are not joined, and are proportionately rather smaller. Little appears to be known about the biology of these genera, nor about the rarely seen single species of the genus *Philetor* of New Guinea.

The genus *Vespertilio*, from which the family takes its name contains only two species, one of which *V. murinus*, is distributed across the whole of northern Europe and Asia from England to Japan. These bats resemble *Eptesicus* but have a shorter and broader ear, and the fur of the back has a frosted appearance owing to the whitish tips to the hairs – hence the name particoloured bat. Their roosts often contain large colonies. *Nycticeius* contains about a dozen species found in the south eastern United States, Cuba, most of Africa, India, Australia and New Guinea. The pararhinal glands are well developed and in some of the species in the subgenus *Scoteinus* from Australia they are comparatively enormous. In these species [259] the glandular pads are naked and so large that they make the muzzle extremely broad and nearly flat above; the lower lip, too, is naked and the area surrounding the eyes is also naked and glandular.

Two genera of this family, *Scotomanes* of central and eastern Asia, and *Glauconycteris* of Africa south of the Sahara, are unusual among vespertilionid bats in having contrasting colour patterns. In the first there is a white stripe along the middle of the russet brown back, a white spot on the top of the head, and another behind each shoulder; the half dozen or so species of the latter are marked with white spots and stripes on a brownish to black ground colour. In the latter, too, there is a fleshy lobe at the corner of the mouth connected by a ridge to the ear. Similar lobes and ridges are present in a more marked degree in the wattled bats of the Australian and New Zealand genus *Chalinolobus* which contains four species. Wood Jones [259] describes them as having a wattle-like lobe which projects from near the angle of the mouth and, together with the short snout and high forehead makes the face somewhat resemble that of a pug dog. Nevertheless he refers to *C. gouldi* as 'this pretty little bat' and 'this beautiful little bat' – and regrets that so little is known of the life of any species. One species *C. tuberculatus* is of particular interest because it is one of the only two mammals native to New Zealand; the other is also a bat, *Mystacina* – both species are discussed below.

Three genera are distinguished by their yellowish colour. Two species of *Rhogeëssa*, a genus from central Mexico and the northern parts of South

America, are pale yellow-brown; they are common in arid regions, and roost in hollow trees and house roofs. The single species of *Baeodon* from Mexico, is similar in colour but nothing is known of its life history, whereas some species of the yellow bats of the genus *Scotophilus*, of which about ten species are known from Africa, southern Asia and Malaya, are common house bats which roost in roofs as well as hollow trees, and sometimes in the open hanging from palm fronds and similar places.

The bats of the genus *Lasiurus* also tend to richness of colouring – the red bat, *L. borealis*, is brick or rusty red, the hoary bat, *L. cinereus*, is yellowish to dark brown with frosting produced by the white tips of the hairs, and the yellow bat *L. floridianus*, is yellowish to orange buff. The genus is widespread in America; the red bat and allied species are found from North America to the West Indies, the hoary bat and subspecies from North America and Hawaii, and the yellow bat and related species from the southern United States to the River Plate. Lasiurines are tree bats and roost in the open in trees and bushes singly or in small groups. It is, perhaps, this open air habit which makes it necessary for them to seek a warmer climate in winter for, like *Lasionycteris*, they are migratory in the northern parts of their range, and as already mentioned, they have been seen migrating in company with that species. A flock of about 200 red bats has been seen flying round a ship about sixty-five miles off the New Jersey coast [69] and on several occasions red bats have been found killed by colliding during their migration with the 1,472 feet high Empire State Building in New York [471]. Although the winter range appears to be the West Indies and Mexico, a certain number of the bats remain throughout the winter in some more northern places, where temperatures often fall below freezing; they fly and feed, however, in the evenings of the warmer days [104]. The bats of this genus are insectivorous, but there is a record [41] of a hoary bat killing and eating a pipistrelle – behaviour that is thought to be abnormal. The lasiurine bats are unique in possessing four nipples on the chest; they can thus suckle the exceptionally large litters of up to four young which they produce – most bats bear a single young, many have twins, but the lasiurines are the only ones habitually to have quadruplets. The mother carries the four clinging to her fur until their combined weights exceed hers; she then parks them in the roost while she is away feeding.

Kerivoula is another genus notable for the bright colour of some of its members – it is found from South Africa through India and southern Asia to the Philippines, New Guinea and north-east Australia. *K. picta*, an Asian species, is orange or even scarlet, with black wings and orange stripes along the fingers. In *K. lanosa* of South Africa the light tips of the dark brown fur are shining bronze in colour. This and some other South African species have often been found roosting in the abandoned hanging nests of weaver birds [406], but elsewhere other species roost in hollow trees, among foliage, or in house roofs. The fur in most species is long and rather curly. *Euderma*, one of the three genera of long-eared bats, also has a peculiar colour pattern – the fur is dark brownish black with a white patch on each shoulder, another above the tail and an almost complete white collar on the neck behind the very long ears, which merges with the white of the under-side. The single species, *E. maculata*, from western North America and northern Mexico is very rare and has seldom been seen. The ears resemble those of *Plecotus*, a genus with five species one of which *P. auritus*, the common long-eared bat, is found over the whole of Europe and temperate Asia, and in north Africa. The other species are North American and Mexican.

When sleeping these bats fold their very long ears down along the side of the body under the wings, leaving the rather long pointed tragus of each projecting like a sharp little horn. When alert the bats hold the ears erect and well forward, but when at rest they fold them, not by just bending them backwards, but by curving the inner edge and at the same time throwing the outer edge into a series of tiny accordion pleats so that the ears are furled before being tucked under the arms. The bats of this genus are insectivorous, but although their flight is rather slow, they can hover when examining foliage in search of insects. The more northern species hibernate. The pararhinal pads are well developed, and in some American species they are so big that they form large swellings on the muzzle, which give them their vernacular name of lump-nosed bats. The desert long-eared bat, *Otonycteris hemprichi*, the only species of its genus, distributed from north Africa to south west Asia, is pale sandy brown in colour, sometimes darker. Its large ears superficially resemble those of *Euderma* and *Plecotus*, but the characters of its teeth and skull show that it is more closely related to *Eptesicus*. The two species of *Barbastella*, covering between them an immense area from western Europe to southern Asia, are closely related to *Plecotus*, though their fairly large ears are not narrowly elongated. The ears of the barbastelles are, however, joined at their bases by a short ridge of skin. The bats are insectivorous, and roost in caves and buildings; they are dark in colour, *B. barbastella* of Europe being almost black, with a frosting caused by lighter tips to the hairs on the back.

The long-fingered bats of the genus *Miniopterus* range from Africa and southern Europe through southern Asia to Australia and some of the

Pacific islands. The genus contains about a dozen species, all characterised by the great length of the second phalanx of the third finger, which is folded back under the wing when at rest, taking the patagium of the wing-tip with it, and by the very long tail and large interfemoral patagium. They are insectivorous, and typically inhabit caves, though they are not confined to them; their roosts are sometimes used by large congregations. The common *M. schreibersi* shows considerable variation in colour, from dark brown through reddish to pale buff with ears and interfemoral patagium almost white; it has an extremely wide distribution, from southern Europe to Japan and the Philippines, Malaysia and southern Australia.

Two genera of tube-nosed bats, *Murina* with eight species, and *Harpiocephalus* with one, have nostrils projecting from the snout in the form of tubes, a condition unique in the microchiroptera but similar to that of the megachiropteran genera *Nyctimene* and *Paranyctimene*. Both genera are oriental in distribution; they are insectivorous, but little is known of their habits. The woolly fur is generally dull in colour, but that of one, *M. aurata*, is bright yellow above.

The fur of the pallid bats of the genus *Antrozous* also is woolly, and in *A. pallidus* of western north America from southwest Canada to Mexico, it is very pale in colour, ranging from creamy to light brown; it is darker in the rarer Mexican *A. dubiaquercus*. The ears are large and rather pointed and have the outer edges pleated. The face is short with a truncated muzzle, and bears a low but distinct horseshoe-shaped ridge above the nostrils, behind which the pararhinal glands form a large rather flat swelling on each side. The ridge above the nostrils is a rudimentary nose leaf, a structure more fully elaborated in the next genus. Some individuals of the pallid bat migrate from the north part of their range in winter, but others hibernate. Pallid bats roost in any suitable shelter available, sometimes in colonies of 100 or more, and fly low when hunting. They come out in the dusk before complete darkness has fallen and, according to Howell [240], 'their large ears give them an odd appearance and their squarish forms are not to be mistaken'. These bats are unusual in taking much of their food, such as wingless crickets, especially *Stenopelmatus* the Jerusalem cricket, beetles and roosting grasshoppers, from the ground – they have even been caught in mousetraps, set by naturalists to catch small terrestrial mammals, when attempting to take the beetles and crickets attracted to the baits [246]. It is also thought that they probably catch nocturnal lizards and geckos, judging from the readiness with which they catch and eat such small reptiles in captivity [135]. These bats, like many others, take prey too large to be eaten in flight to dining roosts, where they settle head upwards and hold the prey in the pouch formed by the upturned tail and interfemoral patagium [48].

At the other side of the world the big-eared bats of Australia and New Guinea – about eight species of genus *Nyctophilus* and one of *Pharotis* – form with *Antrozous* the American pallid bats, the subfamily Nyctophilinae, the only vespertilionid bats with even rudimentary nose-leaves. The ears are long and wide with rounded ends so that they are oval in outline; they are joined at the bases by a transverse ridge of skin. Like the ears of the long-eared bats they can be folded by pleating the outer edges and curving the inner edges outwards. The nose leaf is larger than in *Antrozous* and forms a small naked plate above the nostrils, depressed immediately above them, and with the hind margin standing up from the face. The pararhinal glands form conspicuous swellings on each side of the face behind the nose-leaf. These features together with the shortness of the snout give the bats a distinctly snub-nosed appearance. The bats are insectivorous and have been seen hovering over foliage to pick off insects and their larvae in the manner of the long-eared bats *Plecotus* – the similarities in the ear of both genera may be correlated with their similar feeding habits. It may also well be that the pleated ear of *Antrozous* has some functional relation to the bat's ground-feeding habit, which resembles the foliage inspecting habits of the other genera in taking prey from a solid substrate rather than catching it on the wing. It is possible that the prey in these cases is located by its own faint sounds and not through echolocation by the bat. Although some species have been found in large colonies, these bats are usually solitary, or roost in small groups in hollow trees, behind loose bark, or in crevices in rocks and caves.

A single species, *Tomopeas ravus*, from Peru, is placed in a subfamily of its own because although some of its characters show it is a vespertilionid others resemble those of the family Molossidae. Superficially it resembles a small *Pipistrellus* or *Myotis*, but the shape of the ears is like that of the molossids, and the seventh cervical vertebra is fused with the first dorsal, a feature unknown in any other vespertilionid but universal in the molossids.

FAMILY MYSTACINIDAE

It is often loosely stated that there are no mammals native to New Zealand because the islands were isolated before the mammals were able to spread into them from their points of origin elsewhere. The first part of the statement is not correct, for two species of bat are peculiar to New Zealand – *Chalinolobus tuberculatus*, mentioned above, and *Mystacina tuberculata*, which is assigned to the family Mystacinidae in which it is the only species.

Their presence in New Zealand, however, confirms the latter part of the statement because air-borne mammals could obviously arrive at the islands when terrestrial ones could not. The chances of successful arrival must be exceedingly remote but, theoretically at least, the millionth-chance arrival of a single pregnant female could have been enough to establish each species. *C. tuberculatus* is probably a comparatively recent arrival since it is closely related to allied species in Australia. *Mystacina*, on the other hand, probably arrived in very remote times because it has differentiated so much from all other bats that it cannot be classified in any of the other families; it shows affinities with the Vespertilionidae and the Molossidae, and is probably derived from primitive molossid stock. *Mystacina* is a small bat, about two and a half inches long from snout to tail root, clad in dense velvety fur. The ears and tragi are rather long and pointed and the nostrils are prominent; the short tail perforates the interfemoral patagium and appears on its dorsal surface. The thumb has a well developed fleshy pad, and the soles of the feet and the undersides of the toes are covered with loose wrinkled fleshy skin; the claws of the thumb and toes are very sharp, and each has a small talon or secondary claw at its base. The patagium of the wings is thick near their bases, but very thin in their outer part; the third phalanx of the third finger is bony, and not cartilaginous as in the vespertilionids, and the molossids. These peculiarities of the wing allow it to be folded up and tucked under the thicker part near the body when the bat is not flying [126]. All these characters are correlated with the unusual habits of the bats, for although they can catch insects on the wing, they habitually scuttle about on tree branches and rocks catching insects at rest – the claws, the padded thumbs and feet, and the neatly furled wings allow them to run about freely on the ground and climb in trees and rock surfaces with agility [126]. *Mystacina* – or its ancestor – arrived as a flying mammal, and its evolution has taken it a long way back along the path to becoming terrestrial.

FAMILY MOLOSSIDAE

The Molossidae are the sixteenth and last family of living bats; the family contains nearly ninety species partitioned among eleven genera, and is distributed throughout the warmer parts of America, Africa, Asia and Australia, and extends north to the middle of the United States, southern Europe and Korea. The molossids are free-tailed bats, with the rather thick tail projecting far beyond the edge of the interfenoral patagium; they are small to fairly large bats – some species are as much as six inches from snout to tail root. They are robustly built, and the patagia and ears are thick and leathery; in most species the thick fleshy lips are vertically wrinkled, and the nostrils open on a slightly elevated pad at the end of the truncated snout. The wrinkled upper lips overlap the lower, and in many species, especially the mastiff bats of the genus *Eumops*, the drooping lips at the corners of the mouth give a bloodhound expression to the face. The ears are large but the tragus is small, and a lobe arising from the outer margin of the ear near its junction with the face behind the angle of the mouth, the antitragus, is conspicuous. The wings are long and narrow and, in correlation, the fifth finger is short. The legs are short and stout, and the fibula is well developed, a feature unique among the bats with the exception of *Mystacina*. The feet are short and broad with rather fleshy soles and the outer edges of the first and fifth toes bear fringes of stiff bristly hairs, in many species with spoon-shaped ends. Wood Jones [259] suggests that these special hairs are used in grooming the coat as hair brushes in addition to the comb formed by the claws of the toes. Similar stiff spoon-shaped hairs are present also on the snout of many species but their function there is not obvious.

When the bats are at rest the interfemoral patagium is thrown into transverse folds because it has considerable mobility on the tail over which it can slide up and down. When the wings are folded the second phalanx of the third finger is reflexed or turned forward, taking the third phalanx and the tip of the wing with it. Most molossid bats are agile runners and climbers; Goodwin and Greenhall [175] report that *Molossus ater* of the American tropics 'scuttles over the roost walls and floor, mouse-like with the tips of the folded wings tucked up under the armpits'. On the other hand, the long narrow wings and the structure of the shoulder joint, which is more specialised for flight than in other bats, allows them to fly very fast and with great manoeuvrability. The molossids thus have the best of both worlds for they are highly competent as quadrupeds and exceptionally good as flying machines.

The molossids are insectivorous and emerge from their roosts to forage shortly before dark. They roost, often in large colonies, in many different kinds of natural cavities from caves, rock crevices, hollow trees, cavities behind loose bark, to the shelter of drooping palm leaves, but they also inhabit buildings and are particularly partial to roosting immediately under galvanised iron roofing where the day temperature may reach 130°F. Howell [240] found that *Tadarida brasiliensis mexicanus* was the most ubiquitous bat in California and said,

Not only are its colonies the most numerous but the most populous, and in northern Mexico hundreds of thousands may occur together.... They may be crowded into the corner of an attic, or behind a wooden sign on a building, where one would lay odds that they would be cooked to a crisp by the mid-day sun, but at late dusk out they swarm through a hole that seems too small for a shrew to pass, and away into the gathering darkness.

Some species of molossids are seasonally migratory, but although others may become torpid when roosting in cool weather none of them have a true hibernation. A glandular pouch under the throat is found in many species of molossids; it has more frequently been noted in males than females, but it may well be more universal than has hitherto been recorded because it is conspicuous only when periodically active in the males at the breeding season, and can easily be overlooked at other times and in the females. Some species of tropical and subtropical American molossids have been found to be carriers of rabies but there is no evidence that they normally transmit the disease to man or his domestic animals as does the vampire bat.

The large ears, generally directed forwards when the bats are awake, and in some species looking like fantastic hats, and the voluminous wrinkled fleshy lips give the molossids a very peculiar facial expression. The most bizarre of the family are undoubtedly the two species of the genus *Cheiromeles*, the naked bats of Malaya, the East Indies and the Philippines. They are large bats, the head and body measuring over five inches in length, with fur so short and sparse that they appear to be completely naked. The snout is long and truncated, almost like that of a pig, and the ears, unlike those of most molossids are short and widely spaced. The skin round the neck and under the chin is thrown into loose folds – as is that of the interfemoral patagium and hind end of the body when the patagium is pulled up along the tail if the bat is at rest – so that the animal looks as though it were wrapped in a blanket or loose-fitting robe. A fold of skin, in addition to the wing patagium, extends along the side of the body from wing to leg forming a pouch in which the wings are stowed with the aid of the feet, the large first toes of which are opposable like thumbs, so that the bat can walk about even more freely than *Molossus* with the tips of the folded wings tucked away under the armpits. The pouches are not used, as was formerly thought, for carrying the young – they are present in both sexes. The fold of skin under the throat forms a glandular sac which produces a secretion with a strong musky odour – all the molossids, especially the males, have a similar strong smell. Naked bats are insectivorous, and roost in hollow trees, cavities in rocks and the earth, sometimes in large colonies of a thousand or more.

The genus *Tadarida* (formerly *Nyctinomus*) is a large one containing about sixty species found in the tropics and subtropics all round the world;

it contains a number of subgenera which were once regarded as separate, but are now more logically brought together in the one. The tadarid bats are small to medium in size, and have the ears, lips and long free tails typical of the molossid family. In some species the fur of the back stops short of the tail, leaving a bare patch behind. They roost in hollow trees, buildings or caves often in large colonies. The most spectacular congregations are those of the American *T. brasiliensis* and its many subspecies, sometimes known as the guano bat. Bat guano is rich in phosphorus and nitrates and thus is a valuable fertiliser; where large deposits occur in accessible caves they have been exploited commercially in many parts of the world. The most famous deposits are those in certain caves in Texas and New Mexico where *Tadarida* roosts in almost incredible numbers – the Carlsbad Caverns in New Mexico have a population of about eight million *Tadarida*, and the Ney Cave in Texas has between twenty and thirty million. The number of insects eaten by such huge populations must be staggering. These bats have played a part in at least two human wars; during the American Civil War the Confederates, according to T. Norris writing in 1874 and quoted by Allen [3], extracted nitrate from the guano of the caves for making gunpowder. The most fantastic use of bats, however, is the bat-bomb which was invented in America during the Second World War [335]. Small 1-oz incendiary bombs, with delayed action fuses, were strapped to bats, which were packed in containers holding from one to five thousand. The bats were *Tadarida* from the Ney Cave, and if kept at a low temperature they needed no food or attention for long periods. The container, attached to a parachute was thrown out of an aeroplane and arranged to open at 1,000 feet to let out the bats, which scattered widely and settled in the target area; in one trial a dummy village was successfully burnt down. The perfection of atomic weapons, however, made the use of bat-bombs unnecessary – man's inhumanity to man is matched only by his brutality to the brutes (Plate 9).

Some at least of this species are migratory, for it is estimated that over sixty million of them migrate annually between their winter roosts in Mexico and their summer ones in the southern United States. This species, too, is one of the few mammals in which the young are not necessarily fed on the milk of their own mothers – the naked young are not carried by their mothers but left in the roost during the evening flight, and are suckled indiscriminately by the lactating females on their return.

Some species of the subgenus *Tadarida*(*Chaerophon*) which contains about twenty species in Africa and Indo-Malaysia, are distinguished by the presence of an erectile crest of long hair on the crown of the head in the males. In the allied genus *Platymops*, containing three African species, the skull is much flattened; this feature is thought to be correlated with their habit of roosting in rock crevices. The life histories of the bats in the related genera *Otomops* and *Xiphonycteris*, are little known; the half dozen species of *Otomops* from Africa and southern Asia to New Guinea are distinguished by their large size and huge rounded ears almost like an umbrella over the head – Dorst [117] described *O. madagascariensis* from specimens found in a cave; *O. formosus* from Java has, however, been found roosting in a hole in a tree. Bats of this genus are rare, at least in museums. The single species of *Xiphonycteris*, known by only a few specimens from west Africa, is distinguished by the exceptionally long pointed upper canine teeth.

The genus *Molossus* from which the family takes its name contains about a dozen species from Central and South America – the name refers to the facial resemblance of the bats of this family to dogs of the mastiff type; the Molossian shepherd dogs were famous in ancient Greece. These bats closely resemble those of the genus *Tadarida*, but have only a single lower incisor and a small tragus. Some species, such as *M. ater*, occur in two colour phases, with predominantly red or black fur. This species has large cheek pouches and when 'they are stuffed and crammed to capacity' according to Goodwin and Greenhall [175] 'the bat returns to its roost to chew and swallow its food'. These authors also reported that both sexes have large glands in the mouth over the canines; the orifices of the glands are open when the bat is feeding. The four species of *Promops* have a similar distribution, and are distinguished by having a ridge down the centre of the face from the junction of the ears to the end of the snout. About half a dozen species of *Molossops*, another Central and South American genus, are small to medium-sized bats with the ears not joined across the forehead; they roost in small colonies in hollow trees, and one species has been found actually burrowing into the soft rotten wood of a dead stump. Little appears to be known of the life of the single species of *Eomops* from tropical Africa; it has rounded ears with a small tragus, and a rather long snout.

The eight species of large mastiff bats of the genus *Eumops* are inhabitants mainly of tropical America, though two or three species extend into the southern United States. *E. perotis* has been studied in California by a number of American zoologists. They find that it lives in small colonies in the fissures in cliffs, but that it also often roosts in buildings when available. The bats prefer roosts from fifteen to twenty feet above ground as their long narrow wings necessitate a good drop for easy launching into flight, though they can if necessary spring into the air from the ground. They are, however, peculiarly reluctant to fly except when departing from the roost for a night's foraging. Howell [239] found them very agile when on the ground: 'With wing tips folded above the back so as to be out of harm's way, they scutter across the floor in a sort of gallop, while the action of their "arms" reminds one of nothing so much as an "over-hand" swimmer.' If these bats reach a vertical surface when running on the ground they turn and climb up it backwards, feeling for footholds with the outstretched feet – they share this method of climbing with the horseshoe bats of the family Rhinolophidae, and probably with some others. Krutzsch [281] found that a colony in a cliff roost became active in the late forenoon and early afternoon; the bats moved about and squeaked loudly for some time, but remained quiet from about 4.0 p.m. until they came out to forage after darkness was complete. *E. perotis* has a large mouth which can open to a very wide gape, yet it is believed to feed on small kinds of insects, and Howell asks 'if it habitually feeds only upon small fry, then for what is that huge mouth?' Bats of this genus do not migrate or hibernate, though in the northern parts of their range they may become torpid for short periods during cold spells in winter. The throat gland so frequently found in molossid bats is present in both sexes, according to Krutzsch, and is particularly large in the males during the breeding season. It produces an oily greyish-white secretion with a strong smell – it is suggested that the odour attracts the females, but this function seems to be hardly necessary in bats which roost in colonies of mixed sexes, and even less in species such as those of *Tadarida* which roost in millions. The secretion may perhaps more probably be a sexual stimulant. Howell and Little [242] suggest that the nodular callosity at the base of the thumb may represent a 'modern degeneration of some sort of a suction disc which . . . enabled the distant ancestors of the bats to clamber about in precarious situations. . . .' They base this speculation on the large size of the thumb and pad in the newly born young. They also think that the specialised hairs on the outer sides of the first and fifth toes, which are present in the otherwise naked young, are sensory and have a tactile function when 'the bat is backing into and about dark crannies.'

With the mastiff bats we come to the end of this survey of the families of the chiropterans. Much is known of the biology of many of the enormous number of species, although much more awaits discovery; yet with all our knowledge of them one cannot help feeling how remote is even the familiar little flittermouse on its evening flight after roosting all day in the attic above our heads. The lives of all the peculiar and to us often bizarre and grotesque forms that live in less accessible regions must be at least equally strange; the bats are so withdrawn from our terrestrial world that a real understanding of their lives can be gained only by long and diligent research. To the comparative anatomist, whose studies are less difficult, they present a wonderful example of the infinite variations that can be developed on a single theme – how similar they are in their adaptations for flight, and yet how diverse in all their details.

Chapter 4

Anteaters and Armadillos; Pangolins; Aardvark

IN THE orthodox system of the classification of the mammals the order Primates follows the Chiroptera. Both orders, in spite of their specialisations, have many characters that show their affinity with the Insectivora and the hypothetical generalised primitive mammalian stock. Primates have always attracted particular attention from naturalists because they form the order to which man himself belongs; in the last two decades the attention paid to them has enormously increased. The intensified research has been directed mainly to the behaviour and to the physiology of these animals because of the importance of these subjects to sociology and medicine as applied to man. 'Primatology' has become almost a separate discipline within zoology, and this trend is reflected in the treatment of the order in this Natural History – *The Life of Primates* by Professor Schultz, deals solely with the Primates, which will therefore not be further discussed here.

ORDER EDENTATA

The three orders mentioned above form a logical grouping more closely related to each other than to other orders. We now turn to orders that stand far apart and have no obvious close affinities with them. The edentates or toothless mammals, in which the members of only one of the three living families are toothless, are of obscure origin – Simpson [445] considers that they 'arose from proto-Insectivora, along with several other archaic orders, about the beginning of the Paleocene.' They are now, and always have been, exclusively American; the order as a whole, with the exception of the extinct suborder Palaeanodonta, evolved in South America, and those members of it found in North America either living or as fossils, migrated from the south. Apart from the palaeanodonts the other suborder into which the order is divided, the Xenarthra, contains a large number of extinct genera and species, the most familiar of which include the extinct giant ground sloths, some of which were contemporary with man and persisted almost into historical times, and the enormous glyptodonts protected with domed armour superficially resembling the shell of a giant tortoise. Most of the host of species have long ago disappeared leaving only thirty-one living species which are divided into three families, the anteaters, the sloths and the armadillos. The first two families form the infraorder Pilosa, the hairy ones, and the third the Cingulata or belted ones. The name Xenarthra refers to the presence of extra joint-processes on the arches of the posterior vertebrae. The anapophyses or lateral apophyses of each lumbar vertebra articulate with special processes on the following one, and the pelvis is attached to the vertebral column by the ischia as well as by the ilia, so that the posterior part of the column is unusually rigid. Frechkop [156, 157] correlates this rigidity in the armadillos with the gait, which is plantigrade on the hind feet and more or less digitigrade on the fore, and with the characteristic hunched attitude of some species when sitting.

FAMILY MYRMECOPHAGIDAE

The family Myrmecophagidae contains three genera of anteaters, one containing two species, the others one apiece. The members of all the three genera have long snouts, toothless mouths, long tails and fore feet bearing large elongated claws. All feed on insects, mainly termites and ants, gathered by the long extensible tongue covered with sticky saliva secreted by greatly enlarged salivary glands lying in the neck.

The giant anteater, *Myrmecophaga tridactyla*, is the largest and oddest looking species; it inhabits forests and savannahs from Central America to northern Argentina. It is a fairly large animal, over six feet long from snout to tail-root, covered with coarse shaggy hair except on the head where it is short. The colour is greyish, with a dark diagonal stripe with white edges running from beneath the throat to the back above the shoulders. The tail is about a yard long and covered with long coarse hairs which touch the ground when it is carried horizontally; it is likened to a flag in the Brazilian name of the animal, 'tamanduá bandeira.' The eye lies in front of the small ear at the base of the greatly elongated snout which carries the nostrils and tiny mouth at its tip. The hind feet are plantigrade, but the front ones, which show four fingers – a rudimentary fifth is present under the skin – bear two claws so long that when the animal walks they are turned inwards so that it goes on its knuckles. The anteater tears open the hard mud of termites' nests with these powerful claws and gathers its food with the long mobile and sticky tongue. Giant anteaters are solitary and appear to have no permanent homes; they lie up under the shelter of bush and cover themselves with the tail when sleeping – they do not burrow. It is unlikely however that they are nomads, and probably each has its own territory. If disturbed they make off at a clumsy gallop, but can be overtaken by a running man, and if cornered they can inflict serious wounds by lashing out with the claws in defending themselves. The females, like those of other genera, give birth to a single young, which rides clinging to the back of its mother during infancy. Although anteaters are so specialised for a diet of termites and ants they can be accustomed to eating a mixture of raw egg and minced meat in captivity, and flourish upon it. Attenborough [15] tells of a newly captured anteater that became so fond of its new food that if a dishful was inadvertently left eighteen inches from its cage 'it would stick its nose close to the bars, flick its tongue across the intervening distance and lick up half the contents of the dish' before its keeper was aware of what was happening.

The much smaller tree anteaters or tamanduás, *Tamandua tetradactyla* and *T. longicaudata*, are forest animals and have long nearly naked prehensile tails. The fur, though coarse, is softer than that of the giant anteater and varies much in colour between individuals; it is generally buffish, with a dark collar in front of the shoulders. The snout is similar to but proportionately shorter than that of the larger species. The third digit of the hand bears a large strong claw; the three others have smaller ones. The skin is remarkably thick and tough, but with the thick wiry fur it is not the complete protection against ants that it is sometimes supposed to be. Attenborough [15] saw one slash open an ants' nest in a tree whereupon 'a brown flood of ants flowed out of the gash and swarmed all over the tamanduá which, not in the least dismayed, stuck its tube-like snout into the hole and began licking up the ants with its long black tongue'. Soon, however, it began scratching itself with a hind foot, and then with a fore paw as well, and after a moderate meal descended to the ground and continued scratching the angry ants out of its fur. Enders [134] also records that a tamanduá was attacked by ants when feeding, and found that 'many ants had attached themselves in the tenderer portions of the inguinal region, and under the axilla, clinging on even after they had been killed. All the specimens that Enders came across resisted capture by sitting on the hind quarters, supported by the tail pressed against the ground, and fighting with the fore feet. One fought so well that 'after some time she gained a small tree up which she went in spite of all we could do; it was impossible to shake her hold, so she passed to a larger tree into the top of which she disappeared'. Female tamanduás are the only mammals known to suffer from infestation with parasitic nematode worms in the ovaries. Wislocki

[514] reported two cases from different parts of Central America in which he found filarial nematodes encysted in the ovarian stroma.

The pygmy anteater, *Cyclopes didactylus*, is the smallest of the anteaters, and is no larger than a squirrel. It is nocturnal and arboreal, and though it can walk on the ground like the others, it only does so when going from one tree to another. Its fur is dense and silky-soft, greyish yellow in colour with no dark markings. The snout is comparatively short, the tail long, prehensile and naked on the underside; there are two claws only on the fore feet, that on the third digit is the largest and needle sharp. The hind foot is highly adapted for climbing; Wood Jones [260] showed that the claws of the four toes oppose against a naked plantar pad supported by a tibial sesamoid bone, and that the flexor muscles of the toes are unusually powerful. The animal is thus able to cling to a vertical stem with its tail and hind feet, leaving the fore feet free for tearing open the nests of ants, or for defence against predators. Van Tyne [487] found that even at night the pygmy anteater moves about with slow deliberate movements; in the day it was more sluggish and kept its eyes half shut to mere slits when in direct sunlight, and if undisturbed it soon curled up in a crotch and fell asleep. When startled, 'with a peculiar little sneeze it would suddenly let go with its fore feet and straighten out stiffly, clinging to the perch with only its hind feet and prehensile tail . . . the fore feet are held stiffly on each side of the nose and partly covering the eyes.' It held this defense position for about a minute before relaxing if danger seemed past. Van Tyne reported that it was very gentle and never attempted to claw those who handled it. Others have found that a quick stroke from the very sharp claws when the animal is in the peculiar stance is very painful.

FAMILY BRADYPODIDAE

About half a dozen species of tree sloths form the family Bradypodidae in which there are two genera, *Choloepus* containing the two-toed sloths, and *Bradypus* the three-toed sloths. As they are entirely arboreal they are confined to the forests of Central and tropical South America. They are between two and three feet in total length, with small rounded heads, inconspicuous ears, and eyes directed forwards. The tail is rudimentary, and the limbs long, the arms longer than the legs; the toes and fingers are syndactylous, but the long curved claws with which they are armed are separate. The English vernacular names are misnomers for all sloths have three toes; it is in the number of fingers that they differ. Sloths are exclusively vegetarian, and spend much of their time moving sluggishly suspended below the branches of trees by their claws as they browse. The outer fur, covering a softer under fur, is coarse and shaggy, and over most of the body and limbs is reversed in direction in correlation with the inverted position. The hairs are minutely grooved, giving lodgement in some species to unicellular green algae so numerous that the coat of the animals has a green tinge that is believed to be cryptically protective. The shaggy coat has a rather moth-eaten appearance but although the fur may be inhabited by hundreds of small pyralid microlepidopterous moths it is not apparently eaten by their larvae – the adult moths merely use the fur as a hiding place, and the larvae feed on the scurf shed by the epidermis.

The anatomy of animals of such peculiar habits shows some interesting points. The cheek teeth, five above and four or five below, have open pulp cavities and grow throughout life to replace the loss at the crowns by wear through chewing the leaves which form the bulk of the diet. They have no covering of enamel over the dentine, but are coated with cement. There are no incisors or canines, but in *Choloepus* the front teeth of both series are large and caniniform. The bradypodids are one of the few exceptions to the general rule that mammals have seven neck vertebrae; *Bradypus* usually has nine, and *Choloepus* only six, but the number is not constant in all specimens. *Bradypus* can thus turn its head through an angle of 270°; it has been supposed that this wide movement is an adaptation to the suspended arboreal habit. As *Choloepus* has similar habits, but only six neck vertebrae, the suggestion is not very convincing. Nor has an acceptable explanation been put forward for the widening of the pelvis in the females of *Bradypus* by the presence of an interpubic bone which, according to Frechkop [159] is the ossified interpubic ligament. The function, too, of the retia mirabilia – the plexuses of small arteries and veins – round the joints of the limbs is not fully understood. Similar, but less complicated, retia are present in the limbs of the other edentates, both anteaters and armadillos.

According to Wislocki and Enders [519] the lowest and most labile temperatures recorded in mammals are found in the sloths and the monotremes. They found that sloths, anteaters and armadillos 'all possess extremely low body temperatures subject to wide latitude of fluctuation influenced by the temperature of the environment', and suggested that the sluggish mode of life of the sloths leads to their extremely low temperatures. They record the temperature of *B. cucilliger* as ranging from 27·7° to 36·8°C, and that of *B. griseus* from 24° to 33°C, according to the ambient temperature. They also found that the sluggish pygmy anteater *Cyclopes* has a lower temperature than the other members of the Myrmecophagidae, though higher than the lower end of the range in sloths.

The slow movements and the incomplete temperature regulation of sloths appears to be correlated with the comparatively small bulk of muscular tissue. Britton and Atkinson [52] found that the skeletal muscles weigh only 25 per cent of the body weight whereas those of most mammals are about 45 per cent, and suggested that lack of heat-producing muscle explains in part the animal's thermotaxic inadequacy. These authors report that sloths when cooled do not shiver to increase their temperature, and that when placed in 'ordinary sunshine (35° – 40°C) showed a sharp rise in body temperature to approximately the lethal level (40°C) in about two hours'. The animals died from hyperthermia after panting and showing extreme distress; normally they always sought the deepest shade in their open air pen. On the other hand Morrison [338] found that the temperature of a pregnant *Bradypus* dropped only 6°C when she was placed in an ambient temperature of 10°C, in contrast to the 10° – 13°C fall found in non-pregnant females by Britton and Atkinson.

The female sloth has one young at birth, and carries it about clinging to her fur for a long time, probably for nearly two years, by which age it is nearly fully grown. Enders [134] caught a young *Choloepus* in Panama and kept it as a pet. During his travels the animal was subjected to temperatures from below freezing to 100°F but survived in spite of its own temperature variations. When first taken from its dead mother it clung to its owner's clothes during the day. On her first night she was put in a box with some heated stones, but when they cooled she awoke and 'cried repeatedly until taken into bed by my daughter. Attaching herself to the child's clothing she settled down and slept the night through'. The animal always pressed its underside against its 'mother', to keep warm in cool weather, but the action did not entirely depend upon the ambient temperature, for

At sea level, when the nights were so hot that neither covers nor night clothes could be tolerated by a human sleeper, the sloth continued to sleep with her belly contact with her bedfellow. At lower temperatures when the bedfellow wore clothing, it was comparatively easy for her to keep this contact; her claws grasped the clothing firmly, but she soon learned not to use her claws on the naked skin.

In correlation with the diet of the leaves of trees the stomach of the sloths is comparatively large and complex, and has several caeca or blind chambers connected with it. *Choloepus* can feed on the leaves of many kinds of tree, but *Bradypus* is generally believed to be confined to one species of *Cecropia*, a limitation that makes the successful maintenance of the genus in captivity outside its native land extremely difficult. On the other hand Carvalho [70] found that numerous specimens released in the park of the Museu Paraense in Brazil were not so limited. The park contains many kinds of Amazonian trees but in spite of the abundance of *Cecropia* the sloths were rather seldom seen on them; other trees were liked at least as well, and the animals were seen feeding on trees of the order Euphorbiaceae, Leguminoseae and Bombacaceae. 'The animals settle near the distal end of a stout limb, pulling the leaves with the hands, eating them and dropping large amounts. . . . They are also frequently found on the ground.' They spent the night and rainy spells 'on the forked branches of any available tree, resting on their back and hind limbs and looking very much like a termite nest'. It has been suggested that sloths may spend years or even their whole lives in a single tree; although this appears to be an exaggeration it is true that they are not wide ranging animals and live their lives in a very restricted area, which is a defended territory. A trespasser into another's territory is attacked and driven away; Ingles [250] saw two male *Bradypus* on Barro Colorado Island in Panama 'fall thirty feet to the ground from a palm tree. Although still obviously stunned, they held on to each other with their hind feet. Previous to the fall there had been considerable shrieking. . . . After about thirty seconds both animals released their grips on each other and slowly began to climb the lianas'. This, presumably, was a territorial fight. Sloths are not able to stand on the ground but lie spreadeagled and pull themselves along mainly by hooking the fore claws into irregularities; they can thus move from one tree to another where branches are not close enough for an arboreal passage. Enders' pet *Choloepus* after many experiences of being put to crawl on the floor 'became very proficient, but she never failed to climb off the floor at the first opportunity'. It is perhaps surprising to learn that sloths can swim well – back upwards, not upside-down – and can cross rivers to fresh territory, perhaps when seeking a mate, as they are habitually solitary. Three-toed sloths (*Bradypus*) in the park at Belem easily escaped from the islands by swimming across ten feet of water; Carvalho further records one seen swimming across a river 100 feet wide 'with no signs of discomfort'.

The sloth's digestive and other metabolic processes seem to be as slow as its muscular movements. Britton and Atkinson [52] found that the comparatively large stomach with its contents of chewed leaves varied from 20 to 25 per cent of the total body weight, and even after fasting for several days it was from one-half to three-quarters full and was about 15 per cent of the body weight. Enders' pet sloth emptied its bladder 'at approximately six day intervals, and the colon a few minutes after the bladder'. The food remains in the stomach for seventy to ninety hours, and takes at least a week to pass through the alimentary canal. Honigmann [234] experimenting with animals in the London Zoo, found that the last traces of a meal were not voided for over a month. Krieg [274] found that *Bradypus* descends from its tree to defaecate at a particular spot on the ground, and that several animals habitually use the same latrine – although the animals are normally solitary they thus show at least a simple form of social behaviour which probably keeps neighbours in touch with each other, a necessity when the breeding season arrives.

FAMILY DASYPODIDAE

The family Dasypodidae, containing the armadillos, is the only one with living members in the infra-order Cingulata. The family is richer in species than the others of the order, and contains twenty-one among nine genera. The characteristic armour of these animals consists of a mosaic of small plates of bone developed in the lower layer or dermis of the skin; they are covered with horny epidermis. The plates form a shield on the front of the head that protects the forehead and snout, a shield over the shoulders and another over the rump, separated by separate transverse bands of plates ranging from three to thirty in number according to species. The shoulder and rump shields are distinct in some species, less obvious in others. In all but one genus the tail is covered or ringed with plates, and the exposed outer and lower parts of the limbs and feet are also protected by small plates. The edge of the shell overlaps the side of the body where it joins the soft skin of the underside, and this reduplication is carried to an extreme in the pichi-ciegos. The soft skin between the joints of the shields and between the bands of plates gives flexibility that allows all the necessary body movements; some species can even roll up into a completely armoured ball when disturbed, in spite of the rigidity of the lumbar region of the spine. The underlying panniculus carnosus or skin muscle is well developed in all species, and specially modified in those that can roll up completely. The skin of the underside, and frequently that between the plates on the back, is covered with coarse hair, except in the pichi-ciegos in which it is soft and silky, the amount on the upper side ranging from practically none to a conspicuous coat.

Armadillos range in size from that of a rat to that of a small pig weighing over 100 pounds. They are stout bodied, rather squat creatures, with short muscular legs armed with strong claws, often reduced in number but large in size; in walking or trotting the whole of the sole of the foot is in contact with the ground, but the species with large fore claws usually place only the claws and not the palm of the hand on the ground. The claws are used in digging for food and, in some species, for burrowing; the species that live in burrows generally make nests within them so that in the cooler parts of their range they can keep warm in spite of their rather poor powers of temperature regulation. Animals of this family are distributed from the southern United States to Patagonia, and inhabit both forests and open country. They are described as being insectivorous, herbivorous and carrion-eaters; most species, however, eat any invertebrates they can dig out, vegetable matter, small vertebrates such as lizards, snakes, and mice if they can catch them, and some species are attracted to carrion. A few feed especially but not exclusively on ants, which they capture with their long narrow tongues. The teeth are small and peg-like, without a covering of enamel, although enamel is present on the teeth of a few species at birth; teeth are present only at the sides of the jaw and absent from the front. The number ranges from seven a side to as many as twenty five in the giant armadillo.

As with some other mammals such as some deer, seals and mustelids, delayed implantation is a normal stage in the reproduction of armadillos – in all species save one, as far as is known, only one follicle normally discharges its ovum at oestrus, and fertilisation occurs as usual in the fallopian tube on passage to the uterus. An abnormal double ovulation in a nine-banded armadillo enabled Buchanan [57] to show that the blastocyst can implant only at the fundic tip of the uterus where large blood sinusoids lie close to the uterine epithelium. One blastocyst had implanted in the normal position, but the second which had grown to the enormous size of 1·3 cm, lay free in the cavity of the uterus unable to make attachment elsewhere, and destined to perish from lack of nourishment. The ovum divides to form a blastocyst which does not immediately become attached to the uterine wall at the fundus but remains dormant during two to four months according to the species. When development is resumed, however, what happens is quite unlike what normally occurs in other mammals for, with few exceptions, the single blastocyst gives rise to several embryos – the condition is known as polyembryony. The original embryonic plate divides to produce identical quadruplets or, if more buds are given off by the plate, as many as twelve identical siblings all with the same genetic constitution. The embryos are synchorial and are nourished by a common placenta, and although each is enveloped in its own amnion all the amnions are connected with each other by tubular extensions opening into a separate common amniotic cavity. On the other hand a few species normally have only one young at birth; and the hairy armadillo produces twins from two separate ova, although they appear to be synchorial owing to the fusion of the membranes.

The uterus in the edentates is of the simplex type, a single chamber without paired cornua. A uterus simplex is perhaps more suitable for the production of a single foetus, though it can of course sometimes bear twins or larger numbers in for example the Primates, and conversely the bicornuate uterus of many mammals bears but a single foetus. It has been suggested that all the edentates were originally adapted to bearing a single young at a birth, and to carrying it clinging to the body of the mother, as the homeless anteaters and sloths still do. But when the armadillos took to burrowing and having a fixed home with a nest, it was no longer necessary to carry the young about, and a litter became possible again. Instead of reverting to multiple ovulation, however, polyembryony became established as the means for producing a numerous family. There is, of course, no proof that this attractive speculation is a correct interpretation of the peculiarities of reproduction in the armadillos.

The genus *Dasypus*, from which the family takes its name, contains five species according to Hamlett [200]; they are the nine-banded armadillos, although the number of bands ranges from eight to eleven in *D. novemcinctus* and from six to eight in the other species. They have long tails, and ears proportionately longer than those of other genera. *D. novemcinctus* is the most widely distributed species, found from northern Argentina to the southern United States where it is extending its range and is working northwards to about 33°N, where it is stopped by the occurrence of frosts which it cannot withstand. *D. hybridus* is the common 'mulita' of the Argentine. The dried shells of animals of this genus are commonly made into baskets, with the overarched tail forming the handle, for sale to tourists in Texas, Mexico and the countries of South America.

D. novemcinctus has been more closely studied than any other kind of armadillo owing to its wide distribution, abundance and, particularly, to its common occurrence in the southern United States where it is easily available for the researches of zoologists [263, 264]. In Texas this armadillo commonly digs its dens in the banks of dry streams, and the animals generally make up to four or five dens for occasional use in addition to the home den containing the nest. Dens vary from two to fifteen feet in length, and lie from a few inches to four feet below ground. When bush clearing exposes the stream banks to the trampling of cattle the dens cave in and can become a serious cause of soil erosion when floods follow. The nest is made of dry vegetation and is big enough 'to fill a bushel basket'. Armadillos collect the nest material by raking vegetation with the fore feet well back beneath the body so that it rests on the hind feet. The body is then lowered so that the mass is held against the upper surface of the hind feet, and the animal slowly backs into the burrow tail first. Eisenberg [130], however, saw these armadillos in captivity carry straw to their nest box by hopping backwards and switching the tail from side to side; he saw similar behaviour by *Tolypeutes*, but captive *Priodontes*, *Euphractus* and *Chaetophractus* were not seen to gather nest material, although it is probable that they do so in the wild.

Armadillos dig, according to Taber [465], by using both the nose and the fore feet for loosening the soil, which is pushed into a small pile under the body. The animal then

balances on fore feet and tail, meanwhile placing both hind feet simultaneously over the pile. In this position the body is strongly arched and a sudden tension of the muscles straightens the animal, lifts the hind end off the ground and, as both hind feet are moved backwards in a sudden thrust, throws the pile of soil several feet from the entrance of the burrow. The extremely sharp cutting edges of the claws and the short muscular legs enable the armadillo to dig with amazing rapidity. A captive completely buried itself in two minutes in soil so packed that I required a pick with which to dig it.

Once the animal is dug in it is almost impossible to pull it out; the tail is an excellent handhold, but pulling on it only fastens the animal more firmly because it arches its back and digs the edges of the shell into the earth – Newman [347] says that even if the tail breaks the animal still holds on.

Armadillos remain in their burrows if the weather is cold for a few days, but they do not become torpid although their temperature control is poor [467]. In very dry weather they wallow in mud and ponds; they can swim well and also walk submerged on the bottom of shallow waters. Taber was unable to confirm the story that they swallow air to make themselves buoyant when swimming. In foraging

the nose is pushed into the surface litter as far as possible and when an insect is detected, a conical hole is quickly dug by short, alternate strokes of the fore feet. With each thrust of the feet the nose is pushed deeper into the soil, not once being lifted from the probe while digging is in progress. . . . The ability to locate insects beneath five or six inches of soil is uncanny. I have watched an armadillo shuffle quickly across the forest floor with its nose ploughing a furrow through the litter down to mineral soil. A sudden halt and a sniff initiated a probe resulting in the capture of a beetle larva. Hesitating only long enough to crush and swallow the grub, the animal continued its ploughing.

Similar foraging behaviour was recorded by Enders [133] on Barro

Colorado. Terrestrial insects form the main part of the armadillo's diet, and they are captured by the long protrusible tongue covered with sticky saliva. The salivary glands produce such large quantities of sticky mucous secretion that it is stored in a special salivary reservoir or bladder, as shown by Shackleford [432]. On the other hand armadillos are not restricted to a diet of insects, for Newman and Baker [346] saw one dig up a rabbit nest and devour three four-inch young rabbits in five minutes, and Hamilton [199] found that in August the pulpy fruits of the black persimmon (*Dyospiros texana*) formed 80 per cent of the food near San Antonio, Texas.

Three genera of armadillos are more or less hairy, the three species of *Chaetophractus*, the peludos or hairy armadillos, the single species of *Euphractus* or six-banded armadillo, and the single species of *Zaedus*, the pichi of the pampas of Argentina and Patagonia. These are small to medium sized armadillos with tails of moderate length and small ears; their underparts are thickly covered with coarse hair, and bristly hairs sprout between the plates of the back armour – thickly enough in some specimens of species in the genus *Chaetophractus* almost to conceal the back part of the trunk. Hair is very scanty both above and below, on the giant armadillo, *Priodontes giganteus*, the only species of its genus. This large animal, weighing a hundred pounds or more, is found mainly in tropical South America, and is unique among the dasyposids in its coloration, dark brown on the back with a light buff, almost white, band along each side of the shell. The legs are short and stout, and are armed with large claws, that of the third finger of the fore limb being enormous. The animal commonly sits on the hind limbs supported by the stout moderately long tail, and even when running it is almost bipedal and only touches the ground lightly with the huge claws of the fore feet. It is a powerful digger, and if pursued and cornered it can escape by burrowing with amazing speed – once it has started digging it is almost impossible to hold it back. The five species of *Cabassous*, the naked- or soft-tailed armadillos, a genus widely distributed through Central and South America, are large animals though smaller than *Priodontes*; their colour is dark brown but they lack a distinct light band on the sides. The fore feet bear strong claws, the third being greatly enlarged; the ears are proportionately larger than in *Priodontes*. The comparatively short tail bears only thin widely-spaced plates and practically no hair. The cabassus are able diggers, and are believed to live largely on termites and ants. Ingles [250] watched a captive cabassu on Barro Colorado Island when it was allowed to run at liberty in a clearing, and recorded that it 'searched for ground-dwelling termites and ants by burrowing through the litter and by digging holes, about a foot deep, to some dead root or stump that contained the insects. It apparently located these with the sense of smell by placing its nose close to the ground'. The large sickle-like claw was used to cut roots up to a quarter of an inch in diameter as it dug, and insects were extracted from their tunnels with the long tongue. 'On one occasion the animal entirely buried itself (about eighteen inches) as it worked along an old root infested with termites'.

Figure 7   Three-banded armadillo (*Tolypeutes*), showing the fit of the plates when rolled up.

When alarmed the hairy and six-banded armadillos and the pichi usually cling to the ground with the feet drawn in so that the shell forms a protective shelter, or else wedge themselves in a burrow with the sides of their armour. The nine-banded armadillos can curl up into a ball so that they are fairly well protected, and the giant armadillo and the cabasu also can roll up more or less completely. The two species of three-banded armadillos of the genus *Tolypeutes*, however, are the most highly adapted for this kind of defence. They are small animals with large rigid almost hemispherical shoulder and rump shields separated by only two to three transverse bands; the tail is short and, like the front of the head, is heavily armoured. The edges of the large shields project well over the sides of the body so that when the animal rolls up there is room for the limbs and fleshy ears inside them. When the animal curls up, the edges of the shoulder and rump shields meet, and the notches for the head and tail are exactly filled by the head and tail armour brought side by side; when completely closed the animal forms a ball about the size of a large grapefruit.

Sanborn [413] found that hunting dogs in Matto Grosso ignored *Tolypeutes* but hunted and killed six- and nine-banded armadillos; he thought this was due to painful past experience. He records that *Tolypeutes* when disturbed lay on their sides with the shells partly open.

They held this position until touched on the abdomen or chest and then they snapped together like a steel trap. . . . Any animal that had its paw or nose or lip pinched in this trap would undoubtedly be very careful with this species of armadillo thereafter. What interested me was that an animal, given absolute protection from the largest carnivores of its range, should also be able, and apparently await the chance, to inflict a painful reminder of itself on its annoyer.

I

The underside of *Tolypeutes* is well covered with coarse hair, and the feet of both limbs are peculiar. The second, third and fourth toes of the hind feet are syndactylous, and their claws resemble little hoofs; the claws of the fore feet are large, the central ones being specially strong – when walking the animal, like *Priodontes*, is semi-bipedal, and touches the ground with only the tips of the fore claws. *Tolypeutes* feeds mainly on termites and ants but also consumes some vegetable matter.

Finally the pichi-ciegos, a single species in each of two genera *Chlamyphorus* and *Burmeistria*, inhabiting Argentina and the Chaco region, are the most peculiar of the dasypodids, and differ from the others so widely that they have been placed in a separate subfamily, the Chlamyphorinae. They are small burrowing animals about five to seven inches long, with a banded shell flexible at all the seams between the bands covering the back, and a plate at right angles covering the end of the body. The posterior plate, like the rump shield of *Tolypeutes*, is fused to the sacral bone and the adjoining vertebrae. The dorsal shell in *Chlamyphorus*, however, is attached to the body only along the central line of the back, so that there is a potential space between it and the body over most of its area. The body beneath the shell and on the exposed under parts is clothed with soft fur, which projects as a fringe between the dorsal and posterior plates *Burmeisteria* superficially looks similar to *Chlamyphorus*, but the dorsal shell is attached over a wider area of the back, the posterior plate is less rigid and the head shield is larger. The tails of both are short and bear small plates, that of *Chlamyphorus* being spatulate in shape. The limbs are short, and the claws of the fore legs are large and strong. In *Chlamyphorus* the hair is white and the shell pink; in *Burmeisteria* the colours are duller and buffish. The pichi-ciegos live on ants and other invertebrates, and are said to include some vegetable matter in their diet. Like most armadillos they are rapid burrowers and if disturbed can quickly disappear underground.

Chlamyphorus, according to Minoprio [325] is nocturnal and subterranean, but does not differ greatly in its habits from other species of dasypodids. It is now restricted in its distribution to the centre of western Argentina but was formerly more widely spread – the higher rainfall which would flood its burrows makes the eastern part of the country an unsuitable habitat for it. The story that the animals use the rump shield to stopper their burrows appears to be inaccurate; the shield would automatically protect the animals from any small predator following them in the burrows, but Minoprio points out that the only serious danger of exterminating the species comes from man and his ploughs, dogs and children, although pichi-ciegos are still abundant in their particular habitat. He was the first to publish photographs showing the peculiar posture adopted by *Chlamyphorus* when sleeping – the animal cannot roll up into a ball but curves the body into an inverted U-shape with the nose tucked between the hind feet, and the body supported by the rump plate and the head shield.

ORDER PHOLIDOTA

The Pholidota, the scaly anteaters or pangolins, were formerly included in the Edentata, but are now classified as a separate order. The resemblances between the animals of the two orders are regarded as due to convergence correlated with habits similar to those of the xenarthrous anteaters, and not to close genetic relationships. The similarities in structure and habit are indeed remarkable, but the Pholidota lack the xenarthrous character of the lumbar region of the spine and the ischial articulation of the pelvis, though both of these parts are modified, but in a different way.

Pangolins range in overall length from about two and a half feet in the smallest species to nearly six feet in the largest, the tail making about half or a little more of the total length. The top of the head, the whole of the back and tail, and the outer sides of the limbs of most species, are covered with large overlapping scales produced by the skin as described in Volume I. The unarmoured parts are clothed in short hair, usually sparse. The snout tapers to the tip on which the nostrils and small mouth open. There are no teeth, the eyes are small and surrounded with thick lids, the external ear pinna is rudimentary or absent, the hind legs are longer than the fore, and the tail is broad, convex above, and flat or slightly concave below. The animals are plantigrade on the hind feet, and have five claws on fore and hind feet; those of the central three of the fore are greatly elongated and folded under the palm when the animal is walking so that a pangolin goes on its knuckles. Pangolins are insectivorous and feed mainly on termites and hymenopterous ants; they break open the insects' nests with their powerful claws and capture their prey with the very long tongue copiously supplied with sticky saliva. Some species defend themselves by squirting a jet of strong smelling secretion from their anal glands, and all can protect themselves by rolling into a ball with the tail wrapped outside. The large salivary glands extend far into the neck, and some of the tongue muscles originate not from the hyoid but from the hind end of the elongated sternum which reaches to the pelvis at the posterior end of the abdomen.

Parallels to most of these characters are found in the edentates – the insectivorous habit, the long tail, and large fore claws in the anteaters, the dermal armour and rolled-up defence posture in the armadillos, and the long tongue and sticky saliva in both. But the resemblances do not end here – the tree-pangolins, like the tamanduá and pygmy anteaters, have prehensile tails, and when on the ground both the arboreal and terrestrial species frequently adopt a bipedal gait supported by the tail, as do the armadillos, especially the larger kinds. Frechkop [155] showed that the lumbar region of the spine is strengthened in correlation with this habit by the articulation of the vertebrae, in which the posterior zygapophyses slide in grooves on the anterior zygapophyses of the following vertebrae, and the anchorage of the pelvis is reinforced by strong short ligaments between the ischia and the first caudal vertebra.

FAMILY MANIDAE

The order Pholidota contains only one family, the Manidae, with a single genus, *Manis*, which has been divided into subgenera by some taxonomists to accommodate the seven species, four African and three Eastern. Pangolins are found throughout most of Africa south of the Sahara, India, south-east Asia, Malaysia and the East Indies, chiefly within the tropics. The terrestrial species live in burrows with a den in the terminal chamber, the arboreal species in hollows of trees. Pangolins are common enough over most of their habitat so that their anatomy and systematics are well known, but surprisingly little has been recorded about their general biology and natural history; many casual observations on them are scattered through the literature of zoology, which are summarised in the general treatises, and given in some detail by Mohr [327]. The arboreal species wrap their tails spirally round the branches among which they climb and the larger tailed ones can also suspend themselves by hooking the end of the tail round a branch. The eastern species have small ear pinnae, but the African ones have none; nevertheless Vincent [491], like other observers, found that *Manis tetradactyla* and *M. tricuspis* in the Congo Republic were very sensitive to sudden noises even when of low intensity, and reacted by 'jumping' or rolling into a ball. He also states that they continuously snuffled at the ground when foraging, pausing from time to time to take a deeper sniff.

Pangolins make slow progress on the ground; Vincent records that their gait consists of a slow diagonal walk alternating with periods in which the fore limbs are placed together, with the claws folded under the palm, and the two hind limbs are brought forward together – during the last phase the back is kept strongly arched. He did not see bipedal walking in his specimens, but the giant pangolin *M. gigantea* of central and western tropical Africa is said to be able to outdistance a running man when using this gait.

For quadrupedal walking the greatest speed recorded by Vincent was just under two and a half kilometres an hour over a distance of ten metres.

Most of the pangolins are nocturnal, but *M. javanica* of south east Asia and the East Indies is also partly diurnal, and *M. tetradactyla* of west Africa is more completely so. The scales differ in shape between species, and those of *M. tricuspis* have a subsidiary point at each side of the main terminal one. Rahm [388], however, found that the scales become much worn with advancing age and that the shape of their outline and the details of their surface sculpture were then largely obliterated. Roberts [406] says that *M. temmincki* of East and South Africa is 'a timid animal and when alarmed rolls itself up into a ball that presents only the heavy scales on the outside; its defence appears to be the cutting action of the larger scales, which are worked by powerful muscles and inflict a severe wound on an enemy rash enough to get its fingers or snout between them' (Plate 10).

As pangolins are toothless the insects eaten are swallowed whole, but the almost gizzard-like muscular stomach crushes them with the aid of the small pebbles that it contains – whether the pebbles are swallowed deliberately or incidentally in the course of feeding is not known. In most species the insects are caught in the sticky saliva covering the tongue, but Rahm found that the tongue of *M. longicaudatus*, unlike that of other species, bears 'a surface zone of 4 cm. length, which is covered with light coloured bristles 1 mm. long, pointing backwards' which, presumably, help in retaining and swallowing the prey. The food of pangolins is generally said to be termites, ants and other insects, and indeed the animals do eat these and other insects but there is reason to believe that hymenopterous ants are the main item, and are indispensable for their health. Pangolins are notoriously difficult to maintain in captivity – the smaller species in particular usually die within a month or two, although they can be accustomed to taking an artificial diet of chopped meat mixed with raw egg and milk. Recently, however, the authorities of the Antwerp zoo [12] have succeeded in keeping pangolins of several species in good health on an artificial diet compounded of corn flakes and chopped meat granulated with warm water to which a little wheat-germ oil and a trace of formic acid are added. It is thought that the addition of formic acid is the key to this success, and that it shows ants to be the most important constituent of the diet of the animals in the wild. In captivity, too, it has been found that pangolins enjoy bathing and particularly like being sprayed with a jet of water from a hose.

Little is known about the reproduction of pangolins beyond the fact that the scales of the young are soft at birth and harden in a day or two, and that the young is carried about clinging to the root of the mother's tail.

The scales of the pangolins give them efficient protection, but there is a price to be paid for this defence – they are nearly always heavily infested with ticks, which attach themselves to the skin between the roots of the scales from which they cannot be dislodged by the short legs of the victims. Pangolins are commonly eaten with relish by the peoples of the countries they inhabit; in the east and especially in China the flesh is believed to have aphrodisiac properties, as are the scales which are often to be seen in eastern drug stores of the traditional kind. In Africa, too, the scales are valued as charms and are used in native medicine. In pronouncing the name pangolin, the accent should be placed on the long 'o'.

ORDER TUBULIDENTATA

The Tubulidentata were formerly classified as a section of the Edentata, but are now known to have little affinity with them beyond being mammals. In Simpson's classification they are placed at the end of the superorder Protungulata, the only living order in that group of otherwise extinct orders; Colbert [78] after a close study of one of the few known fossil species, strongly supported the theory that the Tubulidentata are derived from the protungulate order Condylarthra. The Tubulidentata are dealt with here merely for convenience in arranging the chapters of this book, and no affinity with the edentates or the pholidotes is implied.

FAMILY ORYCTEROPIDAE

The order contains only one family, the Orycteropidae, with a single genus and species, *Orycteropus afer*, called by the Dutch settlers in South Africa the aardvark, meaning earth pig – the name is now spelt erdvark in modern Africaans. The aardvark is commonly mis-called 'ant bear' by Europeans in many parts of Africa.

The aardvark is a fairly large animal, and measures up to six feet long overall, including the thick conical tail about two feet long; it weighs up to about 140 pounds. It is indeed much like a pig in general appearance, for the pinkish-brown skin is sparsely covered with coarse bristles, the snout is long and roundly truncated, but it differs in having long ears shaped much like those of a rabbit, a thick stout tail, and in carrying the back arched in a hunched-up posture. The animal is digitigrade, and the strong claws, more like nails or small hoofs, are used for digging – those of the fore limb, four in number, are larger than the five of the hind limb. Like the armadillos and pangolins it habitually sits on the hind quarters with the fore limbs off the ground and, like those animals, it is an ant eater and catches its prey with its long extensible tongue covered with sticky saliva.

Its sight is poor but its powers of hearing and scenting are acute. The unique character of the animal, however, is its dentition. There are no incisors or canines in the permanent dentition, and the cheek teeth, which grow continuously throughout life, are rootless and lack enamel but are covered with a layer of cement. The dentine is arranged in numerous hexagonal prisms surrounding tubular pulp cavities, a structure which gives origin to the name of the order.

The aardvark, which has been classified into several subspecies of doubtful validity by some authors, is a native of the whole of Africa south of 15° north latitude, particularly in open country but also in bush and even forest. It is a fairly common animal, though seldom seen because it is nocturnal, and it is regarded as a nuisance by farmers in some parts of its range owing to its burrowing habits – horses and stock may break their legs by stepping into its burrows, and jackals can creep under jackal-proof fences undermined by aardvarks (Plate 11).

Throughout east and southern Africa the aardvark is notorious as 'the animal that can dig faster than a gang of men' – it is indeed a very fast burrower, but it is frequently dug out by natives who are willing to spend much time and labour on the task because, according to Roberts [406], they are very partial to its fatty flesh, though Shortridge [442] found the meat very poor eating. Shortridge also says that owing to the aardvark's 'amazing "pulling" power' nothing less than a lion trap will hold it.

Aardvarks live in burrows and, although they are solitary animals, Roberts found that the burrows are sometimes close together, probably because the earth of the site is good for digging; he 'counted sixty entrances in one patch of ground not more than 300 yards along by 100 yards wide. The natives informed me that the burrows were intercommunicating underground; traps were kept continuously set at various burrows in the district, but never with success. . . .' Loveridge also says that 'if a trap be set at the entrance of an occupied burrow, an ant bear may remain below ground for a week or more, eventually making its escape by excavating a new outlet. Like an otter, it has an extraordinary flair for detecting and avoiding a trap, however carefully concealed.' The burrows are generally twelve to fifteen feet long with a chamber at the end where the animal can turn round, though Verschuren [489] records one animal seen to emerge from its burrow tail first, pushing out with considerable speed and raising clouds of dust. Aardvarks sometimes shelter in termite mounds which they have opened when feeding, and are said to be indifferent to the bites of the insects from which they are protected by the thickness of the skin. Verschuren also records that aardvarks do not always react to danger by burrowing farther into the ground, but may dig themselves to the surface and try to escape overground. One animal, frightened in its burrow by human disturbance, was heard digging under the surface about eight feet away, and after a few minutes, '*La terre s'ouvre subitment et l'Orycterope surgit littéralement du sol. . . .*' Although speared by the natives the animal escaped into another burrow into which a charge of shot was fired, whereupon the animal '*pousse un cri d'agonie, meurt et est immédiatement extrait du terrier*'.

The aardvark has no layer of subcutaneous fat and its skin is nearly naked; some zoologists have therefore wondered how it is able to maintain its temperature when sleeping underground, as it makes no nest or bed in the burrow. Verschuren measured the temperature and humidity in a burrow and found that both were remarkably constant throughout the day and night so that the animal is isolated from external variations.

When digging the aardvark excavates with the fore feet and kicks the soil backwards with the hind; it folds the ears back to close them, and shuts the nostrils, which are additionally protected by a ring of strong bristles surrounding the muzzle. Aardvark burrows are often used as temporary refuges by wart hogs, and are taken over as dens by spotted hyaenas.

The aardvark feeds mainly on termites, but also takes ants and is said to be partial to wild cucumbers. It takes the insects on the surface but also breaks open termite mounds. Verschuren found that aardvarks open a circular hole about thirty cm. in diameter in the side of a termite mound and dig in for about half a metre, leaving a large pile of debris alongside.

Little is known of the breeding of the aardvark beyond the fact that a single young one is produced at birth. Van der Horst [483] examined a pregnant uterus containing a single embryo in an early stage of development, but found to his surprise that one ovary contained two large corpora lutea, the other three smaller ones, all of similar age.

It is strange that the biology of so widely distributed and fairly plentiful an animal is not better known. The shyness and nocturnal habits of the aardvark make observations difficult, but a careful study of the creature in the wild would no doubt be rewarding to a patient zoologist.

Chapter 5

Rabbits and Hares. Squirrels

ORDER LAGOMORPHA

THE ORDER Lagomorpha contains a little over sixty species of rabbits, hares and their allies, naturally inhabiting most parts of the world except Australasia and South America south of about 30° S latitude where they have been introduced. They were formerly classified as the suborder Duplicidentata of the Rodentia but, although superficially similar to the rodents, the difference in structure and origin of the two groups is so great that their separation is now necessary. Fossils show that the order is an ancient one and that its affinities are with the extinct Condylarthra, which gave rise also to the artiodactyles, whereas the affinities of the rodents are with Simpson's cohort Unguiculata, which contains, with others, the orders of insectivores, bats and primates.

The lagomorphs are small to medium sized animals – the largest are the hares – clothed in soft fur, and tailless or with only a short tail. They are digitigrade, or digitigrade on the fore and plantigrade on the hind feet; the toes – five on the fore and four or five on the hind feet – are provided with claws, and the soles of all the feet are hairy. There are no canine teeth so that there is a wide diastema or gap between the incisors and the cheek teeth, which are high crowned and rootless. The incisors are chisel edged and have open pulp cavities so that their growth is continuous throughout life – these structural adaptations for gnawing led to the former taxonomic confusion of the lagomorphs with the rodents. The members of the two orders are, however, readily distinguished by the presence of two pairs of upper incisors in the lagomorphs. There are two large incisors in both upper and lower jaws, but in addition a small pair of second upper incisors immediately behind the first. These small teeth are peg-like, have no cutting edge, and lie in contact with the bases of the larger teeth in front of them; they appear to be of minor importance, but no doubt serve some function in gnawing – their relation to the teeth in front is similar to that of the back-iron to the cutting-iron of a carpenter's plane, and their function may be similar. The muzzle is 'hare-lipped'; this condition is produced by two hairy lobes covering the naked skin of the upper lip and the openings of the nostrils. The lips also bear hairy lobes which can be brought together behind the incisors, and the lining of the cheeks bears a strip of short hairs inside the mouth.

Most of the food eaten by lagomorphs passes through the digestive tract twice – the faecal pellets passed during the night differ from those discarded during the day, or vice versa in strictly nocturnal species. The night pellets are taken directly from the anus by the leporids, but from the floor of the den in the morning, after they have partly dried, by the ochotonids; it is believed that re-ingestion, or refection, provides the animals with vitamin-B produced by bacterial action in the caecum but not absorbed there – a similar habit in the shrew was mentioned in chapter 2.

The lagomorphs are divided into two families, the Ochotonidae containing a single genus, and the Leporidae containing nine.

FAMILY OCHOTONIDAE

The genus *Ochotona*, the only one in the Ochotonidae, contains about twelve species of whistling hares or rock rabbits, generally known as pikas, a name derived onomatopaeically from their characteristic shrill cries. They are small furry animals – no species exceeds about a foot in overall length and most reach little more than six or seven inches – with short 'bunny' faces and wide rounded ears unlike the long narrow ones of rabbits and hares. The fur is soft and silky and generally reddish-brown or grey in colour, lighter below. The hind legs are little longer than the fore, and there is no tail, although caudal vertebrae are present within. The pikas are typically mountain animals, which live among rocks and in talus screes, though some Asiatic species live in forests, among scrub, or on plains down almost to sea level. They are found over most of Asia north of the southern slopes of the Himalayas (about 30° N) and west of about 50° W, and in the mountains of western America discontinuously from Alaska to California and New Mexico.

Pikas are diurnal, and particularly active in the morning and evening. They live in colonies, which may be due to the availability of suitable ground rather than to any social structure, for they are highly territorial. Kilham [266] recorded that individuals of *O. princeps* in western Montana 'could always be found in the same territories on the rock slide, remaining within boundaries in relation to adjacent territories.' He suggested that owning territory enabled the animals to guard their food stores because, when not harvesting, they kept watch over their territories on a favourite lookout rock where they uttered their cry from time to time; the call apparently serves to advertise ownership and is not always a note of alarm. He also saw the animals marking their territories both by leaving droppings on their fodder-piles and by rubbing their cheeks against stones. The cheek-glands, according to Harvey and Rosenberg [215], are apocrine glands which lie below the ear and are covered with a small patch of yellow-brown hair; their function thus appears to be for marking territory. Kilham saw a pika chase a trespasser back into its own territory and call several times from the trespasser's look-out rock before racing back to its fodder piles. On the other hand pikas at least sometimes show social behaviour, as recorded by Dixon [114] who saw a dwarf weasel pursuing a two-thirds grown pika in the mountains of Colorado. His attention was caught by 'a sudden barrage of alarm notes from numerous pikas on all side of me'. At the first call all the pikas 'immediately ceased foraging and scrambled up hastily to their lookout stations on the crests of slabs of granite where they stood guard and anxiously peered about'. The weasel followed the young pika by scent, 'dodging in and out among the rocks with unbelievable speed' until the pika began to tire. Then suddenly one of the adult pikas left its lookout and cut into the race behind the young one, which soon dodged out. Then a second, a third, and finally a fourth pika joined in the relay, taking each other's places, until the weasel 'showed marked signs of fatigue and gave up the unequal contest'.

Although the mountain-dwelling species of pika generally live among the crevices of rock falls they sometimes burrow into earth slopes, and some Asiatic species habitually dig their own burrows, often in large colonies. None of the species hibernates, and Skinner [447] often saw *O. princeps* sunning itself on rocks blown clear of snow in the Yellowstone National Park on sunny mornings when the temperature was far below zero Fahrenheit. When winter comes, however, pikas generally take to a subnivean life under the snow-blanket whatever their habitat. Their remarkable habit of making hay for winter use is correlated with their inability to hibernate. From about July onwards they start making their winter stores. Martin [313] studied pikas on the Snowy Range in southeastern Wyoming, and found that the pika

besides eating daily from his hayfield, also stores up hay for the winter. He will scamper to the pasture and cut a bundle of plants almost as large as himself which are thus carried by the mouth to the rock slides, often a hundred feet or so away. The freshly cut fodder is placed under the rocks where it is protected from the weather and cannot be blown away. Here it slowly cures into hay.

Kilham often saw pikas harvesting early in the morning 'running to the upper or lower ends of the rock slide to cut stalks of grass or twigs from bushes. The pikas would then race back to their hay piles with the cut vegetation projecting from one side of their mouths. . . . Some animals hid their stores almost completely under rocks while others placed their mounds in the open'.

In Manchuria, Loukashkin [303] found *O. mantchurica* living in the crevices of rocks and under boulders, and hiding its hay stores, weighing fifteen to forty pounds, in caches among the rocks or in holes in trees where they could not be stolen by elk, roe, and mountain hares. *O. hyperborea*, on the other hand, lives in flat places and burrows in the ground, often in large colonies. It makes its hay piles, weighing two to five pounds, in conical heaps on the ground, and turns the piles several times while they are drying, to make well prepared hay. 'Many hundreds and sometimes even thousands of such minute cones can be seen in September, covering an area of two to three square kilometers occupied by the colony'. In the winter, when hardened deep snow covers the steppe 'the Mongolians drive their herds to localities where there are pika colonies and the hungry animals feed upon the pika stocks'. Haga [193] found that the Japanese pika, a subspecies of *O. hyperborea*, lives in talus slopes, and makes hay piles for summer as well as winter use, probably because it prefers dry vegetation to fresh. The animals are very unselective in their hay making, and Haga counted twenty-seven species of plants in their stores. Similarly Beidleman and Weber [28] counted twenty-two species in 5,000 pieces of vegetation from the hay piles of *O. princeps* in Colorado.

Pikas breed during the summer, and the females have litters of two to three young after a gestation of about thirty days.

The pikas show some superficial resemblance to the hyraxes in appearance and in their rock-dwelling habits – both are often called 'cony' in the vernacular. They both use latrines, and produce a thick dark coloured urine which makes incrustations on the rocks.

FAMILY LEPORIDAE

The family Leporidae contains eight genera of rabbits and hares, now found throughout most of the world since some species were introduced into southern South America and into Australia where they were not indigenous. They are readily distinguished from the ochotonids by their long hind legs, long narrow ears, and short tails. They are well supplied with scent glands in the skin – inguinal, anal, preputial, and under the chin – the secretions from which have sexual or territory-marking uses.

The largest genus, *Lepus*, contains over two dozen species, the typical hares, although some American species are known as jack rabbits in the vernacular, and the varying hare is often called the snowshoe rabbit. Hares feed on herbage and the leaves and bark of shrubs; those that live in regions covered by snow in the winter dig out their food with their long sharp claws. Neverthless none of the species digs burrows in the ground either for shelter or to hold nests for their young. They rest lying up among herbage in slight depressions known as 'forms', and the young, which are fully furred and have their eyes open at birth, are born and suckled there. Species that live in cold countries moult into a white winter pelage in the autumn; the arctic hare, north American varying hare or snowshoe rabbit, and the European mountain hare are well known examples. Hares are found throughout most of Europe, Asia, Africa, and North America where there is suitable open country; in the warmer regions their ears are long, but in the colder areas they are much shorter. Some species of hare are among the few mammals in which superfoetation – the simultaneous gestation of two litters of different age – is known.

The best known leporid is the common rabbit *Oryctolagus cuniculus*, the only species of its genus, a native of the Lusitanian and Mauritanian regions, but now dispersed far and wide throughout the world by man. Although rabbits are territorial mammals they dig the burrows in which they shelter close together and form extensive warrens. The young, which are born blind and naked, are suckled in a nest lined with fur plucked by the mother from her chest and placed in a specially dug burrow or stop away from the warren; the mother feeds the young two or three times a day and stops the mouth of the burrow with earth when she leaves it. Rabbits have become a pest partly because they feed on any kind of herbage but mainly because of their great fecundity – they can produce up to six litters a year – and the early age at which they start breeding; yet their potential for increase is often impaired by a prenatal mortality in which a large part of many litters dies during development and is resorbed by the mother's uterus. The virus disease myxomatosis, transmitted mechanically by insects, has killed millions of rabbits, but has disappointed the hope that it would exterminate the species where it is a pest to man.

The genus *Sylvilagus*, containing a little over a dozen species of cotton-tail rabbits is, next to *Lepus*, the largest among the leporids. These rabbits are found throughout America from southern Canada to Argentina, in many different habitats from open land to forests and marshes. They shelter among herbage or in natural hollows, but only one species, the brush rabbit *S. idahoensis*, digs burrows. The young are born in surface nests lined with fur plucked from the mother, but are not precocious like those of hares – they are blind and naked at first. Cottontails can be destructive to agriculture and, like the European rabbit, they are very prolific and flourish in close proximity to man.

The four or five species of red and scrub hares, *Pronolagus*, of South and East Africa are about the size of the European rabbit. They do not burrow but live among rocky places and den up in crevices between boulders. They are the only leporids that use their voices when not hurt or in terror – Roberts [406] on several occasions heard red hares (he does not name the species) uttering a series of loud screams when running away at night, screams that 'cannot be described as anything less than fiendish, and calculated to send a cold shiver down one's back!' The four remaining genera each contain only one species, three of them scarce. The short-eared *Pentalagus* is found only on the Ryukyu Islands south of Japan, the small *Romerolagus* digs its burrows on the slopes of Popocatepetl and neighbouring volcanos in central Mexico, and the short-eared *Nesolagus* inhabits the mountain forests of Sumatra and is the only leporid with a distinct colour pattern – the greyish fur is marked with broad dark brown longitudinal stripes on the head, back and sides. The bristly or black rabbit *Caprolagus* inhabits the *terai* at the foot of the Himalayas, and is difficult to observe because it frequents jungle and long grass among which it is said to make burrows. Its fur is dark grey with a black shading, and the outer hairs are harsh and bristly; the ears and the legs are comparatively short.

ORDER RODENTIA

More than half of the mammals now alive are rodents; according to Simpson [445] the living rodents are probably as abundant individually and in variety as all other mammals put together, and he adds, 'Most of them are small, many are obscure and rare. . . . Their relationships are involved in an intricate web of convergence, divergence, parallelism, and other taxonomic pitfalls.' It is therefore impossible in the scope of this volume to do more than consider their major characters, and to glance at some of the more familiar or interesting genera.

Rodents are small to medium sized mammals, distinguished by the character of their dentition. The incisors in both jaws are a single pair of curved or semicircular chisel-ended teeth with hard enamel only on the front surface so that wear keeps the chisel edges sharp. The pulp cavities are open so that continuous growth from the base keeps pace with the wear at the tips. There are no canine teeth so that a wide gap or diastema separates the incisors from the cheek teeth. The jaw joint is such that the jaws can be thrust forwards to bring the upper and lower incisors into apposition for gnawing, or brought back so that the upper and lower cheek teeth are opposed in an oblique action in chewing. The incisors and lips form a highly versatile mechanism for collecting and preparing food, constructing nests, excavating burrows, and many other purposes. In addition, the fore limbs, generally with five clawed digits, are often used as hands which collaborate with the jaws in these activities.

Many different systems of classification have been proposed for this enormous horde of animals, but the taxonomists agree that a completely satisfactory one has yet to be devised – after much more fossil material has been found and studied. The most convenient, and the one most generally adopted, often with some modifications, is that used by Simpson, in which the order is divided into three suborders and the families within them segregated into superfamilies. This system, although admittedly imperfect in many details, allows the human student of rodents to avoid being overwhelmed by the weight of numbers in this vast aggregation of species.

The rodents are a most versatile order and are adapted to practically all the possible terrestrial and fresh-water habitats – there are no marine rodents – and are distributed over the entire habitable surface of the earth, for they have been introduced by man into the few places where they did not occur naturally, such as some oceanic islands, New Zealand, and some of the subantarctic islands. Their biological success must be attributed to their high fecundity and rapid potential rate of breeding, the quick turn-over of the population, for they are mostly short-lived creatures, and to their small size which enables them to find shelter in and exploit a huge array of minor habitats. They form the prey of a great many carnivorous mammals, birds and reptiles, but their high fertility makes their populations resilient to this constant drain from their numbers. In addition some species are serious pests to man's standing and stored crops, but no human efforts have sufficed to do more than exercise some amount of control over their numbers, and their extermination or local eradication has so far been impossible.

The three suborders of the rodents are the Sciuromorpha or squirrel-like rodents, the Myomorpha or mouse-like ones, and the Hystricomorpha or porcupine-like ones. The premaxillae and maxillae and the lower jaws of rodents are enlarged to accommodate the long roots of the incisors, and to give attachment for the muscles used in gnawing and mastication. These muscles are highly developed, especially the masseters, which are generally larger and more massive than the temporalis. The complex masseter is divided into superficial and deep parts, which may be subdivided into anterior and posterior parts; it is the arrangement of the superficial and deep parts, with the correlated modifications of the skull, that forms one of the most useful characters for distinguishing the three suborders. In the sciuromorphs, in which the arrangements are simplest, and are regarded as most primitive, the superficial part runs from the angle of the jaw to the skull in front of the orbit, and the deep part from the middle of the jaw to the zygoma. In the myomorphs the superficial part is similar, but the deep part runs from the middle of the lower jaw under the zygomatic arch and passes through a hole between the upper and lower anterior origins of the arch to an insertion in front of the orbit. In the hystricomorphs the superficial part is comparatively short and runs from the angle of the jaw to the zygoma, and the deep part runs from the middle of the jaw through a vacuity to a large area of attachment in front of the orbit and anterior base of the zygomatic arch.

SUBORDER SCIUROMORPHA; contains the squirrels, marmots and beavers, in five superfamilies.

SUPERFAMILY APLODONTOIDEA

Family Aplodontidae. This is the only living family in the small superfamily; it contains only one genus and species, *Aplodontia rufa* the mountain beaver or sewellel – which is neither a beaver nor necessarily an inhabitant of mountains. It is a stout thickset animal weighing two to three pounds with short legs and tail, small ears and eyes. The short dense fur is brown, either reddish or greyish, and the guard hairs are sparse. The species inhabits a comparatively narrow strip of country in western North America extending from the south of British Columbia to California, and prefers a wet but not swampy forest habitat. The long claws of the fore limbs are used in digging complicated burrows, with many entrances, close to the surface; a nest of vegetation is constructed in a chamber of the burrow. The animals are mainly nocturnal, and seldom venture far from dense cover; they are vegetarian and feed on the stems of herbage, leaves, twigs and bark. They can swim well, and also climb trees though less skilfully to cut branches for food. They do not hibernate, but gather stocks of food for winter use. It was formerly thought that the tree pruning was done from the surface of deep, frozen, snow and that the animals could not climb, but Ingles [251] saw and photographed them climbing lodgepole pines to a height of twenty feet: 'Twigs up to 30 inches long and $\frac{1}{2}$ inch in diameter are readily cut off in a few seconds from 2–10 inches out from the trunk'. The animal brings the cut branch down in its mouth, descending head first and using the stumps of the cut branches as the rungs of a ladder – the fore feet grasp branches but the hind are merely placed on them. The needles of the pines are eaten, and the twigs are stored in the winter cache. Although mountain beavers dig their burrows close together and form a 'colony' they appear to be solitary rather than social animals, and to have small individual territories. Simpson [445] points out that *Aplodontia* can be considered as the living representative of the most primitive rodent stock. Several families of fossil species are known in the superfamily, one of which includes the oldest known rodent, from the upper Paleocene – a form closely related to the ancestor of all the living rodents.

SUPERFAMILY SCIUROIDEA

Family Sciuridae. This large family is the only one in the superfamily; it contains about 260 species in some fifty genera, the majority in the sub-family Sciurinae containing the tree- and ground-squirrels, the marmots and chipmunks, and the rest in the subfamily Petauristinae containing the flying- or gliding-squirrels. Members of the family are found over a wide area of the world, the whole of Europe, Asia and the East Indies, North and South America as far south as the tropic of Capricorn, and in all of Africa south of the Sahara. Their fur is dense, their toes, four on the manus and five on the pes, carry sharp and rather long claws, their cheek teeth are rooted, not permanently growing, and in many the tail is long and bushy. Most of the sciurines are diurnal and move about in a peculiar jerky manner interrupted with motionless pauses that bring to mind the similar movements of reptiles, particularly of lizards. The eyes of the squirrels are comparatively large, and the eyesight is good – a necessity for animals that take flying leaps from branch to branch among the trees. In addition the squirrels are among the minority of mammals that possess good colour vision – most mammals are more or less colour-blind – and the rich colouration and the patterns of longitudinal stripes in contrasting colours of many species are believed to be correlated with this condition. Tree squirrels not only scamper agilely about the branches of trees, and take flying leaps from one to another, but run up and down the smoothest of trunks with ease, using their very sharp claws to secure a footing. When coming down a trunk they always descend head first.

Although most of the tree squirrels live in woods and forests, some species such as *Rheithrosciurus* of Borneo and *Menetes* of south-east Asia live largely on the ground, and others of the genus *Sciurotamias* live among rocks and mountain cliffs of China. They are primarily vegetarian, eating leaves, fruits, nuts, buds and bark, but many species also eat much animal

K

matter such as insects and other invertebrates, small mammals, young birds, and eggs. The members of several genera, however, are mainly carnivorous – in particular the long-nosed squirrel *Rhinosciurus* of Malaysia lives largely on ants, termites, and other insects which it picks up by using the very long slender lower incisors and the short upper ones in the manner of forceps; the large and very long-nosed *Hyosciurus* from the mountains of Celebes probably has similar feeding habits. Tree squirrels either build globular nests of sticks, twigs and leaves among the branches of trees, or furnish dens in hollow limbs and trunks with bedding of dry plant material. In places where the winter is inclement they collect and store food, not in large caches, but buried in the ground or wedged in the crevices of bark and similar places. The position of the hidden stores is not found by memory, but by random search aided by the sense of smell; none of the tree squirrels hibernates.

The largest genus of tree squirrels is *Sciurus* with about two dozen species, including the common red squirrel of Europe and Asia, the fox- and grey-squirrels of North America – the red squirrels of North America belong to the small genus *Tamiasciurus* with only two species. The species of African striped squirrels of the genus *Funisciurus* are numerous, some fifteen to twenty, and the south east Asian genus *Callosciurus* is as large. The members of the last genus are among the most brightly coloured squirrels; some are marked in strongly contrasting colours, others are white, black, red, pink or yellow, and in one species at least the colouration may differ widely between individuals. There are many smaller genera containing a single species or only a few.

There is a great diversity of size and general appearance among the large number of different kinds of tree squirrel; the smallest is the tiny *Myosciurus* of west Africa, no larger than a mouse, and the largest the species of the Indian and east Asian genus *Ratufa*, marked with rich reds, shining black, yellow and buff, some of which weigh up to about six pounds. The tails of tree squirrels are usually long and bushy, with a tendency to a dorsoventral flattening so that it helps, with the widespread limbs and flattened body, to give some support as a plane in long leaps. When the animals sit on the haunches, generally while holding food to the mouth with the fore feet, the tail is folded up over the back; the members of many genera, too, have conspicuous tufts of long hair on the ears, at least in one phase of their pelage. In many of the ground squirrels also the tail is long and bushy and similarly carried over the back, but in some members of the family, such as the prairie dog and marmot, it is comparatively short and not bushy.

146

The ground squirrels are burrowers, and many of the species are gregarious, digging their burrows in colonies, often of large extent. The animals of many species do not hold individual territories but use the warren of burrows communally, and although there are dominant and subordinate individuals there is little or no hostility between them. The most familiar of the colonial ground squirrels is probably the north American prairie dog *Cynomys*, which makes – or used to make before the American farmers brought it near to extermination – large colonies of burrows, the celebrated prairie dog towns. The distinctive feature of a prairie dog town is not, however, its size, though that can be astonishing, but the construction of the burrows to prevent them being flooded. The shaft goes straight down for several feet before levelling out, and the mouth of the burrow is raised on a mound shaped like a volcano, so that surface rain water cannot trickle in, Smith [451] who studied prairie dogs in Kansas says that contacts between prairie dogs are on the whole friendly and include mouth contacts, anal recognition, huddling, grooming and feeding. The mouth contact or kiss

is effected by one prairie dog turning its head slightly and touching its open mouth (incisors) to that of another prairie dog, after which they may graze together, groom each other, or go in opposite directions. The contact often occurs when a prairie dog is on its mound and another, or others, run to the mound; the first prairie dog 'kisses' the others as they arrive. When prairie dogs are grazing, one may run to another a few feet distant, 'kiss', and return to grazing ... young prairie dogs move about making mouth contact with every prairie dog they see. Animals in the laboratory frequently behave as though they desired to 'kiss'. In such instances the desire seems to be satisfied if a person grasps their incisor teeth by means of his fingers.

Somewhat similar towns are made by one species of the genus *Xerus* in South Africa, a genus in which the outer hair is coarse and harsh and the underfur scanty, so much so that some species are bristly and even named 'spiny'. Several genera and many species of ground squirrel are found in north America, including *Citellus* to which the prettily marked thirteen-lined ground squirrels belong, and *Tamias* and *Eutamias*, the equally handsome striped chipmunks, one species of the last genus being found widely through northern Asia. Over half a dozen species of *Citellus* live in south east Europe and Asia where they are usually known as 'sousliks'.

The prairie dog of America and the spermophile *Spermophilopsis* of central Asia are ground squirrels with short tails, but the marmots *Marmota*, one species of which *M. monax* is the American woodchuck, look very different. They are large stoutly-built rather short-legged animals weighing up to fifteen or sixteen pounds. The ears and eyes are small, the fur dense and thick, and the tail short. Marmots live in burrows which they excavate or in crevices and hollows among rocks, and are active by day – there are about ten species widely spread through North America, Europe and Asia. They are grazers of herbaceous plants, and become very fat during the summer. Like the ground squirrels, they sit upright on their haunches when alerted in order to survey their surroundings.

The ground squirrels that live in cold climates, the marmots, the prairie dog, and the chipmunks hibernate, some of them for very long periods – in their coldest habitats marmots hibernate for as long as eight months. The astonishing internal rhythm that controls hibernation, even when all external stimuli are excluded, which has been demonstrated in the ground squirrel *Citellus* has been mentioned in Volume 1. It is interesting to note that the chipmunks, which generally hibernate for short spells and rouse from their torpidity from time to time during the winter, store food in their burrows and have cheek-pouches in which they carry the food into their caches.

All the tree and ground squirrels and the other sciurines are diurnal and, compared with most small mammals, are surprisingly tolerant of the presence of man; they take cover, of course, on his close approach, but soon peep out to see if the coast is clear. They are, also, rather noisy animals and utter a variety of chuckles and chattering alarm notes in an apparently scolding tone, or signal the approach of danger to each other with loud whistles, such as those of the marmots. The system of vocal signals is most elaborate in the colonial species, especially in the prairie dog; one of its alarm notes is a short bark which has given the species its popular name.

About three dozen species of petauristines, in thirteen genera, make up the subfamily of flying squirrels – the West African flying squirrels, though similar in habits, are basically different in structure, and form a separate family parallel to the sciurids. The flying squirrels glide, for the flight is not powered though the energy needed is not entirely derived from push-off and gravity, because some species at least are said to make use of 'thermals', as does a man in a glider flying-machine. One observer, however, said he once saw a small flying squirrel flap the patagium and gain height of about a yard in a flight of twenty feet. The gliding mechanism consists of a double flap of skin extending from the sides of the head or neck to the wrists and thence to the ankle; in some species a further part extends from the ankle to the basal part of the tail. The area of the patagium is increased by a rod of cartilage at the wrist which is brought forward to strut it out during flight. In many species the tail also increases the area because it is flattened from above downwards, and the long hair at each side makes it featherlike. Once they are airborne flying squirrels have a good deal of control of their flight and are able to avoid obstacles by banking and altering the shape of the aerofoil. On landing on a tree-trunk the body is brought into a vertical position and the limbs thrown forwards so that the body is turned into a parachute which brakes the air speed. When folded the patagium becomes very inconspicuous, and is drawn in to the sides of the body by several modified superficial muscles, unlike that of the colugos described in chapter 2, in which the patagium gives the animal the appearance of being wrapped in a woolly blanket. The large eyes, short faces and snowy fur on the underside make most flying squirrels very attractive creatures to human eyes, an attraction increased by the ease with which the smaller kinds can be tamed to make interesting and gentle pets.

Five species of *Petaurista*, the giant flying squirrels, are found in southern Asia; they are as large as a cat and have long bushy, unflattened tails, the fur generally dark above and light below, but with light colour patterns on the head in some species. The dark-coloured Malaysian flying squirrel of the genus *Aeromys* is rather smaller, and like *Petaurista* is esteemed as food by the natives in some places. The beautiful little flying squirrels of the old world, genus *Pteromys*, are found from northern Europe right across Asia to Japan, and those of the New World, genus *Glaucomys*, throughout America from Canada to Mexico, where suitable woodlands exist. Eight species of *Hylopetes* and seven of *Petinomys*, much like the last two genera, but slightly larger, are found in India and Malaysia. The members of seven other genera, mostly from China and Malaysia, are little known, but two many be mentioned, *Petaurillus*, the pygmy flying squirrels of Borneo and Malaya, scarcely larger than a mouse, and *Eupetaurus* from northern India, with very long thick woolly fur, which is said to live among rocks rather than trees.

All the flying squirrels are nocturnal and, with the possible exception of *Eupetaurus*, arboreal. They spend the day in their nests in hollow trees, old woodpecker holes and such places, but are very active by night. They feed mainly on vegetable matter – leaves, buds, fruits, nuts, bark and seeds, but many species also eat insects and their larvae. Unlike the other sciurids they seem to be rather silent animals. The young spend a comparatively long time in the nest and do not leave it until they are mature enough to fly and fend for themselves.

SUPERFAMILY GEOMYOIDEA

Family Geomyidae. By comparison with the squirrels, and particularly with the elegant flying squirrels, the pocket gophers which form this family are anything but endearing. They are stout-bodied, thickset, bull-necked, blunt-nosed animals with small eyes and ears, and short legs; the species range from about four inches to over a foot in length from snout to tail-root. The tail is short and nearly naked. The incisors are partly exposed even when the mouth is shut, and the fore feet bear large sharp claws on all five digits. Like moles they live entirely underground, and rarely emerge onto the surface. The distinguishing character of the pocket gophers, from which they take their name, is the presence of a large pouch on each side of the head; the pouches open on the cheek behind the angle of the mouth, and extend back under the skin as far as the shoulder – they are lined with fur and can be turned inside out for cleaning, after which they are pulled back into place by a special muscle.

Figure 8 Pocket gopher (*Geomys*), showing the teeth and pocket, the latter shown diagrammatically – the living animal cannot open it as shown.

Pocket gophers are found only in North and Central America, where they are distributed over most of North America west of the Mississippi from south-west Canada to the southern end of Central America, and in Florida, Georgia and Alabama. There are between thirty and forty species, classi-fied into eight genera; the large number of species is believed to have evolved owing to the isolation of populations brought about by the sub-terranean habit and consequent restriction to ground suitable for burrow-ing. The animals feed on roots, bulbs and other underground parts of plants, and sometimes on succulent stems which can be pulled down into the burrow. The feeding burrows run near the surface, but the shelter burrows which contain chambers for storing food, for the nest, and for latrines, are deeper. Some of the excavated soil is pushed out on to the surface where it forms mounds, and some is used to plug the burrow behind the animal, especially in the feeding burrows. Pocket gophers use the incisors as well as the fore feet in burrowing, bringing them into play for breaking the solid ground and for cutting roots. As they dig they push the soil back under the body with the fore feet, and when a pile has accumulated they turn and push it out along the burrow. When using the teeth for digging the lips are pinched in at the diastema so that earth does not enter the mouth; the functional separation of the incisors from the rest of the mouth when gnawing inedible things is commonly found through the rodents.

In regions that are covered in deep snow during the winter pocket gophers come out of the ground and continue their burrows through the snow, and plug them with earth just as they plug their underground ones. When the thaw comes in spring the plugs are left as long cores trailing like tangled ropes over the surface. Marshall [311] found that one species at least makes use of snow cover for travelling, and mentions a snow burrow that crossed a road which would have been an impassable barrier to ordinary burrowing. Pocket gophers do not hibernate, but gather large stocks of food in their store chambers for winter fodder; the capacious pouches are used for carrying the food to the stores. They are solitary animals and aggressively chase any intruder out of their territorial systems of runways. Aldous [2] says that they do not eat all the food in the stores. 'In excavating runways old storerooms full of mouldy plant parts frequently are found'. He also noted that before making a mound 'nearly all the vegetation within a radius of approximately eight inches from the tunnel opening is removed . . . pocket gophers never forage further out than they can reach by keeping their hind feet in the burrow entrance'. Nevertheless they do sometimes emerge, for Sharp [438] found one at night in the rut of a road; when frightened by the lights of a car it 'suddenly ran backwards down the trail,' running in reverse as rapidly as it could move forwards.

Two genera, *Geomys* and *Thomomys*, the eastern and western pocket gophers, totalling about ten species, occur from Canada to Mexico – according to Myers and Vaughan [343] *Geomys* is well adapted to a diet containing a high proportion of grasses, whereas *Thomomys* is largely dependent upon roots and tubers. The other eight genera, containing over two dozen species, are found in Mexico and Central America to Colombia; one species of *Cratogeomys* is found in the southern United States. Some of them, especially *Orthogeomys*, reach a large size, over a foot in length excluding the tail. Species are found from sea level to high on the mountains, and are serious pests to farmers, destroying not only the smaller crop plants but damaging sugar cane, banana plants and coffee bushes. One genus, *Heterogeomys*, inhabits the banks of water channels, natural and artificial, but neither of the two species is truly aquatic. Although the name 'gopher' has a biblical ring because Noah's ark was built of gopher wood, it is etymologically derived from a word referring to the honeycomb result of burrowing – it is indeed applied to other burrowing animals, such as the burrowing land tortoise and the gopher snake – and even to human mining techniques.

Parts of the western prairies of America are covered with closely spaced mounds, some of them seven feet high and containing about fifty cubic yards, which look like European tumuli or round barrows. Their origin has long been a puzzle to geologists, for they are not man-made. They take their name of 'mima mounds' from the type locality, Mima Prairie in Washington State, where the subsoil is glacial gravel. In California, where a stiff clay hardpan lies close to the surface, rainwater stands for some time in the hollows between the mima-type mounds, giving rise to the name 'hog-wallow' country. Dissection of the moulds shows tortuous cores of dark surface-soil extending into the subsoil, forming the so-called 'roots' of the mounds. Scheffer [419] and his partner Dalquest suggest that the mounds were made by pocket gophers, *Thomomys*, over the course of long ages, and that the roots of the mounds are the plugged-up burrows, similar to the surface gopher-ropes left after snow has melted. If this theory is correct, and no other has been put forward which fits the facts so closely, 'the Mima Mounds are amongst the most spectacular – if not the largest – structures created by any mammal'.

Family Heteromyidae. The pocket mice, kangaroo rats and kangaroo mice of the family Heteromyidae, although closely related to the Geomyidae are much more delicate creatures than the pocket gophers, and much more pleasing to human eyes. They are small and mouse-like, and share the possession of fur-lined cheek pouches opening outside the mouth with their more heavily built cousins of the other family. They are exclusively American, and are found throughout the western half of the United States from just over the Canadian border through Mexico and Central America to Ecuador. Most of the species inhabit arid or semi-arid habitats, although a few are found in damper regions; like so many desert mammals they are able to go for long periods without drinking, obtaining sufficient water from their comparatively dry food and conserving it by excreting a concentrated urine. The family contains five genera with no less than some seventy-five species, and consists of two groups, the pocket mice *Perognathus*, *Liomys* and *Heteromys*, which normally walk on all fours, and the kangaroo rats *Dipodomys* and *Microdipodops,* which have small fore limbs and much elongated hind ones, and are saltatorial and bipedal. They all make extensive and often very complicated burrow systems containing store-chambers as well as a nest; in hot places and seasons the mouths of the burrows are plugged with earth or sand so that a more humid and cooler micro-climate is maintained within. The animals feed mainly on vegetable matter, particularly seeds, but some species at least take insects and other invertebrates; they are all nocturnal in habit. All species have long tails, those of the jumpers usually with a terminal tuft of long hairs. The fur of the pocket mice is soft in some species, but in most of them there is an admixture of bristly hair or spines, especially on the rump; the sharp bristly hair is so predominant in the species of *Liomys* and *Heteromys* that their vernacular name is 'spiny pocket mice.' Although the pocket mice are quadrupedal in gait, Bartholomew and Cary [24] watched four species of *Perognathus* and showed that their long hind legs facilitate leap-ing and, together with their long hind feet, allow them to assume a horizon-tal bipedal posture while foraging, so that the front feet are free for searching through the soil for food. The similarity to the stance of the kangaroo rats is then striking. The authors suggest that the evolution of bipedalism in the heteromyids may have been a result of specialised foraging habits, and that the strictly locomotor factors have been of secondary importance. When the mouse is foraging the front feet, equipped with long claws, sift the soil and put the seeds found into the pouches with such rapidity that the action cannot be followed by the human eye.

Many species hibernate or aestivate when the ambient temperature is adverse or food is scarce. Batholomew and Cade [23] investigated these phenomena in the little pocket mouse *Perognathus longimembris* of California, one of the smallest rodents in North America. They found that aestivation and hibernation are the same physiological phenomenon, for even with food available the body temperature may drop to that of the environment, and the animals can become torpid at any ambient tempera-ture between 2° and 25° C. The activity of the animals – feeding, digging, collecting food in the pouches, and speed of movement – is much increased at high ambient temperatures; if they then go into aestivation the body temperature remains high whereas in hibernation it is low. The authors conclude that torpidity is important in coping with both short and seasonal periods of environmental stress, and that the variability of body tempera-tures and 'the range of body temperatures over which normal activities can be maintained make the designation of a single "normal" body temperature of doubtful physiological meaning for this species.'

In the two genera of kangaroo rats the fore limbs are so small that when the animals are walking bipedally they are practically invisible, because they are tucked into the fur, with the hands under the chin. The auditory bullae of the skull are greatly enlarged or inflated in these rats, as they are also in other distantly related rodents such as the jerboas, gerbils, and others which live in similar arid places and have similar jumping progression on elongated hind limbs.

Fitch [145] studied the Tulare kangaroo rat, *Dipodomys heermanni*, in California and found that the length and complexity of the burrows varied greatly according to the nature of the ground. 'In fine, deep soil, the burrows may have a two or three storey arrangement with several entrances and thirty or more feet of tunnel. . . . Even in rocky areas, where hardly any digging is possible, the kangaroo rats may be abundant and find sufficient shelter beneath residual boulders'. They have rather small home-ranges, for few of the hundreds of animals marked and recaptured had gone more than 100 feet. During the winter and spring green leaf material was the main food, but in the dry season the seeds of grasses were the most important item of the diet. Fitch found that widely fluctuating changes in population level from year to year are characteristic of this species.

Bartholomew and Caswell [25] used two species of kangaroo rat, *D. panamintinus* and *D. merriami* from the Mojave Desert in California, for their observations on behaviour, making their records by high speed photography. Their pictures are fascinating, and show not only the details of the rats' locomotion, but a ridiculous resemblance to human boxers when the aggressive little creatures are sparring; the tripod support from the tail when standing on tiptoe to drink from an overhead pipette is charming. When moving slowly both species use quadrupedal hopping, but when going faster bipedal hopping. 'When startled, both species leap vertically into the air, and when attempting to escape both initiate a series of powerful hops during which direction is changed frequently and unpredictably, but during which the animals are in perfect control of their equilibrium.' The erratic movement helps the animals to escape from a pursuing predator until they can reach the entrance to a burrow – the burrows are noticeably larger in diameter than the rats, which may thus be able the more easily to pop into safety. The tail is used as a balancing organ in hopping, and one photo shows the tail swung far forward over the head to maintain balance as one animal leaps over another to avoid attack. The anatomy and the behaviour of these animals is highly adapted to life under exacting desert conditions – the functions of foraging and safety are completely separated. A kangaroo rat leaves the safety of its burrow only to feed, and all activities are such that no hindrance is made to a rapid retreat. 'The eyes . . . are so orientated that is is impossible for the animal to see what it is eating' but they are in the most favourable position for seeing the approach of danger from above.

The development of the cheek pouches reduces the time required for foraging to a minimum by separating the periods of food gathering and eating. Moreover, only the forelimbs and the cheek pouches are engaged in food gathering. Since neither of these structures is at all concerned with rapid locomotion, the animal can leap from feeding to extremely rapid locomotion almost instantaneously. The eyes, ears, and hind legs are not directly involved in feeding. Therefore the sensory mechanisms and the hind limbs can be on continuous alert and ready for immediate use in escape. The value of such a series of adaptations to an animal living in deserts, where the available food is widely scattered over areas offering almost no protective cover or concealment, is obvious.

The intraspecific aggressiveness, especially of *D. panamintinus*, may result in spacing the members of this species 'widely enough so that they do not fully exploit the available habitat' and thus both space and food are sufficient for both species if they avoid direct contact.

SUPERFAMILY CASTOROIDEA

Family Castoridae. This once fairly large family – there are fourteen extinct genera – is now reduced to a single living species, the beaver of the Old World and the New, which some hair-splitting taxonomists still regard as distinct species. *Castor fiber* and *C. canadensis*, however, are probably better treated as subspecies – several subspecies have been described in America, but the different stocks have been so mixed by introductions that such distinctions are of little value today.

The beaver is one of the large rodents, and weighs up to, and sometimes over, sixty pounds, and measures about a yard overall. It is a heavily built, thickset, animal with rather short legs, and has a dense soft underfur beneath the long guard hairs. The eyes and the ear-pinnae are small; the former are provided with a fully developed nictitating membrane, and the latter can be shut to keep water out, as can the nostrils. There are five digits on fore and hind limbs, all with claws, and the hind feet are webbed; the claws of the first and second digits of the hind foot are notched, and are used in grooming the fur. The form of the tail is one of the beaver's most peculiar characters; it is naked, very broad, flattened dorsoventrally, and covered with scales. The scales are oval-hexagonal in shape, and are arranged in about a hundred rather ill-defined rows. The scales do **not** overlap like those of snakes, but form a sort of mosaic; a few short hairs grow in the transverse grooves between the scales. The tail is said to be used in swimming with an up-and-down sculling movement, and is used to slap the water as a warning signal. When the beaver sits in the characteristic hunched-up rodent position so that the fore limbs are off the ground and can be used as hands, the tail extended backwards helps to balance the animal; the tail can also be brought forward between the legs and is held in that position by the female while giving birth. I. & G.B.Steen [456] find that the tail of the beaver plays an important part in thermoregulation. The beaver is unable to loose much heat from the body because of the thick coat of dense fur. When the animals are exposed to heat the naked tail is the main means for losing heat and keeping the temperature of the body from rising. The anus and urogenital opening share a common vestibule or pseudo-cloaca, at the sides of which the large anal glands open outside, and within which the larger preputial glands open inside. The glands are paired, and form large sacs, which secrete a greasy sebaceous material with a strong odour, known as castoreum, that was formerly much prized in pharmacy. The glands, filled with secretion, were removed from the body and dried, and then, according to Martin [312], resembled dried shrivelled pears, and were known as castors; about six pairs of them weighed a pound. The use of the secretion to the animals is probably as a marker for territory; the old fur trappers found that sticks smeared with a preparation containing castoreum strongly attracted beavers, and were the only effective bait that would bring the animals to the traps.

The beaver is widespread in the northern hemisphere – in Eurasia it is found eastwards from the Rhône and Elbe, and from Scandinavia to Siberia as far as the river Lena; in North America from the Arctic circle to about latitude 30° N. It has been exterminated over much of its range by man, who has trapped it for its valuable fur, but its numbers are increasing, particularly in North America, where it is being introduced into many places where it formerly lived.

Beavers have always excited interest because of their social habits and astonishing hydro-engineering works. Where beavers live in large lakes or rivers they usually inhabit burrows which they dig in the banks; there is an underwater entrance, and if the water level falls a new one is made below. Where the ground is soft the living chamber and its ventilation hole are protected above by a pile of sticks and mud. Where conditions are suitable a more elaborate 'lodge' is built of sticks and mud, often entirely surrounded by water, and with underwater entrances. In terrain where small streams and rivers do not provide a wide enough stretch of water, a dam of sticks and mud is constructed to form a pond in which the lodges are built. Cook [81] watched a new dam being built by a pair of beavers, which used some fallen dead wood as well as new-cut aspen and alder; in a couple of months it was 100 feet long and four feet high in the centre.

Numbers of stones and the usual coating of mud were applied to the upstream face. The dam leaked badly until the fall of hardwood leaves, big drifts of which collected along the crest, became waterlogged and presently sank, effectively sealing the dam. A lodge, four feet in diameter and three feet high (from the waterline) was built 250 feet upstream from the dam.

Many aspen sticks and logs were stored underwater for the winter food when the pond, the dam, and all but the top of the lodge, were buried in drifts. It is difficult to realise that these constructional activities are the result of innate behaviour patterns and not of intelligent forethought. The point is nicely illustrated by the observation of Tevis [472] on a colony of beavers living in a man-made pond with a stone built dam: 'the beavers worked on it throughout the summer as though it were one of their own stick and mud dams that had to be continuously repaired.' They covered the outer surface of the dam with sticks, mud, stones, water plants and so on but 'although such material would have strengthened a beaver dam it had no effect on this particular structure'. In some places beavers dig canals, often extending for many hundreds of yards, along which they float cut timber to their ponds.

The social relationships of the beavers are peculiar; they are colonial and gragarious but not co-operative. A colony consists of a number of lodges each occupied by one family – parents, kits and yearlings – and the colony prospers only if its numbers are maintained. The apparent collaboration in building the dam, repairing the lodges, and collecting and storing food, is an illusion. Each animal works independently and, apparently, only for itself. Tevis reported of his colony,

Each animal acted as though it alone were the owner and caretaker of Lincoln Pond and the environs. One never assisted another in dam maintenance, lodge maintenance, tree felling, tree sectioning, or food procurement. Although two or more might feed side by side, the reason was not sociability, but the fact that the most easily obtained food for one was the most easily obtained for all.

Nevertheless they lived amicably together and rarely showed any emotion, and even the old male had little social dominance. Their daily routine, 'repeated with monotonous regularity', was to sleep or rest in the lodge most of the day, and come out in late afternoon to feed, collect food, work

157

on the lodge or the dam during the night, and to play or loaf until returning to the lodge in the early morning. Tevis inferred from the sounds coming from the lodge that some activity went on during the day, feeding, scratching, shaking the fur and so on, but the main sounds whether the beavers were asleep or awake were rumbling and belching noises from the movement and expulsion of intestinal gases. 'Having spent most of the daylight hours crowded into the low, stuffy chamber of the lodge, they seemed to delight in the nightly opportunity to separate, when each could go its own way in its own time.'

When swimming in the pond beavers normally dive quietly so that the surface is scarcely rippled; if suddenly frightened they crash dive abruptly, but if the danger is not immediate they do the warning dive. 'Lifting its great flat tail high into the air, it would bring it down hard against the surface of the pond, whip it up again, and dive beneath with a geyser-like eruption of water – a sound as though two boulders had been thrown into the pond'. Not all the beavers that hear the sound respond; those in deep water ignore it, those in shallow water move to deeper, and those on land hurry to the pond.

The industry of the beaver is proverbial; Shadle and his team [435] have given its results in figures. A colony consisting of a single pair and its eight young cut, in forty-nine months, 1,584 trees between one and seventeen inches in diameter. This and five other colonies of about the same size, in an average period of slightly over two and a half years cut 5,424 such trees, of which seventy-six were wasted because they fell so that the beavers could not use them. Over 52 per cent of the trees were only an inch in diameter, and no more than 7·6 per cent were over five inches. The species preferred is the poplar (cottonwood) followed by willows, maples, cherries and others, including some conifers. The bases of trees are often stripped of bark but the trees left standing.

Osborn [356] found that the litter size of the beaver is most commonly two to four, that there is often resorption of some of the embryos, and that the peak of the season of births, in Wyoming, falls in May.

The work done by beavers affects not only their own comfort and livelihood, but the whole animal and plant ecology of their habitat. The ponds are reservoirs which slow down the drainage of the river basin, conserve summer ground water, and raise the level of the water table. As Cornwall [83] says, 'The entire region benefits. Forests flourish and water-loving plants colonise the ponds and swamps. Four-footed game, water birds and their predators immigrate and multiply. . . . The northern wilderness is thus filled with natural resources, valuable to man whether primitive or civilised'. When the dams finally silt up and are colonised by grasses, herbs and shrubs, extensive beaver-meadows fill the valley bottoms. These amenities are so desirable that much work has been done in America and Russia on re-introducing the animals to places from which they had long been exterminated by ruthless trapping. In many remote places the difficulties of transporting the living animals to their proposed new quarters are so great that ingenious techniques have been devised for dropping them by parachute from aircraft – the cages open automatically on reaching the ground, and the animals are liberated where they can form new colonies. This method has been very successful, and beavers are consequently increasing in numbers in Russia, Siberia and America; on a smaller scale they have been re-introduced into Switzerland and tributaries of the Rhône. All this work has stimulated research on beavers so that the literature about the animals is very large – it is unfortunate that space does not allow further consideration of the many fascinating aspects of beaver biology here, beyond wondering how the landscape must have been affected by the works of *Castoroides*, the extinct giant beaver of the Pleistocene of North America, which was 'as large as a black bear'.

SUPERFAMILY ANOMALUROIDEA

Family Anomaluridae. The scaly-tailed squirrels are found only in the tropical and subtropical forests of west and central Africa; although they superficially resemble the sciurid squirrels they are not closely related to them phylogenetically. There are four genera and nine species distinguished by the presence of two rows of overlapping horny scales on the underside of the base of the tail – the scales are said to be used as a sort of ratchet in climbing the trunks of trees by pressing the tail against the bark while the sharp claws of the hands and feet grip surface irregularities. All species except one are gliders; those that are, have a patagium from neck to wrist, thence to the ankle, and from the ankle to a short distance along the tail. As in other gliders the patagium is strutted by a rod of gristle but, unlike that of the flying squirrels and phalangers, it is attached to the elbow and not to the wrist. Most of the flying scaly-tailed squirrels are nocturnal, and spend the day sleeping in hollow trunks and branches; some species feed mainly on leaves, others eat fruits and seeds in addition, and some of the genus *Anomalurus* take the nuts of the oil palm in west Africa.

Little is known about the lives of these animals, but Sanderson [415] has recorded some interesting notes about them in the north Cameroons forest area. He found that *Anomalurus fraseri* rests in hollow trees during the day, and emerges just after sunset. When leaping from tree to tree the patagium is curved so that the animal has 'the shape of a small umbrella' and the animal usually lands on the trunk rather than the branches of a tree. Just before landing the head is lower than the tail, but at the last moment 'the fore legs are thrown back over the shoulders, the head comes up, and the tail sweeps up to meet it over the back'. This brings the animal into a vertical position to land on the trunk and run upwards, using the two fore feet together 'then pulling up the two hind feet and arching the back like a giant loop-caterpillar', holding by the hind feet and the scaly tail while the fore feet are moved ahead. The animals can thus gallop up the smooth trunk of a giant forest tree at an astonishing speed. Sanderson found that this species eats flowers as well as the fruits of various trees, and that the minutely masticated contents of the stomach have a pinkish-purple colour. Chalmers [74] records that a captive *A. jacksoni* from Kenya not only ate various fruits and leaves but also enjoyed rose petals, and seemed to prefer pink ones. Sanderson states that the single young stays in a hollow near the top of a tree until it is nearly full grown, and is fed by both parents, which bring it chewed food carried in the mouth so that the cheeks 'are swollen with a wad as large as a tangerine'.

The habits of the single species of the allied genus *Anomalurops, A. beecrofti* the West African flying scaly-tail, are quite different, according to Sanderson. He found that it lives in the higher layers of the forest and sleeps among dense bunches of leaves and twigs at the tops of trees, never in hollow trunks. It is also diurnal and fully active at noon. Sanderson suggests that the bright leaf-green colour of the back, a colour which fades to dull olive green after death is correlated with daytime activity. The preference for tree-top roosting is probably the reason for the occurrence of *Anomalurops* in deciduous mountain forest where there are no hollow trees suitable for *Anomalurus*.

The genus *Idiurus*, the pygmy flying scaly-tails, contains four species, two of which were studied by Sanderson. They are gregarious and live in hollow trees, often in large colonies of up to a hundred animals, frequently of the two species together – *I. macrotis* and *I. zenkeri*. These tiny squirrels, unlike *Anomalurus*, fly very slowly with the patagium taut and more or less flat, and can steer up and down and to the sides, and land with a graceful upward curve; on landing they run, using the fore and hind legs diagonally. Although they are nocturnal, when disturbed during the day they

seem to be in no way disconcerted by the light, but issue forth from the holes, career about the surface of trunk and branches, faster than mice on a level floor, take to the air in clouds, floating away among the neighbouring trees like bits of soot from a chimney, steering themselves over and around obstacles with the facility of birds, and landing silently, hundreds of feet away, without any apparent loss of height.

The single species of the genus *Zenkerella*, the flightless scaly tail of Cameroon and Spanish Guinea, closely resembles its flying cousins except that it has no patagium. Nothing is known of its life and habits, for only seven specimens had been collected up to 1957. The suggestion has been made that as it does not fly it may be diurnal like the tree squirrels; the collectors [1] however, say that it roosts in hollow trees like *Idiurus*, and is nocturnal.

Family Pedetidae. This family contains only one genus, *Pedetes*, the jumping hare or springhaas of eastern and southern Africa, with at most two species, which are considered by some systematists as merely sub-species of one. Jumping hares are found in Africa from Angola on the west and Kenya on the east, to Cape Province in the south. *Pedetes* is about the size of a large rabbit, and in general form resembles a small kangaroo; the hind legs are long and are used for progression by bipedal hopping. The head is rather heavy and blunt-nosed, the eye large and dark, the ears long and shaped somewhat like those of a kangaroo. The fore limbs are short; the tail, as long as the body, ends in a bushy tuft of hair. The fur is soft but rather thin, and buffish brown in colour; the tail tuft is nearly black. When sitting on its haunches, with the soles of the long feet on the ground and the tail extended behind, the hunched attitude of the body is much like that of a kangaroo. The first digit is lacking in the foot, and the third is longer than the others; the four digits are armed with short triangular claws. The fore limb bears five digits all armed with long sharp claws, and the palm carries well developed thenar and hypothenar pads. The thenar pad, lying over the radial sesamoid bone, is capable of some movement through the action of slips from the palmar muscles attached to it; it is covered with thick hardened epidermis resembling a nail. Wood Jones [259] thought that the sporadic occurrence of a radial sesamoid resembling an extra digit in the mole, jumping hare, and giant panda 'is a very remarkable fact'. Be that as it may, the structure of the hypothenar pad on the ulnar side of the hand, overlying the pisiform bone, is certainly remarkable: the pad is elongated in line with the fifth digit near the edge of the hand and 'is fringed on the whole length of its free margin by a growth of stiff hairs'. The pad and its fringe are a hair brush used for brushing the face and the long facial vibrissae – a young animal in captivity used the toilet apparatus in this way almost immediately after it was born.

Jumping hares live in dry country where the soil, often sandy, is suitable

L

for burrowing. The animals dig burrows with the long claws of the fore feet, and spend the day underground, emerging to feed only at nightfall. The burrows are generally not very complex, although there are usually several escape holes opened from within and thus not betrayed by a spoil heap; when at home the animals plug the burrows with loose sand some way from the opening. One observer quoted by Roberts [406] said that the jumping hare leaves its burrow at night with a mighty spring 'apparently with a view to avoiding lurking enemies'. When unhurried and foraging, jumping hares walk on all fours 'like a rabbit', but when alarmed they bound away hopping on the hind feet and holding the fore feet clenched under the chin – they can clear eight or ten feet at a hop. When hopping the tail is held in an upward curve behind, and the animals follow an erratic course, jinking like hares. They are often picked out by the lights of motor cars at night, and even people familiar with the creatures find the sight of these peculiar hoppers, apparently with no fore limbs, wildly bounding about, more than a little ridiculous (Plate 12).

Jumping hares feed on the roots and rhizomes of grasses and other herbage, which they dig up with their fore claws and incisor teeth; the buccal pads at the corners of the mouth are well developed, and can be pinched in to isolate the mouth cavity and keep the soil out when the incisors are thus used. In cultivated country jumping hares can be a great nuisance to farmers because they dig up young cereals to eat the seed and roots, and are particularly destructive to maize, eating both ripe and green cobs.

Pedetes is not a social animal, each burrow is inhabited by one animal or at most a pair, and although the burrows sometimes communicate they are not used communally; where many burrows are dug close together they do not form a warren, but are associated because the soil there is suitable for burrowing. The animals sleep in a chamber within the burrow, sitting with the head and fore limbs bent forward between the thighs and the tail curled round, but no nest is made, even for cradling the young – although the females have four nipples on the chest they rarely give birth to more than a single offspring. When digging or underground they can fold back the ears to keep out the sand, closing the meatus with the elongated metatragus.

Some naturalists have stated that jumping hares do not go far from their burrows when feeding; indeed one would expect animals with homes in burrows to hold territory, and to have a restricted home range. On the other hand some have said that they travel distances of several miles to feed, and in severe droughts to go twenty miles in a night to find water;

this is probably exceptional and due only to the stress of unusual conditions.

Shortridge [442] says that the flesh of *Pedetes* is white and firm but rather tasteless: 'most Colonists have a prejudice against eating it'. On the other hand it is one of the main sources of meat for the bushmen of South-west Africa, and he goes on, 'Easily caught, much hunted by natives, and only as a rule producing a single young, Springhaas, nevertheless, seem to maintain their numbers, even in populated areas'.

The family Pedetidae brings us to the end of this review of the sciuromorph rodents, but before leaving them we should notice that although the first five families dealt with form a reasonably homogeneous group, the last two, the anomalurids and the pedetids, are of doubtful affinities. There seems to be good reason for placing the two families close together, but their position relative to the first five is much more distant. Indeed, they have at one time or another been associated with numerous different groups by various students of the rodents, and are here placed among the sciuromorphs in accordance with Simpson's classification, in which he points out that the arrangement is tentative and really only a marriage of convenience; the possible illegitimacy of the union is therefore stressed by including them in the suborder under the heading: ? Sciuromorpha *incertae sedis*.

Voles, Hamsters, Mice and Rats

THE MYOMORPHS are by far the largest suborder of the rodents, and run to over 1,100 species in some 225 genera, out of the rodent total of nearly 1,700 species and over 350 genera; the equivalent figures for the sciuromorphs are about 400 species in seventy-one genera, and for the hystricomorphs about 180 species in fifty-eight genera. Two families contain the greater part of the myomorph species, the Cricetidae with about 567 species and the Muridae with nearly 460. Large groups of genera in both these families contain species that resemble one another pretty closely, so that superficially they show the same theme with little variation. The species in the smaller families, with comparatively few genera, are particularly interesting because they mostly differ conspicuously from those of the two large families, and are generally much more specialised in their habits and adaptations. This does not mean that the members of the great host in the large families are without interest, but merely that they tend to be less eye-catching, and that consequently less is known about the biology of many of them – all species turn out to be of equal zoological interest when we have discovered enough about them. In this chapter, however, the sheer weight of numbers forces us to examine the two large families less closely than the smaller ones.

FAMILY CRICETIDAE
Turning first to the Cricetidae, the largest family in species, although it is surpassed in genera by the Muridae, which have ninety-eight as against ninety-seven, it is convenient to deal with it under the headings of five subfamilies and, within the largest of them, the Cricetinae, three tribes, as shown in the following table.

Family Cricetidae
Subfamily Cricetinae
Tribe Hesperomyini

Tribe Cricetini
Tribe Myospalacini
Subfamily Nesomyinae
Subfamily Lophiomyinae
Subfamily Microtinae
Subfamily Gerbillinae

TRIBE HESPEROMYINI

By far the largest tribe in the subfamily Cricetinae is that of the Hesperomyini, the rats and mice native to the New World. They are found throughout the Americas from north to south, living in a great diversity of habitats; although they superficially resemble the Old World mice and rats they are classified as cricetids on characters of the cheek teeth and skull. Most of the species are typically mice or rats, depending upon their size, but a small number show special adaptations to unusual ways of life.

Numerous species of rice rat, genus *Oryzomys*, are semi-aquatic and live in coastal marshes; others are terrestrial. They are found from the United States through Central and most of South America. In swampy places they make nests above ground in tufts of vegetation or low shrubs; elsewhere, on the ground among grasses or even in burrows. Although they dive and swim well they have no special adaptations, such as fringed digits, for aquatic life. They are docile in captivity; Hamilton [198] who studied them in the tidal swamps of the Virginia coast found that they spend

much time on their meticulous toilet. They wet the back of their hands, then draw them over the ears and down over the face. The bottom of the feet are carefully licked, the tail is run lengthwise through the hands, while the rat nibbles gently on it. One hind foot is held by the fore paws while each toe is slowly licked and cleaned. The moistened claws are then used to preen the fur of the back, groin and shoulder. I have never seen any other small mammal so solicitous of its toilet.

The food is mainly vegetable matter, though animal food is also taken, and where the animals are plentiful they are sometimes pests in rice fields.

The giant rice rats, *Megalomys*, of some West Indian Islands, nearly as large as rabbits, but with long tails, are probably extinct – the volcanic eruption of Mont Pelée on Martinique in 1902 seems to have destroyed their last habitat. They were scarce before that date because they were killed as plantation pests as well as for food – their flesh was relished in spite of the animal's musky smell. Only a few stuffed specimens remain. The spiny rice rats, *Neacomys*, of tropical America, resemble *Oryzomys* but have spiny bristles mixed with the fur of the back. Two neotropical genera, *Nectomys* and *Scapteromys*, each with several species and numerous sub-species, are semi-aquatic – some species have fringed tails or hind feet, typical adaptations for swimming. Hershkovitz [224], however, considers it most probable that 'their swimming modifications are, primarily, adaptations for meeting the exigencies created by the periodic floodings of their habitats' because they are often found away from streams or lakes, and do not appear to be 'dependent upon them for foraging or refuge'. The large South American genus *Rhipidomys* contains the climbing mice, which live in trees and bushes; the fur is soft and velvety, and the tail, unlike that in the genera already mentioned, is hairy and carries a terminal tuft.

The rather similar paramo and tree mice of the genus *Thomasomys*, with many species throughout most of South America, have thick soft fur and long tails; some species live on the ground but others are arboreal. The large tree rats – about half a dozen species of *Tylomys* – of tropical America superficially resemble the common brown rat, but are adapted for climbing by their large hands and feet; the rather smaller species of the genus *Ototylomys* are distinguished by their large naked ears. The vesper rat, the only species of its genus, *Nyctomys sumichrasti*, has a large eye enhanced by a ring of dark hair surrounding it, and is unusual among rats in that it builds nests like those of squirrels in trees – it is nocturnal and seldom comes to the ground. The American harvest mice, about fifteen species of the genus *Reithrodontomys*, though much smaller, also make nests above ground – balls of grass and fibre usually placed in tall herbage of low shrubs They are found from southwest Canada to Ecuador.

Over fifty species of deer mice of the genus *Peromyscus*, are found throughout North and Central America; their eyes and ears are large, and their underside and feet are usually white. These mice are the American counterpart of the Old World field- or wood-mice of the genus *Apodemus*. Their abundance and ready availability have led to them being the most intensively studied of all the American small rodents both in the field and in the laboratory; in consequence much of their biology is known in great detail. These mice have small cheek pouches, described by Hamilton [197], who showed that although small when empty they can be greatly distended when a mouse is collecting food for storage.

The females of most small rodents will not tolerate the presence of another adult in, or even near, the nest containing her young, but in one subspecies of *Peromyscus* males have been seen sharing the care of the young, at least in captivity. Horner [237] records that male *P. maniculatus gracilis* assisted the female in nest building, brooding and washing the young, and in removing the young to a new nest if the old one were disturbed. This co-operation has not been seen in other species of *Peromyscus*, in which the female's behaviour is similar to that of other mice in that she is hostile to male intrusion into her nest. Similar co-operation has, however, been observed by Blair [42] among captive specimens of the nearly related pygmy mice, three species of *Baiomys* from the southern United States and Central America, which are only two to three inches from nose to tail root, and are thus among the smallest of mammals. The two species of grasshopper mice, *Onychomys*, found in various arid areas from Canada to Mexico, are unusual in their diet; they are insectivorous and carnivorous, and get their name from their favourite article of food.

Over sixty species of South American field mice make up the genus *Akodon*; superficially they resemble the Old World voles but have comparatively long tails. Their habits appear to be similar to those of voles, though some species enter houses and become pests. They inhabit many different habitats throughout South America even to the subantarctic forests of Tierra del Fuego. The seventeen species of *Zygodontomys*, less widely distributed in South America, also resemble voles in their way of life – they are active during the day as well as the night, and live in nests and runways among dense grass. A like number of species of burrowing mice, genus *Oxymycterus*, is found from eastern Brazil to northern Argentinia; as their English name implies, these mice live in burrows. The snout in this genus is long and pointed – the first species known was collected by Darwin during the voyage of the *Beagle*, and was named *O. nasutus*; both the generic and specific scientific names are taken from this character. The mole mice of southern South America, about ten species of the genus *Notiomys*, have stout bodies, short tails and ears, and long claws in adaptation to fossorial life. They are found in a wide range of habitats, where they burrow through the soil and seldom emerge on the surface.

Two or three species of marsh rats of the genus *Holochilus*, found from Venezuela to Patagonia, live in swampy places, but can become serious pests to cane and rice fields. The population periodically builds up to huge proportions to produce a 'ratada' or plague of rats – these rats make feeding platforms of cut vegetation on which they sit at night eating the food cleared from the surrounding areas, and at such time take so little notice of human disturbance that they can be caught by hand [480]. Similarly several species of cotton rats, genus *Sigmodon*, of the southern United States and Central America, often build up enormous populations which do great

167

damage to crops; they are medium sized rats which are not restricted to a vegetable diet but also eat invertebrates and other animal food. Over twenty species of pack- or wood-rat, genus *Neotoma*, are found over most of North and Central America from British Columbia southwards; they build large houses of sticks and twigs on the ground, as described in Volume I – their habit of picking up any bright object from men's camps to work into their nests, and of dropping in its place the stick they happen to be carrying, has given them their alternative name of 'trade rats'.

The pericots, about fifteen species of the genus *Phyllotis*, have soft fur and large hair-covered ears, as does the single species of *Chinchillula*. Some species of the first are found at sea level, but others at high altitudes in the Andes, where with *Chinchillula* and some species of *Akodon* and others they form a characteristic small mammal fauna; although they are not adapted to life at great heights by a high level of haemoglobin in their blood, they have, according to Morrison and Elsner [339] faster breathing and heart rates than lowland mice. The South American highlands are the home of a number of other genera of endemic mice. Pearson [370], when studying the mountain viscacha, a rabbit-sized hystricomorph described in the next chapter, at an altitude of 16,000 feet in Peru came across many of them, including the vole-like *Punomys*, active by day and night. He mentions *Chinchillula sahamae*, 'most beautiful and silken-haired of Andean mice; and, closest associate of all, big-eared *Auliscomys boliviensis flavidior*'. *Auliscomys* is one of the subgenera of *Phyllotis*: he continues,

While watching viscachas in the morning sun, you may see the little head of one of these *Auliscomys*, topped by ridiculously long ears, pop out of a crevice. Then the whole mouse darts out to some sunny rock where he seems to melt in the relaxing rays of the sun, half closes his eyes, and sits contentedly among his sun-bathing viscacha neighbours. . . . With eyes adjusted to viscachas it is always surprising and amusing when one of these little elves appears instead.

When the viscachas flock out to feed in the afternoon, 'I have seen *Aulomyscus* follow along to scratch and nibble among the "grownups", the late sun fairly glowing in its golden ear patches'. Another genus of big-eared cricetines contains the two species of *Reithrodon*, the rabbit-rats of the plains and open country of southern South America. Finally, five genera of South American water rats, such as *Ichthyomys* and *Rheomys*, are known from their habits as the fish-eating rats. They have rather large feet and tails; some species live in small mountain streams with fast-flowing currents, others (*Daptomys*) in the rivers of tropical forests.

TRIBE CRICETINI

The tribe Cricetini contains the Old World hamsters, together with two other genera. Hamsters are stout-bodied blunt-nosed rodents with short legs and tails; they live on the ground and dig extensive burrows in which they make their nests. They have large cheek-pouches used in collecting food which is stored in the burrows; in many places they are serious pests to agriculture. When the cheek pouches are crammed with seeds or other food the hamsters, like some other pouched rodents, use the fore feet to press behind the cheeks and squeeze out the contents when they are to be discharged. The stores are eaten during the winter, whenever the animals interrupt their hibernation. The common hamster, *Cricetus cricetus*, the only species in the genus, is found from Europe to Siberia both on wild steppe and cultivated land. The fur is thick, brown on the back, but black below, with white patches on the cheeks and flanks. The single species of *Mesocricetus*, *M. auratus*, the golden hamster, is similar though smaller, but has a white or cream underside. Its burrowing and food storing habits are also similar. The species is found in arid places from southeast Europe to Asia Minor; all the hundreds of thousands now in captivity throughout the world are descended from a female and her litter of twelve young found in Syria in 1930.

The genus *Cricetulus* contains seven species widely distributed from eastern Europe to China. They have rather longer tails than the other hamsters, hence their name of rat-tailed hamsters – some are also known as rat-headed hamsters. The largest species, *C. triton*, is about eight inches long with a tail of about half that length. Loukashkin [304] studied it in Manchuria, where it is a serious crop pest, and found that like other hamsters it lives in extensive individual burrows and is intolerant of intrusion by its neighbours. He described it as extremely fierce and savage, and recorded that it attacks and kills other small rodents and that, unlike many other hamsters, it is quite untamable in captivity. He found as much as five kilograms of food, mainly soy beans and millet seeds, in a single store chamber of a burrow. Two species of dwarf hamsters of the genus *Phodopus*, with short tails, are found in arid parts of Siberia and northern China. The single species of *Calomyscus* from central Asia is unlike the hamsters, and resembles a wood mouse; it has a long tufted tail, and large ears like those of a mouse. It is limited in distribution, and is considered by Osgood [361] to be a late Pleistocene immigrant from North America, now on the way to extinction. The white-tailed rat of southern Africa, *Mystromys albicaudatus*, the only species of its genus, is the other cricetinid that looks nothing like a hamster, and is the only member of the tribe found south of the equator. It is about the size of a common rat, with grey upper fur and white belly, feet and tail. Roberts [403] found white-tailed rats living in burrows and coming out at night to feed on seeds and plants – in captivity they are 'tame and playful'.

TRIBE MYOSPALACINI

The tribe Myospalacini contains a single genus, *Myospalax*, with five species of mole-rats, found from Russia to northern China. The animals are cricetids and are quite distinct from the mole-rats of the separate family Spalacidae and the hystricomorph Bathyergidae dealt with below. They are stout-bodied, short-legged animals with short tails and soft fur; the largest species is as big as a common rat. The eyes are very small and there are no external ears; the four claws of the hand are large. *M. psilurus* was studied in north China by Zimmerman [530]; the animals are subterranean and live in long burrows which are marked at intervals by mole-heaps of excavated earth – they feed on roots and other vegetable matter. Loukashkin [302] found *Myospalax* plentiful in an extensive area of sand dunes with scattered willow bushes and pine trees in northern Manchuria – cone shaped heaps of wet sand thrown up by the mole rats were found along the track of the Chinese Eastern Railway, 'sometimes like a chain reaching hundreds of meters in length'.

SUBFAMILY NESOMYINAE

The subfamily Nesomyinae consists of seven genera peculiar to Madagascar, where they have probably differentiated to fill the different ecological niches available to them, although some systematists think that they may not all be of common origin. The two species of *Macrotarsomys*, one of them very small, superficially resemble the European wood mouse *Apodemus*, but have very long and slightly tufted tails. They are burrowers, whereas some of the five species of *Eliurus*, which also have tufted or even bushy tails, are arboreal. The three species of *Nesomys* look much like house-mice with white undersides and hairy tails, and the rather more vole-like *Brachytarsomys albicauda*, the only species of its genus, has a fairly long tail the distal half of which is white. Two further genera, *Gymnuromys* with one species and *Brachyuromys* with two, are more rat-like in appearance; in the first the tail is long, in the second it is short. The most striking member of the subfamily, however, is *Hypogeomys antimena*, the only species of its genus, a large-eared mouse as big as a rabbit. It lives in dense forest, now rapidly being destroyed, and digs long deep burrows. Although the animals of this family have been closely studied by the systematists, it is unfortunate that little has been recorded about their lives and habits.

SUBFAMILY LOPHIOMYINAE

The subfamily Lophiomyinae consists of a single species, *Lophiomys imhuasi*, the maned or crested rat, found in eastern Africa from the southern Sudan to Kenya. The crested rat is a large thick-set rat with a blunt head, small ears, short legs and long tail; its apparent size is increased by the long fine fur, bushy tail and erectile crest of long coarse hair on the back. The general colour is grey, with lighter patches on the face, extending to behind the shoulder, surrounded by darker ones – the colouration is diffuse and variable. This rat is nocturnal and, although it lives in burrows or natural crevices, it is arboreal; the first digit of the foot is opposable to the others. In the skull the temporal fossa is closed by an expansion from the frontals and parietals which reaches the jugal, a unique character in rodents; the upper surface of the skull is also peculiarly granulated. The diet of the crested rat consists of all kinds of vegetable matter.

SUBFAMILY MICROTINAE

The subfamily Microtinae contains the numerous species of voles and lemmings – about a hundred species in eighteen genera. Members of the subfamily are found throughout Eurasia and North America to northern Central America. Among the several genera of lemmings *Lemmus*, containing the Norway lemming, *L. lemmus*, is the most widely known, from the spectacular migrations that follow the periodic explosions of its population, as discussed in Volume I. Similar migrations are sometimes made by other species, such as the brown lemming of northern Canada and Alaska, and the Siberian lemming. The lemmings, about the size of a small rat, are stout-bodied, blunt-nosed animals with short tails and small eyes, and the ears hidden in the rather long fur. The claw on the first digit of the hand is wide and flat, and is enlarged during winter when the animals burrow through or under the snow. Soper [453] found the brown lemming *L. mucronatus* the most abundant animal during the winter in Baffin Island, and for that reason he says, 'it holds for even the most casual wanderer in those solemn lands of the north a peculiar appeal. For many months of the year when nature becomes as inert and lifeless as is possible upon the globe, a fleeting glimpse of even such an inconspicuous creature, or its tiny trails, provides a passing stimulus and pleasure that the average person can hardly be expected to appreciate'.

The five species of collared lemmings of the genus *Dicrostonyx*, found throughout the arctic, are even more highly adapted to facing the harsh winters of the north, for their dense fur turns white in winter, and the claws of the second and third digits of the hand are enlarged, and develop two points during winter, in adaptation to burrowing in hard snow. None of the lemmings hibernates; all species find their vegetable food beneath the snow. Banfield [18] points out that they wander not only over the ice of frozen lakes but over coastal shore-ice to reach the offlying islands. Two species of *Synaptomys*, the bog lemmings, are sporadically distributed over North America, and range well below the arctic; they look more vole-like than the other lemmings, and are found in marshy places. The single species of wood lemming, *Myopus schisticolor*, of northern Eurasia is the smallest species, and like all the others, it is subject to periodical fluctuation in numbers.

Figure 9 Collared lemming (*Dicrostonyx*), feet showing the bifid claws developed in winter, and the single pointed ones of summer.

The numerous species of voles or meadow mice are classified into about a dozen genera, of which *Microtus* with nearly fifty species is the largest. Voles are mouse- or rat-sized animals, with blunt snouts, small eyes, and the ears partly or wholly hidden in the fur; in most of them the tail is rather short. Species of *Microtus* are found throughout North America and the northern middle parts of Eurasia. Most of them live among dense vegetation, and make surface runways under the mat of grasses, or shallow burrows below it. They feed mainly on grasses, and make their globular nests from shredded grass stems. Species of this genus are subject to very large periodical increases in population leading to the vole plagues and population crashes discussed in Volume I. The five species of *Clethrionomys* the red-backed or bank voles, have an equally wide distribution between them, but are less dependent upon grasses; they spend much of their lives climbing among shrubs and low bushes of which they often bark the twigs and stems during winter. Several genera of Asiatic and far eastern species include *Alticola*, the thick furred mountain-voles of the high mountains of central Asia; the species are generally light in colour, and have the soles of their feet partly furred. The single species of *Arvicola*, the water vole, is distributed over most of Europe and Asia to Siberia and Asia Minor, and is divided into many subspecies. It is one of the largest voles – it is about the size of a rat – and swims and dives well, but not all of its races are aquatic and in many parts of its range the local race is terrestrial. For this reason the scientific name of the water vole is, paradoxically, *A. terrestris*. As with all voles the diet is mainly vegetarian, but the aquatic forms eat considerable quantities of water invertebrates as well.

The spruce voles of North America, five species of *Phenacomys*, are very similar to bank voles, but are even more arboreal and make their nests in trees, where they feed on leaves, bark and pine needles. The genus *Pitymys* contains eleven species of pine voles, found in North America and most of Europe and Asia Minor, in various habitats ranging from open country to forests. Some species are rather mole-like, and live a mainly subterranean existence; they sometimes damage the roots of crops and of fruit trees. Their fur is velvety, the external ears and the eyes are small, and the claws of the fore limbs are rather large. The general appearance of the long-clawed mole-vole, *Prometheomys schaposchnikowi*, the only species of its genus, from the Caucasus mountains, is similar, but the fore claws are even longer; this species throws up the earth from its burrows like mole heaps. On the other hand, the still more mole-like species of *Ellobius* have short claws, although the fore and hind feet are broad, and use mainly their incisor teeth in digging – as also probably the last two genera do to a lesser extent. There are two species in the genus, which is found from central Russia to Mongolia and Asia Minor; *E. talpinus*, from the Russian steppes, makes very long and complicated burrow systems. Steppe lemmings, three species of *Lagurus*, are found in the west of the United States, and from the Ukraine to Mongolia, where they inhabit rather arid regions such as sandy semi-deserts with scattered plants of *Artemeisia* – hence their American name of 'sagebrush' voles. They are small fossorial voles which thrive in captivity; they are prolific breeders and have been used as laboratory animals. The numbers of steppe lemmings, like those of most voles, periodically reach plague proportions; and, like most voles, the animals do not hibernate but lay up stores for winter use.

Two genera of comparatively large animals, *Ondatra* and *Neofiber*, the musk rats of America, are among the least vole-like of the subfamily. The musk rat *Ondatra*, found over much of North America, is aquatic in habits and structure; it is rather larger than the common rat, and has thick fur, the hind feet partly webbed and fringed with stiff hairs, and a long scaly tail

173

flattened from side to side. It inhabits marshes, ponds, lakes and rivers, and in its way of life it much resembles the beaver. Like the beaver, it digs extensive burrows in the banks of rivers – the entrance is below water – or, in swampy places, it builds lodges of reed stems above the surface, and makes feeding platforms nearby; it does not, however, fell trees or build dams, but it does clear channels through thick growths of water plants. The dense, soft, underfur is the musquash of commerce, a fur so esteemed that the animals have been introduced into Europe to be farmed for their pelts – escaped animals have multiplied so that they have become pests to agriculture and a threat to the stability of river and canal banks in many places. The round-tailed muskrat, *Neofiber alleni*, the only species of its genus, is found in Florida and southern Georgia, where *Ondatra* is absent. It is slightly smaller than *Ondatra*, and less aquatic, although it lives in marshy places and along the banks of rivers and other waters. The hind feet are not webbed, though they bear fringes of stiff hair, and the tail is not flattened. These animals make their nests in mounds of vegetation which they heap up on the surface, and like *Ondatra*, they also make floating feeding platforms.

SUBFAMILY GERBILLINAE

The Gerbillinae, the last subfamily of the Cricetidae, are small rodents inhabiting desert and arid places throughout Africa, and from south east Russia and Arabia to China. About eighty species are known; most of them are mouse-sized animals with soft silky fur, large dark eyes and moderately prominent ears. The tail in many species is long, and in all it is hairy, often with a tuft at the end; the hind legs and feet tend to be long, and some species are saltatorial. The skull in all the members of the subfamily has inflated auditory bullae, a peculiar character commonly found in rodents of desert and arid regions.

The gerbils, about thirty-five species of the genus *Gerbillus*, are found throughout Africa and southern Asia to India. They are among the longest legged of the subfamily, and have some resemblance to jerboas. Harrison [210] found that the Mesopotamian subspecies of *G. dasyurus* is colonial and lives in numerous burrows in sandy hillocks at the edge of the desert. 'The mice were crepuscular and nocturnal in habit, appearing, as the light faded, to dart from one system of burrows to another.' About two dozen species of naked-soled gerbils of the genus *Tatera*, have between them a similar distribution to that of the genus *Gerbillus*. The hind legs of these animals are shorter than in *Gerbillus* but the tail is long and the general appearance is more rat-like. These gerbils, also, are colonial and live in complex burrow systems with many entrances. The genus *Meriones,* the jirds or sand rats, from southeast Russia through southwest Asia to north Africa, runs to about a dozen species. They are even more rat-like, with comparatively shorter tail though with a terminal tuft; they live in very arid places in colonies with labyrinthine burrows. Some species are nocturnal, but Harrison [210] found *M. lybicus syrius* in Iraq to be mainly diurnal; in a large colony he found 'the animals running about and feeding in the open and in the full sunlight at midday. When running fast the tail is held erect at right angles to the body'.

Of the nine other genera, most of them containing only a few species, the fat-tailed gerbil, *Pachyuromys duprasi,* the only species of its genus, from northern Africa and Egypt, is the strangest in appearance; it is a mouse-like animal with buffish, silky fur, large eyes and a short very thick club-shaped tail. The short-eared gerbil of southern Africa, *Desmodillus auricularis,* also the only species in its genus, is stoutly built and runs like a rat; Shortridge [442] found it 'not noticeably saltatorial', and that it is insectivorous as well as graminivorous – 'the wings of quantities of locusts were often observed almost choking the entrances of some of the burrows. . . . Still more frequently the burrows may be found packed with feathery grass heads' from which the seeds had been 'carefully picked off and eaten'.

FAMILY SPALACIDAE

The small family Spalacidae contains only one genus with three species, the mole rats of southeast Europe, southwest Asia, Egypt and Libya. Members of the genus *Spalax* are among the most peculiar of the rodents in appearance. They resemble moles in size and shape, and have short soft velvety fur which, however, is yellowish brown; on the other hand, the snout is not pointed but rounded and blunt, the tail is vestigial and invisible externally, and the fore feet are small and relatively delicate – Reed [397] who observed mole rats in Iraq calls the claws 'merely blunt round nubbins'. Mole rats live like moles, permanently underground, though, like them, they sometimes come out on to the surface by night – they are not, however, carnivorous but feed mainly on roots, bulbs and tubers, and in some parts of their range are pests to agriculture. They use their huge permanently exposed incisor teeth for digging, and kick the loose earth back with their hands and feet; in hard ground they push the excavated material up to the surface and make mounds like mole heaps or pocket gopher mounds, but in softer and damper ground they tamp it into the burrow walls. The rounded head is flattened and the snout which is broad, padded and horny, is said to be used for tamping the earth in the burrow and for pushing it along the tunnel when mounds are made. Watson [500], who studied a living specimen in Greece, found that in fairly loose soil the animal tunnels largely by thrusting the wedge shaped head into the earth and lifting it to compact the soil overhead. The snout bears a row of stiff bristles, running from the nostril towards the ear, which are almost certainly tactile in function.

The species of *Spalax* are the only rodents that are blind; Reed found 'that there is no external indication of an eye and the animals are obviously completely blind. In preparing a skin, a tiny black pinhead of an eye is found, however, adherent to the inner surface of the dermis, but no traces of ocular muscles or optic nerve could be seen'. The external ear, too, is rudimentary and the ear orifice is pushed back to the region of the shoulder by the great development of the jaw joint and muscles; as Reed says, the jaw joint has become particularly specialised, and the fore-and-aft movement of the jaw condyle in the glenoid fossa, which is universal in rodents, is greatly accentuated so that the lower jaw can be pulled further back than in other rodents. As would be expected, the burrow systems of *Spalax* are extensive, and reach a depth of nine or ten feet; rounded chambers contain the nests made of dry grasses, and others are used as food stores and as latrines – even in cold mountainous regions the animals do not hibernate but are active under the snow during winter.

Reed [397] who accompanied an archaeological expedition to Iraq says, '*Spalax* obtrudes itself upon the archaeologist's consciousness, for the rodent is usually a prior excavator in the archaeologist's site and had often been mixing things up for thousands of years, to the exasperation of man, who thus finds small cultural artefacts displaced ... and out of stratigraphic sequence'. He also refers to a *Spalax* released in a room, whereupon it at once began 'tearing the plaster off the wall'. Harrison [209] collected *Spalax* in Kurdistan in the extreme north of Iraq, and found that the animals 'uttered quiet grunting noises quite unlike the normal rodent squeak'. Grassé [183] records that *Spalax* are '*Très méchants, ils essaient de mordre quiconque les approche*'.

FAMILY RHIZOMYIDAE

The family Rhizomyidae contains the bamboo rats, *Rhizomys* and *Cannomys* of Malaysia and south east Asia, and the African mole-rats *Tachyoryctes* of tropical east Africa. All are mainly subterranean in habit, but they are much less specialised anatomically than *Spalax*. The skull, though broad, is less flattened than in *Spalax*, the limbs and tail are longer, and the fur is thick and soft but not velvety. The eyes and ear-pinnae, though small, are present in the rhizomyids, and the claws of the feet are well developed although not particularly long or specially modified. The incisor teeth are large and permanently exposed but are not generally used for burrowing, which is accomplished with the claws – the hind feet kick back the soil loosened by the fore. The three species of bamboo rat, *Rhizomys*, found from southern China to Sumatra, make their burrows in bamboo thickets and feed upon the roots and shoots of bamboos. The lesser bamboo rat *Cannomys badius*, the only species of the genus, found from Nepal to Thailand, feeds more on the surface on a wide range of plant materials, leaving its burrow at night to do so. This species is said to use its incisors as well as its claws for burrowing more constantly than the members of the other genera of the family.

Tachyoryctes, the African genus of the family, contains fourteen species. The genus is unusual because its range overlaps that of other mole-rats belonging to a distantly related family, the Bathyergidae, which are not even myomorphs. The specialised subterranean ecological niche exploited by herbivorous rodents is generally limited, so that it does not support more than a single group, and even species within a group seldom overlap in range and habitat. No doubt the habits of the members of the two families differ sufficiently to avoid direct competition for food and territory, but the facts have yet to be discovered. The African mole-rats are often abundant where the soil structure is suitable to them, and their tunnels then honeycomb the ground; the excavated earth is thrown up as mounds. Mole-rats of this genus, as well as the bamboo rats, are highly esteemed as food by the natives of the regions they inhabit – Walker [495] relates that they are regularly eaten by the Wanderobos 'who pour water into their holes, thus making them surface and easy to catch', and Harrison and Traub [213], who collected mammals in Selangor, said that one of the bamboo rats, *Rhizomys sumatrensis*, 'is considered such good eating that most specimens caught go to the pot rather than to the museum'. The members of this family, like *Spalax* and those of other families of subterranean rodents, appear never to drink but to obtain all the water they need from their food.

FAMILY MURIDAE

We now come to the Muridae, the enormous family of Old World rats and mice – their numbers are so great that we can glance only briefly at some of its more interesting members. They are distinguished from the other myomorph families, and particularly from the cricetids, mainly by details in the characters of the teeth and skull. In Simpson's opinion they probably arose from the cricetids at a relatively late date, most likely toward the end of the Miocene, in the tropics of the Old World. He adds, 'If ability to survive and multiply and to adapt quickly to a great number of local situations is taken as the criterion, the murids are incomparably the most successful and dominant mammals, not excepting man'. The murids naturally inhabit the whole of Europe and Asia below the arctic circle, Australia, and Africa, but not Madagascar or New Zealand; they are now cosmopolitan because some species such as house mice and rats have been universally, though unintentionally, introduced by man throughout the world.

The division of the 475 living species into subfamilies goes some way towards giving a clear view of this host – but not far, because the first, the Murinae, contains sixty-eight of the ninety-eight genera, whereas of the other five one contains two genera and another one only.

FAMILY MURIDAE

Subfamily Murinae
     ,,   Dendromurinae
     ,,   Otomyinae
     ,,   Phlaeomyinae
     ,,   Rhynchomyinae
     ,,   Hydromyinae

SUBFAMILY MURINAE

In the subfamily Murinae the genus *Mus* contains about twenty species, of which the most familiar is *M. musculus*, our commensal house-mouse which has accompanied man to the farthest corners of the earth – many of its great number of races or subspecies, however, are 'wild' and do not associate with man. Some subspecies of *M. minutoides*, found in many parts of Africa, are among the smallest of mammals, being less than two inches in length from snout to tail root. Two species of the genus *Rattus* which are commensal with man are equally familiar – the brown rat *R. norvegicus*, and the black rat *R. rattus*, many races of which are not black. *Rattus* is the largest genus in the subfamily; it contains at least 120 species and, counting the subspecies, it has more than 560 named forms. The habits and habitats of the animals in this genus, the largest among the mammals, range widely – some species are arboreal, some terrestrial, most are nocturnal but some diurnal, many are burrowers, all are primarily eaters of plants but most also eat animal matter and are practically omnivorous. In some species of both *Mus* and *Rattus* the hair is coarse and inter-mixed with bristles or spines. A similar tendency to spininess is shown by the several species of *Arvicanthus*, the African grass-mice, which resemble the field vole *Microtus* in their habits – some of them are diurnal, and some have become commensal with man. The single species of *Golunda*, the Indian bush rat *G. ellioti*, and the numerous species of *Pelomys*, the swamp rats of Africa, also have bristly coats; but the greatest degree of spininess is shown by the spiny rats and mice.

In the spiny rat *Tokudaia osimensis*, the only species of its genus, found in the Riu Kiu Islands, coarse spines stick out above the fur over most of the head and body; and the spiny- or porcupine-mice, several species of the genus *Acomys*, found in arid and rocky places from India through the middle east to most of Africa, have hard spines projecting above the fur over most of the back – some forms are commensal with man and destructive to stored products. The spines are evidently of practical importance, for Hanney [203] found that in Nyasaland 70 per cent of the diet of the barn owl, *Tyto alba*, consisted of ten species of mice belonging to several genera, but that the spiny mouse *Acomys cahirinus*, which was plentiful, 'was the only species of rodent found to be generally avoided by owls'. The field- or wood-mice, seven species of *Apodemus*, inhabiting the whole of Eurasia north of the tropics, live in burrows but, like most murines, can climb and often spend much time foraging among the branches of low shrubs, using their tails as balancers. The single species of harvest mouse, *Micromys minutus*, found in most of Europe and part of Asia, is a more habitual climber among tall vegetation, and has a semi-prehensile tail. These tiny mice make globular summer nests among the stems of tall plants up to a yard or more above the ground, but retire to burrows in the winter. The climbing mice *Vandeleuria oleracea* and *Hapalomys longicaudatus*, from south east Asia, are more strictly arboreal and grasp twigs between the opposable great toe and the other digits on the hind foot. The forest mice of Africa, about four species of *Thamnomys* and eight of *Grammomys*, and several rare genera of the Far East, also live arboreally although they are less specialised anatomically; the African forest mice make nests in hollow trees or among branches – they have long tails which exceed in length that of the head and body. The acacia rats of Africa, the several species of *Thallomys*, like a number of African small rodents, carry their new born young permanently attached to their pelvic teats for a period ranging from a week to a fortnight, much like some of the small pouchless marsupials.

The general colour in murines is some shade of brownish or greyish on

179

the back, lighter, sometimes white, below, but there are species in many genera in which the dorsal colour is darker along the centre of the back and in some is a definite stripe. In the striped mice of Africa the pattern is further elaborated; one of the two species of *Hybomys*, *H. trivirgatus*, bears three dark longitudinal stripes; the single species of *Rhabdomys*, *R. pumilio*, has five dark stripes separated by lighter ones, and some of the six species of *Lemniscomys* are longitudinally striped with six or more alternating light and dark stripes on each side – in some forms the light stripes are broken up into rows of more or less confluent spots. All these mice are diurnal; they live among grass and dense vegetation, and generally make nests of vegetable fibres above ground.

Only four species of murines have cheek pouches, which differ from those in some other rodents by opening inside the mouth. They are all African; two species of *Beamys* from East Africa are little known, but the pouched rat, *Saccostomus campestris*, is common from the tropics to the Cape – it is a stout-bodied rat with short limbs and tail, and lives in burrows with chambers for storing food for the winter. The giant pouched rat, *Cricetomys gambianus*, is a large animal with long tail and large ears, and cheek pouches of great capacity. When the pouches are full the animal seems unable to empty them by muscular action alone and uses its hands externally to press the contents back into the mouth. This handsome nocturnal murine lives in burrows, and is found throughout most of tropical Africa. Recent (August 1968) press photographs have shown the bodies of this species exposed for sale as food in Biafra, and the claim is made that they show that starvation as a result of the civil war has brought the natives to the necessity of eating rats. This is nonsense based on ignorance; the giant pouched rat has always been prized as food in west Africa, together with many other species of small mammal such as the rufous nosed rat *Oenomys hypoxanthus*, and the fat mice of the genus *Steatomys* in the subfamily Dendromurinae, all of which are regarded as 'good beef' (Plate 13).

Some specimens of *Cricetomys* are nearly as large as a rabbit, and species of two genera from the Far East approach it in size, as they are well over a foot in length from nose to tail root. In one of these, *Hyomys goliath*, of New Guinea the tail is covered with overlapping scales which, it is thought, may be used in climbing as are those on the tail of the flying squirrel *Anomalurus*. The five species of giant naked-tailed rats of the genus *Uromys*, about three times the size of an ordinary rat, are arboreal but the scales of the tail form a raised mosaic and do not overlap; the tail although not prehensile is said to be used in climbing by curling round and gripping twigs with its roughened surface. Their large size, like that of *H. goliath*, is commemorated in the names of some species – *U. rex* and *U. imperator*. The genus inhabits New Guinea and neighbouring Pacific island groups, and at least one species, *U. caudimaculatus*, is found in Queensland.

This species introduces us to the interesting group of murine genera native to Australia – the marsupials are so characteristic of the Australian mammal fauna that the large number of native placental mammals (Eutheria) is sometimes overlooked. The centre of origin of the murines is probably south east Asia, and so it is not surprising that these versatile mammals, together with some members of the subfamily Hydromyinae, have reached the Australian continent and evolved diversely there – at least seven genera are peculiar to Australia, and two have reached Tasmania The young of many small rodents cling so tightly to the nipples of the mother when suckling that if she is disturbed they are carried by her as she escapes; it is peculiar that in Australia, the land of the marsupials, this habit seems to have been noticed in the native murines more frequently than elsewhere. One species of *Conilurus*, the rabbit rats, so called from their large ears, formerly inhabited southern Australia, and others are found in the north and New Guinea. Many species of native rats and mice are known – diurnal, nocturnal, arboreal and burrowing; the largest genus, *Pseudomys*, contains over twenty species including one from Tasmania where, also, the broad-toothed rat *Mastocomys fuscus* of south eastern Australia lives. The three species of *Leporillus*, the stick-nest rats, now greatly reduced in numbers by human settlement, are big rats with large ears like those of the rabbit rats. They build houses somewhat like those of the cricetine pack rats of America; the houses are piles of sticks up to three feet high and four in diameter and, as Wood Jones [258] says,

in the making of them a vast amount of material is employed. Often the nest is constructed round some central object such as a dead bush, and the whole mass of sticks is so securely interplaced that it is a matter of considerable labour to pull one to pieces. Within the nest a small community of rats is established, some half dozen individuals living in even the smaller nests. In places from which the rats have been exterminated for some considerable time the remains of these massive nests are still to be seen; and the strange part of it is how the animals ever managed to collect and interlace the great mass of dead wood that has proved so resistant ... it is often a great puzzle how the rats could have shifted the large branches and pieces of wood entering in to their construction.

Another genus of particular interest is *Notomys*, containing over a dozen species of hopping mice. These mice have a superficial resemblance to small jerboas; the hind feet are greatly elongated, the long tail bears a tuft at the end, the ears are large, and the facial vibrissae long. In the foot the tarsus and the central three digits are elongated, but the first and fifth digits are present though they are small – the two interdigital pads between the long digits are enlarged. These mice inhabit lightly timbered sandy country and dig complicated systems of burrows; Wood Jones [258] considers them to be typical desert animals. He kept several species in captivity and, speaking of *N. cervinus*, he says,

> In captivity it is extremely gentle and makes a very beautiful pet, its activity being astonishing. It is a very remarkable thing to see this animal as dusk comes on, for although it is content to spend its time during daylight curled up in its nest, when evening sets in it becomes transformed. It will stand erect on its hind legs beneath a bench three feet from the ground and next moment, and seemingly without effort, it is sitting on the bench cleaning its whiskers with its delicate hands.

Some species of *Notomys* live in the same habitat as the marsupial *Antechinomys*, and even inhabit the same burrows, thus providing a very remarkable instance of two entirely different animals whose evolutionary paths have converged, occupying the same ecological niche. Some species of *Notomys* have a conspicuous glandular pouch beneath the chin; its function is unknown, and it is only a rudiment in other species (Plate 2).

There are many other genera of murines, too numerous to mention by name, inhabiting Malaysia, New Guinea, the Phillipines, and other western Pacific islands, some of them rare or little known monotypic genera of which few specimens exist in museums. Before leaving the subfamily, however, we may remember that the abundant bandicoot rats, two species of *Bandicota*, widely spread throughout India and south East Asia where they are pests to agriculture, are the original bandicoots, and that the name transferred to the Australian marsupials has become so firmly established that the word 'rat' is now added to the name of the donors.

SUBFAMILY DENDROMURINAE

The subfamily Dendromurinae contains some thirty species in seven genera; in spite of the subfamily name only some of the genera are arboreal, and even those species that do climb trees make their nests in burrows or on, or close to, the surface of the ground. The subfamily is confined to Africa, mainly to its southern half. About ten species of climbing mice of the genus *Dendromus*, which is found over most of Africa south of the Sahara, are arboreal and have long prehensile tails, only three slender digits on the fore limb, and a dark stripe along the back; they are nocturnal and more or less omnivorous. Although they build nests in long grass or low trees and sometimes use bird nests, they can, according to Roberts [406], burrow despite the slenderness of their fore legs and toes. Two other little-known genera of tropical climbing mice, with long tails which may be prehensile, *Deomys* and *Prionomys*, are arboreal, and another, *Delanymys*, discovered in Uganda only in 1962, lives in swamps. The three remaining genera are all terrestrial; *Malacothrix* and *Steatomys* are burrowers, but the two species of *Petromyscus*, the pygmy rock mice of the dry western mountains of South and South-west Africa, with long silky fur, are reported by Shortridge [442] who discovered the genus, as nocturnal and hiding by day 'in crevices and under rocks on hill-slopes where there is plenty of outcrop'. The single species of gerbil-mouse *Malacothrix typicus* of South Africa, with a very short tail and enormous ears, makes long burrows in sandy soils and throws up mounds of earth like moleheaps, plugging the burrow with the earth excavated from the nest chamber, and making a new entrance from below with no tell-tale heap to betray it. The thirteen species of fat mice, *Steatomys*, a genus found throughout Africa south of the Sahara, also throw out earth in mounds from their burrows. They are among the few murids that hibernate; there is a layer of fat under the skin, and it becomes greatly thickened before hibernation, which in South Africa lasts from April to October. Fat mice are nocturnal, and feed mainly on the seeds and bulbous roots of grasses, but are known also to eat insects – as already mentioned, they are highly esteemed as a bonne-bouche by the natives of many parts.

SUBFAMILY OTOMYINAE

The small subfamily Otomyinae contains only two genera. *Otomys*, with eight species of swamp rat, extends from Abyssinia and the Cameroons to the Cape; swamp rats have blunt snouts and short legs and tail, and thus look rather like voles. The species that live in swamps are good swimmers and make shelters of grass and herbage among matted vegetation; those that live in drier places among scrub vegetation make piles of sticks as shelters over shallow holes in the ground. The two species of Karoo rat, *Parotomys*, of South Africa, differ in having enlarged auditory bullae; one lives in colonies of burrows generally in open ground with sparse vegetation and the other makes dome-shaped surface nests of interlaced sticks and grass up to two feet high at the base of matted bushes, with a burrow beneath – Roberts records that 'it has a habit of uttering a melancholy, not very loud, note from its retreat when an intruder approaches close to the nest'. The diet of all the species of the subfamily consists of grasses and herbs; some species carry their food to the nest, round which the uneaten remains are scattered.

SUBFAMILY PHLOEOMYINAE

The subfamily Phloeomyinae contains seven genera of arboreal mice and rats of south-east Asia and the East Indies; some of them are large in size such as the giant tree rats of Celebes (*Lenomys*), New Guinea (*Mallomys*), and Flores (*Papagomys*), the biology of which is little known. The tree rats of the genus *Pogonomys*, from New Guinea and neighbouring islands, are smaller and have long prehensile tails; and the half dozen species of tree mice of the genus *Chiropodomys* found from Assam and southern China to Borneo and the Philippines, have a slight tuft at the end of the tail and like all members of the subfamily, very long facial vibrissae – although these mice are arboreal their tails are not prehensile. The most striking members of the subfamily are, however, the cloud rats of the Philippines, two species of *Phloeomys* and one of *Crateromys*. They are large animals – *Phloeomys* is the largest of the murids – with thick bushy fur that makes them appear even larger. The *Phloeomys* species are rather diffuse in colouration, brownish grey with darker areas round the ears and on the back and shoulders; the long tail is well haired and dark in colour. *Crateromys*, the bushy-tailed cloud rat is very variable in colour, dark brown or nearly black above and lighter below, but some specimens are marked with white or light brown patches over the shoulders and fore limbs and the front part and underside of the body. The soft fur covers the long tail which is thus quite unlike the 'rat tail' typical of the murids as a whole. The bushy tail, dense fur, short snout and small ears give this species a strong resemblance to the dormice, especially *Glis* the fat dormouse. All the species in these genera are arboreal, and feed on vegetable matter of various kinds. *Crateromys* inhabits the mountains of Luzon, but *Phloeomys* occurs in the lowlands as well as in the mountains.

SUBFAMILY RHYNCHOMYINAE

Another peculiar murid of the Philippines, the shrew-rat *Rhynchomys soricoides*, is the sole representative of the subfamily Rhynchomyinae. Unfortunately nothing is known of the biology of this animal; very few have been seen, and there are apparently less than a dozen specimens in the museums of the world, all captured high up on Mount Data, Luzon. The general form much resembles that of a shrew; the fur is dense and velvety, the ears are short and rounded, the eye is small, and the elongated snout is pointed. The skull is remarkably shrew-like, though there is of course a zygomatic arch; the nasal part is elongated and the upper incisors are very short; the lower jaw is slender and its incisors are procumbent. The cheek teeth are reduced to two small flattened molars on each side of both jaws. The action of the incisors must resemble the forceps-like action of the corresponding teeth in shrews, and the small simple molars show that the diet of the animal is probably soft-bodied invertebrates – it is a pity that little is known of this interesting creature.

SUBFAMILY HYDROMYINAE

Two genera, *Chrotomys* and *Celaenomys*, both containing a single species, of the subfamily Hydromyinae, also inhabit the mountains of Luzon; They too are rare animals, known from only a few specimens. Both resemble *Rhynchomys* in their shrew-like form and pointed snouts, but their skulls are not so extremely modified as in that species; there are three molars in *Chrotomys* but only two in *Celaenomys*. Both genera are believed to be herbivorous rather than insectivorous. *Chrotomys* is handsomely marked with a dorsal stripe, from face to tail-root, of buff or orange flanked by black bands which merge into the brownish grey of the flanks. *Hydromys* the genus from which the subfamily takes its name, contains three species of which *H. chrysogaster*, the Australian water rat or beaver rat, is the most widespread, extending from New Guinea and neighbouring islands throughout continental Australia to Tasmania. It is a large species often reaching a foot in length from snout to tail root, and it is highly specialised for aquatic life. The fur is dense and 'seal-like', the eyes and ears are rather small, and both fore and hind feet are partly webbed. The colour ranges from dark to light brown on the back, and from golden orange to yellowish white on the belly; the tail is well clad with dark hair except towards the tip where it is white. The head is rather flattened from above down so that the muzzle is broad, and the vibrissae grow from thickened pads on each side – pads which resemble those of other aquatic mammals such as otters, seals, desmans and others.

Australian water rats are nocturnal and rest by day either in burrows in the banks of streams or rivers, or in hollow logs and similar places, where they make nests of shredded vegetation. The food consists mainly of crayfish and molluscs, but fish, birds and their eggs, and some vegetable matter are also taken. The rats bring their food to eating platforms such as a rock or log; Troughton [478] records that freshwater mussels too large to be opened by the rats' teeth are exposed in the sun on such platforms until the shells gape and the rats are able to extract the meat. He also tells that when, in 1937, the supply of muskrat skins fell off, the Australian fur trade turned to water rat pelts, and that suddenly these inconspicuous and comparatively harmless rodents were claimed to be the cause of serious economic damage. The much smaller false water rat *Xeromys myoides* of Queensland lacks the aquatic specialisations but nevertheless feeds upon freshwater molluscs – it has rarely been collected and few specimens exist in the museums. A similar paucity of specimens and lack of knowledge about their biology applies to nine genera, mostly with but a single species, from New Guinea, some of which are aquatic, others terrestrial. Of these *Crossomys monktoni* from the mountains of eastern New Guinea is the most aquatic. Its dense fur is said to be water repellant, the ears are small, the hind feet large and webbed, and the tail is bordered with two rows of hairs, confluent towards the tip, and thus resembles that of some aquatic insectivores. The brownish grey colour of the upper surface is abruptly separated from the white of the underside – although the animal is obviously aquatic nothing seems to be recorded of its habits.

After skimming over the immense family of murids and noticing some of its peculiarities we now turn to the remaining families of the myomorph rodents.

FAMILY GLIRIDAE

The seven genera of dormice, only one of which contains more than a single species, form the family Gliridae. They are medium sized to small rodents with thick soft fur, bushy tails generally as long as the body, rounded ears, and large dark eyes surrounded in most species with a dark ring; their vibrissae are long and abundant. Nearly all are nocturnal; they are good climbers and live in trees and bushes or among rocks. All species hibernate, and the common European species have been much used in studies on the physiology of hibernation. Members of the family are found throughout most of Europe south of latitude 60° N, Asia Minor, western Asia and north Africa, and one genus occurs in Japan; in Africa the genus *Graphiurus* is widely distributed throughout the continent south of the Sahara.

The fat or edible dormouse, *Glis glis*, the largest species of the family, much resembles a small squirrel though it has not the squirrel's sinuous bounding gait and does not carry the tail over the back when at rest; it is found over most of Europe and Asia Minor and beyond. The diet, like that of most dormice, consists mainly of seeds, nuts and fruit, but also includes insects. The fat dormouse makes its summer nests in holes in trees, but its winter hibernating nests are sometimes made underground. It often enters buildings and may be destructive to stored fruit, and even makes its nests in lofts and sheds. It becomes very fat in the autumn before hibernation –

so fat that it was valued as a luxury article of diet by the Romans, who fattened this dormouse in captivity for their feasts. The species is not native to the British Isles but was introduced into Hertfordshire at the beginning of the century; it has become established in the counties north of London, but has not spread far, and has occupied an area of only about a hundred square miles during some sixty-five years.

The common dormouse, *Muscardinus avellanarius*, is much smaller and more mouse-like, the length from snout to tail-root being three to three and a half inches, that of the tail slightly less; its dense light brown fur gives it a compact stocky appearance. It is found throughout Europe to western Asia where it lives in woods of low trees and bushes, hedgerows and coppices. The limbs are rather short; the first digits on all feet are rudimentary but the others are long and clawed, and the hands and feet are rotated outwards when grasping a twig. Dormice make globular nests of shredded bark and fibre, often among branches a foot or two above ground, but their hibernation nests are usually on or below the ground. Hurrell [248] noted that captive dormice showed bursts of activity at night and 'would chase each other surprisingly rapidly through the maze of thin twigs and branches ... sometimes making shrill squeaking noises'. Occasionally one mouse met another travelling in the opposite direction on the same twig when 'quickly and agilely it would swing under the branch and pass the other upside down. It was characteristic of their behaviour that these activities would come to a sudden abrupt halt, and they would sit hunched up, frozen and immobile, for anything up to three-quarters of an hour before moving again'. This freezing may help to save them from discovery by owls, which hunt by sound.

The garden dormouse, *Eliomys quercinus*, with a similar distribution but extending also into the western parts of north Africa, is rather larger. The upper parts are greyish brown, the underside cream or white, the long tufted tail is black, often with a white tip, and the dark area round the eye extends backwards on the side of the head under the ear giving a pleasing expression to the face. Garden dormice live among shrubs and dense vegetation, but in places lacking such cover they are more terrestrial and live among rocks and make their nests in the crevices.

The tree or forest dormouse, *Dryomys nitedula*, slightly smaller, is rather similarly marked, but has a bushy tail which is not black but brownish, as is the back. The species is found throughout eastern Europe to central Asia and parts of Asia Minor. In the colder parts of its range the forest dormouse hibernates for some months, but in the evergreen oak maquis of Israel Nevo and Amir [345] found that hibernation was intermittent and involved only part of the population. These authors record that in Israel it is typically arboreal, though in some parts of its range it lives among bushes or even bare rocks – 'when frightened *Dryomys* leaps up to 2 m between adjacent branches of trees. Occasionally it jumps diagonally towards the ground, where it proceeds relatively slowly and unskilfully, compared with its swiftness in its arboreal domain'. It builds globular nests of leaves, lined with finer material, among branches up to seven metres from the ground, often in groups of two or three nests in the same tree. *Dryomys* is omnivorous, and feeds on nuts, acorns, fruits, insects, and sometimes birds' eggs; the food is eaten on feeding platforms, usually old bird-nests in Israel, especially those of blackbirds and turtle doves. In Israel *Dryomys* breeds from March to December and the females have two or three litters, generally of three young, during the season.

The Japanese dormouse, *Glirulus japonicus*, is a little smaller than the European and Asiatic *Muscardinus*; it is found only in Japan, far to the east of the habitats of the rest of the family, isolated by nearly fifty degrees of longitude. It resembles the common dormouse in form, but has a flattened tail, and a dark stripe along the centre line of the back. *Myomimus personatus*, known only by a specimen from the neighbourhood of the Russian-Iranian border and some bones from owl pellets, appears to be the sole dormouse that is not arboreal, according to Dukelski [120], who based her conclusion on the form and arrangement of the pads on the vole-like feet.

The genus *Graphiurus*, containing the African dormice, consists of several species, from three to nearly twenty according to different systematists; it is found throughout Africa south of the Sahara. Some species are arboreal and make their nests in hollow trees, but others live in rocky terrain devoid of tree or bush cover. The latter is the habitat of the largest species, *G. ocularis*, the Cape dormouse of South Africa, a prettily marked dormouse, grey above and white below, with a black streak running from the nose to surround the eye and extend behind the ear above a white shoulder-band confluent with the white of the underside. It lives, like some other species, in the crevices of rocks, and eats both vegetable matter and insects. Another and smaller species, plainer in colouring, which Roberts [406] places in the subgenus *Claviglis* containing arboreal species, *G. (C). kelleni*, sometimes makes its nest within that of a South African social spider. The spiders' nests, according to Roberts' source, varied

from a single chamber about the size of a hen's egg to a mass of very tough, felted silk, as large as a man's head, and intersected throughout by passages and chambers. In this the dormice hollow out a chamber of suitable size, which they line with feathery grass heads, the downy seeds of various flowers and even a few stray feathers. The dormice drive out the spiders when occupying their nests . . . the spiders gathered in numbers outside, while the nest contained young dormice.

Another species, *G. (C). woosnami*, nests in hollow trees, but in the dry Kalahari region Roberts found it often lived and reared its young in the nests of weaver birds.

FAMILY PLATACANTHOMYIDAE

The family Platacanthomyidae contains only two genera, each with a single species. *Platacanthomys lasiurus*, the spiny dormouse of southern India, is the size of a small rat – about six inches from nose to tail root, with a tail which, with its terminal tuft, is nearly as long. The colouration is brownish above and white below, but the most prominent feature is the coat of the back which is thickly set with sharp flattened spines. Spiny mice inhabit hilly regions, and make their nests in hollow trees. Although they are abundant in some places little seems to have been added to our knowledge of their habits since the notes made by their discoverer Baker in 1858 were published. They are quoted by Jerdon [256] who includes the observation that 'Large numbers of the Shunda palm (*Caryota*) are found in the hills, and toddy is collected from them; these dormice eat through the covering of the pot as suspended, and enjoy themselves. Two were brought me in the pots half drowned'. *Typhlomys cinereus* of south eastern China is mouse-like, with a tail longer than the head and body together. As the generic name indicates, the eyes are small though the animals are not blind; the ears, however, are prominent and nearly naked. There are no spines in this genus, and the fur is dense and close. The animal inhabits mountain forests; there appears to be no record of its habits or life history.

FAMILY SELEVINIIDAE

Selevinia betpakdalaensis is as unusual in appearance as its specific name – it is the sole species of the family Seleviniidae, and was discovered only in 1938. It is a small mouse-like animal with a long tail; the fur is dense and long so that the feet are almost invisible as the animal stands, looking like a barrel-shaped mechanical toy. It was found in the vast Betpak-dala desert, one of the most barren and sterile parts of the Kirghiz steppes, without human inhabitants, and forming a flat waterless plain with salty soil sparsely covered with wormwood and saltgrasses. Bashanov and Belosludov [26] record that the animals were not common and were seen only before sundown, moving by short jumps on the hind legs. Although these authors say that no definite burrows of the rodents were found, they add that the observers believed that a deserted burrow noticed near one belonged to it.

It was a small vertical cul-de-sac with two horizontal short lateral holes ... wherein was found a small store of *Salsola laricifolia* leaves and wormwood shoots. As the remains of the same plants were present in the mouth cavities and in the stomachs of dissected animals, it is assumed that the stores kept in the burrow belonged to the rodent in question and that these plants constitute its principal food.

Walker [495], however, states that the diet consists solely of invertebrates such as insects and spiders, but this information is derived only from captive specimens. It is also suggested that the animals shelter under stones or other objects during the summer, and burrow only for hibernation during the winter.

FAMILY ZAPODIDAE

The small family Zapodidae consists of two subfamilies, the Sicistinae containing the birch mice, and the Zapodinae containing the jumping mice. All have long tails, and the jumping mice have elongated hind feet. About six species of birch mice, genus *Sicista*, which is found from northern Europe through central Asia to China, are the only members of the sub-family. The European species, *S. betulina*, is brown with a narrow black stripe along the back. Birch mice live in birch scrub and open woodland, and in meadows and steppes; they are agile climbers and can support themselves on very thin twigs, helped by their semi-prehensile tails. They are nocturnal, and live by day in shallow burrows in which they make their nests. They feed on seeds, leaves and insects, and hibernate in winter. Hanström [204] gives a photograph of a birch mouse swimming well and buoyantly in a pool in Sweden where, however, it is common only in the central districts.

The subfamily Zapodinae contains three genera, one, *Eozapus* from the high mountains of western China, the other two from North America where between them their ranges cover the continent from Canada to New Mexico and California; the single species of *Napaeozapus*, *N. insignis*, occupies the most northern parts, and the three or four species of *Zapus* cover the remainder. These jumping mice have elongated hind feet and very long tails, and when alarmed they escape by making long leaps of up to six feet or more, though in slow progression they go on all fours. The meadow jumping mice (*Zapus*) are found in widely varied habitats from alpine meadows to deciduous woodlands but, whatever the terrain, they prefer damp, swampy places or the banks of watercourses. Krutsch [280] reports that *Zapus* is an agile climber and that it also swims well. All the species hibernate; Krutsch quotes an observation that they will awaken at times 'famished for water and will drink and drink before going back to sleep'. *Zapus* lives in short shallow burrows during the summer, but digs deeper ones in drier places for hibernation. These mice are nocturnal, and their food consists of seeds and berries, with a large proportion of insects; the mouse when eating sits on its hind quarters and holds the food in its hands. The long tail of the jumping mice is important as a balancing organ to control the trajectory of the body during leaping. Svihla and Svihla [464] record that a *Z. trinotatus* which had lost its tail in a trap 'made desperate attempts to escape. These jumps were about a foot high and for a distance of three feet. It could not land on its feet properly, but instead turned somersaults and landed on its back every time'; and Miller [322] noted that one which lost its tail in a mowing machine 'turned end over end so that it was as likely as not to strike the ground facing the direction from which it had come'.

The habits of *Napaeozapus insignis* appear to be similar to those of *Zapus*. Snyder [452] watched a female collecting nest material by night in Ontario, and carrying it to her nest burrow. The nest chamber was less than six inches from the surface in a short burrow, and 'the material was carried in the mouth and seemed to be supported by the front feet. While the animal was laden its jumps were short and slightly labored, being not more than two feet in length, but after it emerged from the hole one leap usually carried it to the end of the tent, which was seven feet long'. Several times the work was suspended while the mouse executed a curious 'dance': 'There did not seem to be the slightest motive for this performance. Its movements gave one the impression of a rubber ball fastened to an elastic band being bounced carelessly up and down, here and there'. Five naked and blind young were born in the nest in the burrow, the mouth of which was blocked with earth during the day, so that no sign of its presence was visible. When the adult was alarmed it escaped by rapid jumps covering from ten to twelve feet.

FAMILY DIPODIDAE

The family Dipodidae contains the jerboas, small to medium sized animals, and the most specialised of the jumping rodents, with very long hind and short fore limbs. The family is widely distributed through northern Africa, Arabia, Asia Minor, south Russia and Central Asia to China. Most of the twenty-seven species fall into the seven genera of the subfamily Dipodinae. The jerboas of this subfamily are medium sized, soft furred animals,

sandy-brown or grey above and white below. They have large eyes and
ears, and a long tufted tail: the hind limbs are greatly elongated, and the
fore limbs short and carried under the chin where they are almost invisible;
the vibrissae are long – nearly as long as the body in some species. They
inhabit arid, but not necessarily hot, regions with sparse vegetation and
live in burrows that they dig themselves. The jumping or hopping gait is
closely correlated with the habitat; it is only possible to a small mammal
living on fairly open ground clear of close vegetation, it makes a large area
of territory available for foraging in a comparatively unproductive environment, and its speed and opportunity for erratic jinking are said to help in
avoiding capture by predators. Jerboas are nocturnal and crepuscular, and
most species hibernate – a few are said to aestivate – but they do not lay up
stores to be eaten in the intervals of wakefulness. They are highly adapted
to living in arid deserts, and are able to live for long periods without drinking. They obtain their necessary water from the moisture content of their
food, which includes insects and underground bulbs and roots as well as
the seeds, stems and leaves of plants. The species inhabiting north Africa
and the Near East are the best known – naturalist are not plentiful in the
deserts of Central Asia and Mongolia where a number of species flourish.

In most species there are only three toes, the first and fifth being entirely
lacking, and the three elongated foot- or metatarsal-bones are fused into a
single cannon bone. The cervical vertebrae are fused, and the auditory
bullae are inflated, a character found also in other desert rodents.

Dipus sagitta, the feather footed jerboa, the only species of its genus, is
found from the Caucasus to northern China. The fringes on the toes
together make a hairy 'boot' for hopping on loose sand. Feniuk and
Kazantzeva [144] who studied the species among the sand hills of west
Kazakhstan, where the air temperature near the surface of the ground can
reach over 50° C by day, state that the 'conditions of life for the majority
of animals are very hard, but jerboas can live under the conditions of a
relatively favourable micro-climate, owing to their nocturnal and fossorial
habits of life'. The animals plug their burrows with sand except during
periods of low temperatures, and spend the day in a moister and cooler
micro-climate; the early morning winds 'rapidly erase all trace of the
animals and their holes, and re-establish the ripple marks characteristic of
aeolian sands'. The burrows usually have several emergency escape holes
that stop short of the surface; the nest, made of shredded vegetation, is
placed in a larger chamber. The Russian authors found that in feeding
the jerboa passes slowly from one place to another, with back bent and legs
strongly flexed, while digging for insects and seeds. In order to procure the ears

of *Elymus giganteus*, they rise high on their hind legs and gnaw the stalks. After having satisfied its first hunger, the jerboa runs around its territory, visits its temporary burrows, meets another jerboa, and bathes in the sand. When meeting the jerboas sniff and then chase each other, making bounds of about 2 to 3 metres; at an unhurried pace, however, their jumps average 30, 40 or perhaps 60 cm. in length.

These jerboas hibernate for about four and a half months from November to March, and breed on emerging; the females produce two litters generally of three young, during the summer.

The comb-toed jerboa, *Paradipus ctenodactylus*, so called from the fringes of stiff hairs which increase the surface area of the foot applied to the ground, is similar to *Dipus* but has longer ears; it is the only species of its genus. Vinogradov and Argyropulo [492] who studied it in the deserts of Russian Turkestan, its only habitat, found that although in many ways its habits resemble those of *Dipus*, its burrows are not usually plugged with sand, and do not have emergency escapes.

The genus *Jaculus* contains four species, and is found from northern Africa to Turkestan; it contains two species well known to western zoologists, the greater and lesser Egyptian jerboas, *J. orientalis* and *J. jaculus*, which have been used in experimental researches on the physiology of desert animals. They do well in captivity and, although they are naturaly nocturnal, Laurent [288] reported that the numerous specimens of *J. orientalis* which he studied readily adopted diurnal habits – Dickson [110], on the other hand, found her pet *J. jaculus* remained strictly nocturnal. She adds that to the

children of Arabia these small mammals take the place of the mechanical toy of the west. The boys learn at a very early age, when tending camels or sheep, how to dig these little rodents out of their holes and catch them. Usually before they hand one over to some smaller brother in the tent the hind legs are broken, so that it cannot escape. Later it will be killed, by cutting its throat, cooked in the hot ashes of the fire, and eaten by the children. These jerboas are purely nocturnal in habit and are seldom seen except in the headlights of a car after dark in the desert.

Allactaga, with ten species, and *Alactagulus* with one, spread throughout Asia from west to east. They are the five-toed jerboas, although one species, *Allactaga tetradactyla*, which lives in Egypt, has only four. In these jerboas the three central metatarsal bones form a cannon bone, and the first and fifth are delicate splints alongside it – their toes do not touch the ground with the three central ones in hopping. The habits of these species are like

N

those of the other genera, and the animals are easily accustomed to captivity; Hutton, quoted by Sterndale [457] notes that *A. elater*, the small five-toed jerboa of Afghanistan, 'is easily tamed and lives happily enough in confinement if furnished with plenty of room to leap about. It sleeps all day, and so soundly that it may be taken from its cage and examined without awaking it; or at most it will half open one eye in a drowsy manner for an instant, and immediately close it again in sleep'. The single species of *Stylodipus*, *S. telum* of Russia and west Asia, and the two species of *Pygeretmus*, *P. platyurus* and *P. shitkovi*, from Turkestan, are distinguished by their tails – *Stylodipus* by the long hairs giving it a feather-tailed appearance, and *Pygeretmus*, which has five toes, by its fat tail, which resembles that of some other fat-tailed species of myomorph rodents.

Three species of *Salpingotus* and a single one of *Cardiocranius* both Asiatic genera, form the subfamily Cardiocraniinae; the first has only three toes but the metatarsals are not fused into a cannon bone, and the second has five toes with no fusion of metatarsals. They are small, especially the tiny *Salpingotus* species, and the tympanic bullae of the skull are large, especially in *Cardiocranius* in which the inflation is more extreme than in any other jerboa. Little is known of the biology of these species, and the same must be said of the long-eared jerboa *Euchoreutes naso* from the Central Asian deserts, which is the sole member of its subfamily, Euchoreutinae. It is a five-toed species, and has very long ears; it is probably plentiful enough in its native deserts, but there are few specimens in the museums of the world.

Hooper [235] who made detailed studies on the middle ear in the microtine rodents discusses and reviews the work of other authors on the function of the inflated auditory bullae found in so many desert rodents, 'the large amount of space afforded by the cavities reduces the damping of the tympanic membrane by increasing the size (thus decreasing the resistance) of the airy cushion behind the membrane. . . . Parts of the middle-ear system may act as acoustic filters, selecting on the basis of frequency or direction of the sound'. He points out that Petter [375] suggests that the enlarged bulla is one of the most important adaptations that allow the desert gerbils to communicate in spite of the low population density resulting from the aridity of their environment. 'Species or races of *Meriones* in which the populations are low and the individuals well spaced have larger middle-ear cavities than *Meriones* with denser populations'. Species with large bullae and sparse populations are thought to be able to perceive feebler sounds and to communicate with each other over greater distances. Hooper concludes, however, that much more experimental, ecological and behavioural research is needed before an understanding of these peculiar anatomical structures is possible.

The jerboas, which are among the most specialised, and the most charming, of rodents – delicate little creatures that can withstand the rigours of an environment so severe that it would be fatal to many a more robust animal – bring us to the end of this survey of the myomorph rodents.

Chapter 7

Porcupines, Cavies, and Agoutis

THE HYSTRICOMORPHA, the third and last suborder of the rodents, contains fourteen families, nearly all of them found only in South America. Two further families of uncertain affinities, which probably belong to this suborder, are included at the end of the chapter.

FAMILY HYSTRICIDAE

The family Hystricidae, the Old World porcupines, with fifteen species in four genera, is the largest and most familiar of the three families found outside the New World – the others contain the cane rats and dassie rats of Africa. The members of all the genera in this family have some at least of the hairs of the back modified as spines, and in many the spines are formidable quills. They are ground-living, burrowing animals, and are nocturnal. The family is distributed over the whole of Africa and throughout southern Asia, and one species extends into southern Europe; the genus *Hystrix* containing the common porcupine and others enjoys this wide distribution. In all members of the family the quills are grown only on the hinder part of the body and tail; the fore part bears shorter spines, bristles and coarse hair.

Most of the species of *Hystrix* are heavily built animals weighing up to fifty or sixty pounds; they are short-tailed, plantigrade and rather slow-moving. They dig deep and very long burrows, sometimes close enough together to form a warren. They are primarily vegetarians, feeding on bulbs, tubers and roots, but they also habitually gnaw bones, which they collect and bring to their burrows and often leave scattered near the entrance. Their behaviour is correlated with the safety provided by their array of quills. Although they are nocturnal they scuffle about noisily, snuffling and grunting, and if disturbed or alarmed erect the quills and the crest of long bristles on the head and neck and rattle the tail quills. The latter differ from the others in being tubular, open-ended and not pointed; they are attached by thin stalks so that when the tail is vibrated they make a sound something like that of a rattlesnake. The quills themselves act as a warning not only by their sharp points but in their conspicuous colour pattern of black and white bands. The animals thus give both visible and audible warning that they should be left alone. If they are actually attacked they run backwards very quickly and jab their quills into the enemy, and those that strike home come out of the porcupine's skin and are left in the victim. This habit is the origin of the erroneous story that porcupines can shoot their quills at an enemy. Porcupines need these defences, for carnivorous mammals from man downwards much relish their flesh – only the larger predators, which can kill them by a heavy blow on the unprotected head, are successful in dealing with them if they can avoid the backwards charge – but many an unfortunate leopard and tiger are at this moment limping about with porcupine quills in their paws, or with faces like pincushions. Like other animals, Hillaire Belloc's tuptophile small boy, weeping with his palms stuck full of quills, learnt that:

> To strike the meanest and the least
> Of creatures is a sin;
> How much more bad to beat a beast
> With prickles in its skin!

The Indonesian porcupines of the genus *Thecurus* are smaller, and have shorter quills but more numerous spines. The quills have comparatively thin stems and expanded ends so that they rattle as the animal moves about. The long-tailed porcupines of the genera *Atherurus* and *Trichys* are more slender-bodied, and look like large spine-covered rats, some of the spines on the back being large enough to be called quills. Their most prominent character, however, is the tuft of bristles at the end of the rather long tail, from which they take their name of brush-tailed porcupines; the base of the tail is spined but the part between base and tuft is nearly naked. In some species the bristles of the tail tuft are flattened and expanded at intervals along their length. The spines are arranged in longitudinal bands easily seen in the new born young, in which they are only rudimentary points projecting from the skin; when the spines grow longer they overlap each other and the arrangement is obscured, as shown by Rahm [390]. Brush-tailed porcupines are forest animals and most of them are good tree climbers, and much more agile on the ground than the larger species of *Hystrix*. They are nocturnal and live by day in burrows; the diet consists mostly of vegetable matter, but insects also are eaten. Several species of *Atherurus* are found in south-east Asia, and one in equatorial Africa from

197

coast to coast. The two species of *Trichys* from Borneo, Sumatra and Malaya are very similar in appearance, but have rather longer tails.

FAMILY ERETHIZONTIDAE

The tree porcupines forming the family Erethizontidae are exclusively American; one genus, *Erethizon*, is found over most of North America from the extreme north to Mexico, but not in the south-eastern United States; three more between them cover Central and South America east of the Andes to northern Argentina. They are all arboreal, and are confined to the regions covered with forest or woods. Although some species are almost entirely covered with spines, none of them have long quills like the members of the family Hystricidae.

The North American or Canadian porcupine, *Erethizon dorsatum*, is the only member of its genus, and has been studied in considerable detail by American zoologists, some of whom have even established it as a laboratory animal. It is a fairly large species, ranging up to a yard in length overall, and in weight from fifteen to twenty pounds, though examples weighing as much as forty pounds and more at the age of five years have been recorded. The legs are short, the tail is thick and the snout rounded. The body is clothed in rather long coarse guard hairs that give a shaggy appearance, and hide the shorter quills scattered among them, except on the tail and lower back. The colour is dark brown or blackish with a grey overcast, and the sparsely haired face is black. Tree porcupines are nocturnal and crepuscular, and lie up during the day among rocks or tree roots, in hollow logs, and similar places; by night they travel by well worn paths to favourite feeding trees. Batchelder [27] found that porcupines, when climbing a tree trunk, reach forward with the fore feet and then move both hind feet up together, supporting the body by the tail and the fore claws. Among branches the animals move slowly, but are astonishingly sure-footed, and go out on to branchlets that can barely carry their weight, standing up to pull down twigs above them. They feed on the leaves, buds and bark of many kinds of tree, but in the winter they prefer conifers, especially the hemlock. They are rather wasteful feeders, and let many small branches fall to the ground; in winter the dropped clippings are eaten by deer and rabbits, which are attracted to this easily obtained fodder. Shapiro [437], however, who found that deer are attracted to porcupine trees, considered that the amount of food they get from this source is not of great importance.

Porcupines have a strong smell derived from their urine, which plays an important part in their lives. They urinate at their den entrances, along their trails, and particularly at the foot of the feeding trees. This habit is doubtless a marking of territory, though the animals do not appear to defend territory – perhaps the marking suffices, for in general the animals are solitary, though Struthers [461] records colonies of up to fifty porcupines living in contiguous dens. He also recorded up to nineteen in a single large den, but found that the animals feed in their own trees after following the trail to the feeding grounds. The urine is also used in the courtship display or stimulation; Shadle and his partners [436] state that when a male meets a female in advanced oestrus both animals stand up on their hind legs and tail, facing each other a few inches apart, whereupon 'the erected penis of the male begins to discharge urine in frequent, short spurts which in less than a minute may thoroughly wet the female from nose to tail. . . . If near the point of mating, she may offer only perfunctory objections to the urinary shower. The quantity of urine used and the force with which it is discharged are surprising'.

Porcupines chatter the incisor teeth, pushing the lower jaw forward and striking the inside of the lower against the outside of the upper teeth, making a sound used in mutual recognition, as a reaction to something strange, or as a challenge between rival males. They also make many whining and grunting sounds, often with no apparent communication content; some resemble the human voice, and Batchelder [27] quotes Dr J.B.May who told him that when walking through the trailess woods of New Hampshire, 'I was surprised to hear, a little to one side, an old man mumbling to himself. As I was more than a mile from the nearest road or farm, in pretty rough ground, I turned aside to investigate. In a small alder swamp I found a good-sized porcupine feeding near the top of an alder and talking to himself very much as you describe the sound, *déahp-déahp*'. Porcupines are not liked by foresters because they damage and even kill trees by eating the bark, particularly the sweet cambium. Curtis and Kozicky [92] say that although the damage to individual trees is often spectacular the overall damage in forests is not great; the damage to conifers produces stag-headed trees, but few of them die. Shapiro [437] points out that the animals return year after year to the same tree to feed on the callus produced by previous damage; the feeding range of individual porcupines may extend to about thirteen acres, and each animal eats about a pound of foliage and bark a day.

The spines or quills of the tree porcupine are particularly cruel, for their very sharp tips are covered with microscopic barbs, and their bases are loosely attached to the skin so that they readily come out and remain in the victim of quilling. The barbs are overlapping scales and cause the quill to work forwards with every movement of the victim's muscles, so that it travels through the flesh remarkably quickly; when thus drawn into the tissues it can travel an inch a day. If cornered a porcupine backs towards its adversary and swipes with its tail, leaving a plentiful supply of quills as a souvenir. Shadle [433, 434], however, who handled many porcupines says that the popularly supposed horrible pain and discomfort from being quilled are greatly exaggerated; though not pleasant, there is merely a pricking and itching feeling with sometimes a slight inflammation which subsides in twenty-four to forty-eight hours – he found quills much less troublesome than splinters if pulled out at once. Quick [383] also found that there was scarcely any sensation other than a tingling of the flesh after some fifty quills were struck into his arm – 'However, the extraction of the quills was somewhat more painful and caused minor bleeding'.

Shadle also found that the porcupine itself experiences the same discomforts that other animals feel when quilled; they scratch and rub the affected area, and pull out the quills with the forepaw when not too deeply embedded, or use the incisor teeth like pincers. One animal that had been quilled

> sat on its haunches, lifted the quilled arm, and taking one of the foreign quills between his upper and lower incisors, he pulled it out. The quills were so well buried in his flesh that it took a hard pull to dislodge them, but apparently none broke off in the process. When a spine was pulled out he would take it from his mouth with his left fore paw and drop it to the floor.

A writer in *Life Magazine* [11] told of a farmer in Nebraska who had a tame porcupine; 'He has only slapped me lightly with his quills once or twice, and he didn't mean it even then for he immediately helped pull the quills out of my hand with his sharp teeth'.

In spite of the protection given by the quills porcupines are often eaten by predators, by pumas, wolves, coyotes, bobcats and lynx, but particularly by fishers and foxes. Quick [383] who examined the bodies of large numbers of fur bearing carnivores, found quills in many different parts of the body embedded in all tissues from the stomach wall, through which they protruded into the peritoneal cavity, to liver, lungs and kidneys. Daniel [99] similarly found quills in most of the viscera of fishers, and numerous perforations of the stomach and intestines. He remarks that the porcupine is the favourite food of the fisher and is also eaten by many other carnivores. It is astonishing that predators survive in apparently good condition with quills working their way through the vital organs. The quills usually drift about the body and eventually work their way out unless they come to rest against bone. Quick found that the body-fluids soften and erode the quills if they are retained for some time, and concludes that ingested or injected quills are not necessarily lethal to mammals which prey upon porcupines. These findings do not wholly confirm Struthers' conclusion [461] that

speed of flight, cunning, keen senses of smell, hearing or sight, have been sacrificed for two great specializations, a coat of needles and a set of teeth which are capable of rendering fit for consumption the bark of trees that no other animal wants.... They have nothing he wants and with his coat of spines very few of them can do him harm.

On the other hand, as Struthers points out, the low fertility rate – a female seldom produces more than a single offspring at a birth once a year – and the long adolescence – sexual maturity is not attained until the third autumn – do suggest that the expectation of life among tree procupines is high, and that food supply and predation are not pressing problems.

Figure 10 South American Tree-porcupine (*Coendou*).

South America is the home of three genera of tree-porcupines, of which the genus *Coendou*, containing five species, is the most widespread, covering the whole of the continent from Mexico to the north of Argentina. The largest species of coendu is eighteen to twenty inches from snout to tail-root, the others are a few inches smaller; in all the tail is about as long as the body and is hairy and prehensile. The tip of the tail of the coendus curls upwards, unlike that of other prehensile-tailed mammals which curves downwards. In the widespread *C. prehensilis* the body except the underside is covered with short stout quills, speckled black and white, the base only of the long tail carrying spines. In other species the quills are hidden under the long hairy fur which is said to conceal the animals among branches by its resemblance to tufts of moss. The limbs are short and the four digits on fore and hind feet bear stout claws; the first digit of the hind foot is represented by a fleshy pad supported by a bone of the tarsus so that the foot is broadened. The head is rounded, and the naked snout without spines gives the animals a ridiculous resemblance to a blobby-nosed clown; Durrell [125] speaks of the 'great swollen nose whiffling to and fro and on each side a small, cunning, and yet somehow sad little eye brimming with unshed tears'. Coendus are arboreal, and climb slowly and deliberately among the branches of trees, using feet and prehensile tail, in search of leaves, fruits, buds and juicy twigs. They sleep in hollow limbs or among dense tangles in trees, and are generally said to be nocturnal, but little is recorded of their habits. Cabrera and Yepes [65], however, said that a captive specimen was active only when seeking food, and that it regularly fed at nine in the morning and four in the afternoon every day; it was only once noticed to be active at night. This animal was fed on a large variety of vegetable foods; when eating it picked up the food with the teeth and then sat up and held it to its mouth with both hands. In cultivated regions coendus are said sometimes to damage crops. The number of young is variously stated as up to four, or only one – the latter perhaps the more likely to be correct. The biology of the ouriço preto, *Chaetomys subspinosus*, of north-eastern Brazil is not well known, although the animal is common enough. It is much like the coendu but although the spines on the head and shoulders are sharp, over the rest of the back, sides, and limbs they resemble coarse bristles. The bristles do not extend on to the tail, which in spite of its appearance is said to be not prehensile. Another genus, *Echinoprocta*, with a single species, lives in Colombia at considerable elevations in the mountain forests. The back and sides are entirely covered with spines, brown to black and not speckled. The tail is short, hairy and non-prehensile. Little is known of the life of the species beyond that it is arboreal and plentiful in its habitat.

FAMILY CAVIIDAE

The family Caviidae contains the guinea-pigs or cavies, and the mara or Patagonian hare. It is distributed throughout South America with the exception of the Guianas, north-east Brazil and Tierra del Fuego. All members of the family have four toes on the fore and three on the hind feet.

The genus *Cavia* contains six species, though many more have been named, the best known being *C. porcellus* which is the worldwide pet and laboratory guinea pig, and was domesticated to be used for food by the Indians of Peru before the Spanish conquest. Wild cavies – six species of *Cavia* and three each of *Galea* and *Microcavia* – are stout-bodied, short legged animals the size of large rats, with short ears and no external tail; their hair is soft and rather long, especially on the hind quarters. The colour is a brownish pepper-and-salt. Cavies are nocturnal and live in burrows – the writer has found one species living in the gardens and under house foundations in the suburbs of Rio de Janeiro, where cavies are known as *preás*. The food consists of all kinds of vegetable matter. The number of young in the wild is usually two – the female has only two teats – but may be more in domesticated varieties. The young are peculiarly precocious and have already cut their teeth when they are born and, although they are suckled, they can eat solid food almost from the start – they are able to run about nearly at once. The genus *Cavia* contains several subgenera, formerly considered full genera, but the moco, *Kerodon rupestris*, of north-eastern Brazil, is usually considered as generically distinct. It is larger than the other cavies, and lives in arid rocky areas, coming out from its dens among the rocks to feed on vegetation at night, sometimes climbing trees to obtain it.

Figure 11 Patagonian hare (*Dolichotis*).

The two species of mara or Patagonian hare, of the genus *Dolichotis* are much bigger than the cavies – large specimens weigh up to about thirty pounds. They have long legs, fairly long ears, blunt rounded muzzles and short tails. The fur is fairly long, dense and crisp, grey above and whitish below, with a white patch on each thigh connected to that of the opposite side by a narrow band above the tail. The vibrissae and the eyelashes are long and black; the short thick tail is sparingly covered with hair at the base but the tip is naked. These animals inhabit suitable places throughout central and southern Argentina, and formerly extended much further towards the southern border of Patagonia. They inhabit arid and almost desert regions with scattered vegetation of coarse grasses and shrubs on which they feed. They shelter in burrows, sometimes dug by themselves and sometimes taken from other animals. Maras in many ways look very unlike typical rodents; when standing or walking they are digitigrade and then superficially resemble small ungulates; when squatting they sit on their haunches with straight fore legs like a dog; and when reclining at ease they loll half on the side with extended fore and hind legs, in an attitude resembling that of a kangaroo basking in the sun. When they run, however they hop with a gait much resembling that of a rabbit – this gait, and the long ears and legs have earned them the misnomer of Patagonian hares.

During the voyage of the *Beagle* Darwin made a journey over the country between the Rio Negro and Rio Colorado which, he said, deserves scarcely a better name than that of a desert, where he saw many maras, which he called 'agoutis'. This country

is inhabited by few birds or animals of any kind . . . but the agouti (*Cavia patagonica*) is the commonest quadruped. This animal here represents our hares. It differs, however, from that genus in many essential respects; for instance, it has only three toes behind. It is also nearly twice the size, weighing from twenty to twenty-five pounds. The Agouti is a true friend of the desert; it is a common feature of the landscape to see two or three hopping quickly one after the other in a straight line across these wild plains. . . . Where the Biscacha lives and makes its burrows, the Agouti uses them; but where, as at Bahia Blanca, the Biscacha is not found, the Agouti burrows for itself.

The elegant long-legged build of the body, the soft colour of the fur, the rather large dark eye shadowed by long lashes, and the readiness with which maras become semi-tame, make them very ornamental animals in captivity. In one of the German zoos they have free run of the grass plots and make a beautiful exhibit unconfined by fences and undisturbed by visitors who keep to the paths and obey the 'verboten' notices.

FAMILY HYDROCHOERIDAE

The family Hydrochoeridae contains a single genus, *Hydrochoerus*, with two species, the capybaras or carpinchos of South America. *H. hydrochaeris*, found throughout South America east of the Andes south to the river Plate, is the largest living rodent, but *H. isthmius* of Central America is considerably smaller. Capybaras are heavily built stout bodied animals with short legs, blunt rounded snouts and small ears and eyes – their generic name, meaning water-hog, well fits them. The larger species weighs up to a hundred pounds and more. The long hair is coarse and rather sparse, dusky brownish in colour, rather lighter below; the digits, four in front and three behind, are partly webbed at the bases and bear short broad nails. The tail is represented merely by a horny protuberance. Like the mara and some other related genera, the capybara is digitigrade, and when resting often sits up on its haunches. The adult males have a large oil-secreting gland on the upper surface of the muzzle, as noted by Waterhouse [498] in 1848, which just over a hundred years later was shown by Rewell [401] to be a mass of hypertrophied sebaceous glands (Plate 14).

Capybaras live near water – lakes, rivers and streams – and are found not only in marshes and swamps and the dense vegetation of wet places, but also on the banks of streams and lakes in more open country. The diet consists both of water plants, and grasses and herbage grazed from the banks. When alarmed capybaras take refuge in the water, in which they swim with only the top of the head exposed and the eyes, ears and nostrils just above the surface. Although many authors note the resemblance of capybaras when feeding on land to pigs, the writer has been struck by the resemblance, when seen at a moderate distance, to small hippopotamuses when they stand in the shallows at the edge of a lake. Capybaras live in small herds of from three or four up to thirty animals; they are diurnal where they are not disturbed by man, and they do not burrow. Although they are so large as to be safe from attack by smaller predators they are much preyed upon by the jaguar on land and, it is said, by caimans in the water; they are also harassed by man in both media, although the flesh is little esteemed as food and the hide is of small commercial value.

Darwin, one of the first scientific naturalists to record field observations on the animals of South America, came across capybaras in the south of Uruguay. He notes in his journal,

The largest gnawing animal in the world, the *Hydrochaerus capybara* (the water hog), is here also common. One which I shot at Monte Video weighed ninety-eight pounds: its length, from the end of the snout to the stump-like tail, was three feet two inches; and its girth three feet eight. These great rodents occasionally frequent the islands of the mouth of the Plata, where the water is quite salt, but are far more abundant on the borders of fresh-water lakes and rivers. Near Maldonado three or four generally live together. In the day time they either lie among the aquatic plants, or openly feed on the turf plain. When viewed at a distance, from their manner of walking and colour, they resemble pigs: but when seated on their haunches, and attentively watching any object with one eye, they reassume the appearance of their congeners, cavies and rabbits. Both the front and side view of the head has quite a ludicrous aspect, from the great depth of their jaw.

He found the animals very tame because jaguars had been exterminated from the region, and no one thought them worth the trouble of hunting. He watched four of them

from almost within arm's length (and they me) for several minutes, when they rushed into the water at full gallop with the greatest impetuosity, and emitted at the same time their bark. After diving a short distance they came up again to the surface, but only just showed the upper part of their heads. When the female is swimming in the water, and has young ones, they are said to sit on her back.

It is not only the depth of the jaw that gives the capybara its ludicrous appearance; the position of the eye, high up and far back, produces a supercilious cast of countenance. Durrell [125] tells of a captive capybara which, housed in a cage, awoke him at night. The animal was 'sitting by the wire looking very noble', and plucking the wires with its huge incisor teeth so that 'the whole cage vibrated like a harp'; when the noise died away it then 'raised his large bottom and thumped his hind feet on the tin tray, making a noise like stage thunder'. Capybaras, if caught young, readily become tame and almost domesticated. The smaller *H. isthmius* is equally amenable; Trapido [477] records that it takes well to captivity, becomes docile, and feeds well. At the Gorgas Memorial Laboratory in Panama the capybaras are 'permitted to graze at will during the day on the grass lawn of the laboratory grounds. At night they readily enter a large outdoor cage where they are locked up'. The largest male recorded weighed fifty-eight pounds and the largest female sixty-two pounds, about half the weight of the larger species. Trapido found the gestation period to be between 104 and 111 days, a little less than that for *H. hydrochaeris*, and that the three or four young weigh between two and three pounds each at birth.

FAMILY DINOMYIDAE

The family Dinomyidae contains a single genus and species, *Dinomys branicki*, the pacarana, found on the slopes of the Andes in the region of the headwaters of the Orinoco and Amazon rivers and their tributaries. The pacarana is a scarce animal, and little is known of its life in the wild. It is a large rodent, weighing up to thirty pounds; the body is thickset, the legs rather short, and the tail, unlike that in the closely related families is comparatively long – about six inches in length. The eyes and ears are small and although the digits have strong claws the animals are said not to use them for digging. On the other hand Tate [468] reports that an Ecuadoran collector said that pacaranas live in burrows 'which are usually placed under rocks. Usually there is a male and female and two young. The animals make no trails or runways but wander about as they will. When prodded with a stick while they are in their burrows they set up a tremendous growling. They are so stupid that they will not run away in the daytime'. The coat colour is dark greyish brown with longitudinal rows of white spots, the spots of the two rows nearest the centre of the back tending to become confluent into stripes. The head is rather large and, according to Crandall [87] gives the pacarana 'a somewhat formidable appearance' which, however, 'is not borne out by its nature, for in captivity it is calm, almost phlegmatic, and non-aggressive'. Pacaranas are slow-moving, and inhabit forested and rocky places in mountain valleys, where they feed on fruits, leaves and the stems of plants. When eating they sit on the haunches and hold their food in the forepaws like the dasyproctids. They are the prey of ocelots and coatis, and their flesh is relished by the natives – according to Hodge [233] they are nearing extinction. The pacarana is said to have been discovered accidentally by a person who fired blind in the direction of a disturbance outside his house at night, and in the morning found that he had killed the first specimen of the animal to be known to science.

FAMILY HEPTAXODONTIDAE

An allied family the Heptaxodontidae, containing several genera and species, is known only by fragmentary bones from Puerto Rico and Hispaniola in the West Indies. Some of these animals were as large as the capybara and were eaten by the natives, but all are now extinct; they were killed out either before or shortly after the Spanish conquest.

FAMILY DASYPROCTIDAE

The family Dasyproctidae contains four genera, two of agoutis and two of pacas. The agoutis are rabbit-sized animals weighing up to about eight pounds, but pacas are larger and reach about thirty pounds. They are digitigrade and cursorial, and when running bear some resemblance to small ungulates; when resting they sit or crouch on their haunches. The seven species of agouti in the genus *Dasyprocta* are brownish or reddish yellow, often almost orange, black or white on the hind quarters; the ears are small and the tail very short and hairless. The hair is rather coarse and is particularly long on the rump – the name *Dasyprocta* literally means 'shaggy bottom' – and when alarmed or excited the animal erects it just as the related porcupines erect the quills on the same part of the body. Agoutis are found from Mexico to southern Brazil in many different habitats, from forests and thick bush to savannahs and cultivated regions. Their food consists of all kinds of vegetable matter. Enders [133] found that in Panama *D. punctata* wandered about the forest floor when feeding, smelling here and there, but he was not able to identify the food exactly. He says, 'when mildly surprised, they run at a very fast trot that is very unlike a rodent's gait, and break into a gallop when frightened. On being disturbed suddenly they give voice to one or two loud, squealing barks as they bolt for cover'.

Agoutis are diurnal except where they are harrassed by man and his dogs; where they are much distrubed they lie up in their burrows by day. Ingles [250] saw them coming out in bright sunshine on Barro Colorado Island to feed on breadfruit dropped from a tree by coatis: 'Sometimes they would approach within four feet of the feasting coatis without being molested'. Enders thought that agoutis must fight much among themselves, 'for the majority of those taken had torn or nicked ears, and some had long parallel scratches that might have been made by the claws of others'. Agoutis have a strong smell, produced by the secretion of the anal glands; they are nevertheless much hunted for the pot; Enders remarks that the flesh is firm and white when cooked 'and is good eating if one can forget the strong odor so much in evidence in skinning'.

The name acouchi is given to the two species of the genus *Myoprocta*, but it is merely a convenience, for the words agouti, acouchy, cutia, cotiara and others are variations of Amerindian names and are used interchangeably in South America. The acouchis are similar in build to the agoutis but are much smaller, about the size of cavies. The two species, the red acouchy *M. acouchy* from the east, and the green *M. pratti* from the west, may be no more than a cline stretching across South America. *M. pratti* is the only member of the Dasyproctidae that has been carefully studied in captivity; its behaviour has been fully recorded by Morris [337]. He found that acouchis are intensively active diurnal animals which sleep in surface nests, not burrows, by night. They live in forests, especially near rivers and marshes, and at the slightest sign of danger they flee into dense undergrowth so that little is known of their habits in the wild. Like the agoutis, they freeze when alerted, and often stand with one foot raised while listening. They wear paths radiating through their territory from the sleeping cavity, by trampling the vegetation. The tail is much longer than that of the

1 Long-beaked
Echidna,
Tachyglossus
(p. 18).

2 A rodent and a
marsupial from
arid parts of
Australia:

(a) *Notomys
cervinus*, a rodent,

(b) *Antechinomys
spenceri*, a
marsupial. They
were once thought
to show similar
adaptations for a
hopping gait.
Antechinomys
does not hop
(pp. 25, 182).

3 Cuscus (*Phalanger*) (p. 33).

4 Kangaroos drinking at a waterhole. Although they are gregarious they avoid contact (p. 41).

5 Tenrecs:

(a) *Centetes ecaudatus* (p. 49),

(b) *Hemicentetes semispinosus* (p. 49),

(c) *Geogale* sp. (p. 51),

(d) Stridulating spines of *Hemicentetes nigriceps* (p. 49).

6 A young Colugo
(*Cynocephalus variegatus*)
(p. 70).

7 A fisherman bat (*Noctilio*)
and its reflection as it dips the
toes of its left foot at the start
of a sweep (p. 82).

8 The face (a)

and profile (b) of a
lattice-winged bat
Centurio senex
(p. 95).

9 The evening
flight of bats
(*Tadarida
brasiliensis*)
leaving Eagle
Creek Cave
(p. 113).

10 A pangolin *Manis temmincki* digging for ants in Kenya (p. 133).

11 A young aardvark (p. 135).

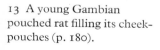

12 A spring-haas (*Pedetes*) photographed at night. The flashlight is reflected through the widely open pupil of the eye (p. 162).

13 A young Gambian pouched rat filling its cheek-pouches (p. 180).

14 A capybara, showing the gland on the nose (p. 205).

15 Trained striped dolphins (*Lagenorhynchus obliquidens*) jumping together (p. 254).

16 The commercial fishery for porpoises in the Little Belt, Denmark (p. 242).

17 African hunting dogs *Lycaon pictus*; the colour pattern differs individually (p. 259).

18 Fur coats of the big cats: jaguar, leopard, leopard, cheetah, cheetah, clouded leopard, leopard, jaguar. The nearer to extermination the higher the prices (p. 287).

19 Sexual difference in the fur seal; a bull with two cows and a pup (p. 290).

20 Bearded seal (p. 297)

(a) dry with the ends of the moustache bristles curled,

(b) wet, with the bristles smooth and uncurled.

21 A manatee breaks the surface with its snout as it rises to breathe (p. 307).

22 Rock hyraxes sunning themselves (p. 326).

23 Savanna forest destroyed by elephants; Tsavo National Park (p. 319).

24 Indian wild ass (*Equus hemionus*) (p. 348).

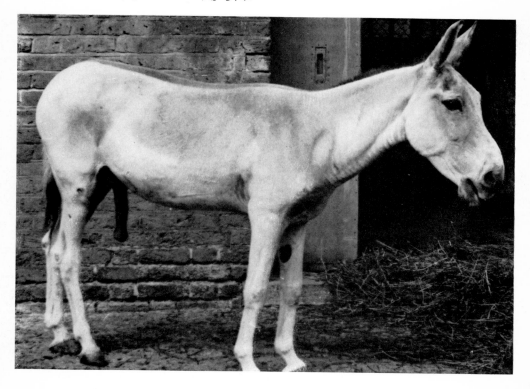

25 A camel with the palu
partly blown out of the corner
of its mouth (p. 366).

26 Giraffes necking
(p. 382).

27 A tree trimmed to hour-glass shape by browsing giraffes (p. 381).

28 Pronghorn buck (p. 385).

29 Pronghorns; white signal patches on the rumps (p. 386).

30 The white-tailed gnu, or black wildebeest; long hair adorns the face, chin, neck, and chest (p. 393).

agoutis – an inch or so in length – and bears white hairs at the tip; it is erected and rapidly wagged by the males when pursuing the females during courtship. In the pursuit the males, like those of the cavies, maras, tree porcupines, rabbits and others, squirt jets of urine at the females. The feeding habits are peculiar, for any food object with a covering of peel is elaborately skinned before being eaten. The animal squats on its hind quarters and holds the food to its mouth with the hands. In peeling a potato, strips of skin are removed from end to end and the potato is rotated until the removal of successive strips completes the peeling. Potatoes are 'stripped of their peel with great accuracy and with more precision than by most human beings'. When the peel is removed, the potato 'is then moved round in the front feet and examined for any small areas that may have escaped peeling. These are then dealt with. Only at this point does eating begin'. And then, after the potato has been completely devoured, the animal often picks up the peelings and eats them too! In the wild the food no doubt includes things of which the peel is distasteful or even toxic, and which is not eaten. The pattern of food preparation is so stereotyped that the peeling routine is followed in captivity even when the peel, like that of a potato, is edible.

Acouchis, like many other rodents, also bury superfluous food for future use; they do not, however, make larders, but tend to bury objects in places where they have not previously buried anything. This scatter-hoarding appears to have evolved 'in connection with increased mobility within the home range. By having food buried all over their range, the acouchis are not heavily dependent on one special site that might become endangered in some way'. Acouchis, like agoutis, are agressive towards each other, and fight among themselves until a social hierarchy is established; during fights 'the rump hair is spectacularly erected. . . . After a severe fight, a victor who has not won easily is liable to gnash his teeth ferociously and audibly'.

The two species, the paca, *Cuniculus paca*, and the mountain paca, *Stictomys taczanowskii* are much larger animals, with stout bodies and comparatively shorter legs and rudimentary tail. The fur is reddish brown with longitudinal rows of white spots, those on the flank tending to coalesce into white stripes. The former lives in forests, the latter in the more open uplands and thus has a thicker, softer coat. Pacas are nocturnal, and lie up by day in burrows that they dig themselves. Their flesh is much esteemed in Brazil; I have seen them hunted with packs of mongrel dogs on the island of Itaparica in the Bahia de Todos Santos – the dogs brought the pacas to bay usually in a stream or in a burrow, where the hunters shot them. The zygomatic arch is immensely developed in the pacas, and extends down-

o

wards and backwards as a broad convex plate. Two peculiar cheek pouches lie on each side under the arch, and have been investigated by Howell [241] and by Hershkovitz [225]. The external pouches are rather small but the internal ones fill 'the relatively large chamber formed by the medial concave surface of the inflated maxillary root of the zygomatic arch'. The pouches are not used for carrying food but for making a peculiar rumbling sound by blowing air from the external into the internal pouches, which act as resonators. Pacas, like acouchis, squirt urine at each other during courtship, and also occasionally as a gesture of aggression or defence; Pearson [370] relates that some natives of Panama believe that the paca also does this when cornered, 'and that the urine will blind humans'.

FAMILY CHINCHILLIDAE

The family Chinchillidae contains the chinchillas and the viscacha. The viscacha, *Lagostomus maximus*, the only species of its genus, is a medium sized animal weighing up to about fifteen pounds. The head is large and the muzzle blunt and rounded so that it has an almost clumsy appearance which is heightened by the peculiar colour pattern of a dark band from cheek to nose under a light band below the eye. The eyes lie well back near the short ears, the tail is long and well haired, the hind legs long, and the colour grey above, white below. The black stiff vibrissae are long. The middle of the three toes on the hind feet bears a swelling towards its end thickly beset with stiff curved white bristles; this arrangement is believed to be a toilet digit, used in grooming the fur. The viscacha is or was found widely throughout the lowlands of Argentina – its numbers have been much reduced because it is regarded as vermin from its burrowing and grass-eating habits.

Viscachas dig extensive burrows – their long vibrissae and closable nostrils are correlated with burrowing – and have the peculiar habit, well described by Darwin, of

dragging every hard object to the mouth of the burrow: around each group of holes many bones of cattle, stones, thistle stalks, hard lumps of earth, dry dung, &c., are collected into an irregular heap, which frequently amounts to as much as a wheelbarrow would carry. I was credibly informed that a gentleman, when riding on a dark night, dropped his watch; he returned in the morning, and by searching in the neighbourhood of every biscacha hole on the line of road, as he expected, he soon found it. This habit . . . must cost much trouble. For what purpose it is done, I am quite unable to form even the most remote conjecture.

It is, however, probably an extension or part, of the burrow sanitation routine, for they carry rubbish out of their burrows and add it to the pile.

The burrows, dug in sandy or clay soil, form warrens some of which have been known to be occupied for decades. In the evening the animals come out and sit at the entrances; Darwin noticed that they are then very tame and 'a man on horseback passing by seems only to present an object for their grave contemplation. They run very awkwardly, and when running out of danger, from their elevated tails and short front legs, much resemble great rats'. On the other hand Llanos and Crespo [301] recorded that when being chased they can touch forty kilometres per hour, make huge leaps and jink with right angled turns and reverses, for more than 600 metres to reach a burrow. At the breeding season the males frequently fight, and the two young, born after about four months' gestation, are precocious and start eating long before they are weaned. Perhaps the most striking point in the life of the viscacha is the astonishing amount of soil dug out by the animals in the excavation of their burrows, amounting to many tons for each warren.

The three species of *Lagidium*, the mountain viscachas, live in the region between the timber line and the snow line of the Andes from Bolivia and Peru to Patagonia. They much resemble rabbits with long bushy tails carried curled upwards towards the tip. The colour varies much between individuals but in general is greyish above, lighter below; the fur is deep and soft but of little commercial value because it wears badly. Mountain viscachas live in arid places with little vegetation and have their dens among rocks, especially the tumbled boulders of rock falls; they dig little and then only where the earth is soft. The animals are diurnal; Pearson [370, 371] who camped among the viscachas (*L. peruanum*) 16,000 feet up the mountains, records that

They get up with the sun, or sometimes a bit before it, and go to some favourite perch on top of a boulder or cliff where they dress their fur and sit on their haunches soaking up the welcome heat of the early morning with eyes half closed in contentment. In the evening they usually feed until after sunset and return to their burrows before dark.

When sun-bathing one 'often sat Buddha-like on his ischia with the sides of his back feet approximated and his paws dangling on his fat belly'. Mountain viscachas are gregarious; a colony may contain up to eighty animals living two to five in each burrow. There are no territories, and the members of a family group like to cuddle up to each other as they bask in the sun. Pearson saw no real fighting, and even in the breeding season aggression was minimal and sometimes diverted by a squirt of urine, although the aim might not be perfect.

Mountain viscachas have few enemies apart from man, and show little fear of intruders when they are among the rocks and near the entrances of their dens – when alarmed they utter a series of high-pitched, mournful warning whistles which puts the whole colony on the alert. Although the Andean fox (*Dusicyon*) is plentiful, and Pearson found fox tracks after a snowfall leading unerringly from one hole to the next he never saw traces of a successful catch: 'Considering the abundance of viscachas, however, the foxes were surprisingly unsuccessful in catching them'. When not hurrying viscachas move in short hops, but if alarmed the hops reach six feet or more; Pearson records their most characteristic method of locomotion is 'by reckless, carefree bounds from rock to rock, or ledge to ledge'. Viscachas have not a great variety of food available to them, and feed mainly on coarse grasses and some lichen, but they are not choosy and eat whatever is at hand. Pearson found that captives were remarkably slow and inefficient eaters, chewing grass one blade at a time and taking twelve seconds to eat each grain of barley. 'At this rate it would take three and one half hours of continuous chewing to eat one day's requirement (about 130 calories) of this comparatively calorific cereal. It is no wonder that wild viscachas go about their evening feeding on meagre vegetation with intense, almost desperate seriousness'. They do not, however, store food or even take it to their burrows for consumption.

There is one main breeding season each year, but occasionally there is a post partum oestrus so that females can sometimes produce up to three young a year – only a single young is produced at each birth after a gestation of about three months. The young are precocious at birth and can eat solid food from the first, although they are suckled for some time as well. Pearson found that, although there is no histological difference between the ovaries of the two sides, the right ovary is always the functional one, and the pregnancy is always in the right horn of the uterus – unless the right ovary is surgically removed, whereupon the ovary and uterine horn of the left side become functional. This phenomenon is unexplained. Although there is no delayed implantation, the early growth of the embryo is extremely slow. Pearson showed that the fertilised ovum reaches the uterus, becomes attached, and begins the formation of the extra-embryonic structures 'at a rate comparable with that found in many other mammals (eleven days) but growth of the embryo itself is so slow at these early stages that is is still a 1-mm filmy oval at 23 days (when a rat would already have produced an entire litter)'. After the embryo is three millimetres long growth becomes quicker, and the birth weight of about half a pound is reached at the end of a total period of about 100 days. A similar slow embryonic growth rate in the early stages is known in other hystricomorphs such as the chinchilla and the domestic guinea pig.

In adaptation to life at great altitudes the blood of viscachas has a greater affinity for oxygen than that of lowland mammals, and the soft fluffy coat protects the animals in the low prevailing temperatures, although the scarcity of guard hairs allows the fur easily to be wetted – 'and a wet viscacha does not survive long in this climate'. Under the best of conditions, however, their survival is short, for Pearson concluded that few live more than three years. Viscachas are of negligible economic importance and enjoy the security

of living in a barren land of few guns and have the good fortune to occupy cliffs and rockpiles which man does not covet. . . . Viscachas are so easily shot and trapped that if man so wished he could readily reduce them from one of the most abundant animals of the high Andes to one of the rarest. Meanwhile, their thin whistle and cheering presence will be welcomed by shepherds and travellers in a barren realm.

The two species of *Chinchilla* have not been so fortunate; their extremely soft dense, silver-grey fur is the most valuable of the South American furs, and consequently they have been hunted almost to extermination, for as the prices paid for their skins rocketed the scarcer they became. Laws to protect the animals may help to save them, but they have not yet recovered in numbers to their former abundance. They have however been brought into captivity and semi-domestication on chinchilla ranches where they are bred for their pelts; the biggest money, however, seems to have been made not by producing pelts but by selling stock to would-be ranchers, with no technical knowledge, anxious to get rich quick. The establishment of chinchilla ranches in many parts of the world has led to a fall in prices of the fur. Chinchillas are much smaller than mountain viscachas – many are little bigger than large rats – and have rather long ears, large eyes, short legs, long whiskers and bushy tails. They are found in Peru, Chile and Bolivia, those with the most valuable fur coming from great heights in the Andes. They live in colonies, formerly of a hundred or more, and seek shelter in holes among rocks and boulders. Like mountain viscachas, they eat the coarse vegetation of their arid habitat. The young, like those of so many hystricomorphs, are precocious and can run about within a few hours of their birth, and eat solid food within a few days.

FAMILY CAPROMYIDAE

The family Capromyidae contains the coypu and the hutias, the latter name spelt 'jutia' in Spanish, in which language 'j' is the aspirate. The coypu is widespread in South America south of the tropic, but the hutias are confined to the West Indies and Bahamas where most species are rare, all are becoming rarer, and some are extinct. Hutias are stout bodied, short limbed, rat-like rodents, some species of the genus *Capromys* confined to Cuba and the Isle of Pines being as large as or larger than rabbits. The tail of *Capromys* is long and hairy, and in *C. prehensilis* it is prehensile. Bucher [58] described the slightly smaller *C. melanurus* as nocturnal and very squirrel-like, running along branches and jumping from tree to tree and using the prehensile tail as a balancer. At least one species, *C. pilorides*, is still common, for Angulo and Alvarez [8, 10], who studied the internal anatomy of the animals, during sixteen months examined 227 specimens in their laboratory in Havana, and between three and four hundred wild caught ones exposed for sale in the Havana market. The short-tailed hutias, three species of *Geocapromys* of Jamaica and some of the Bahamas, are all now very scarce. The Jamaican *G. browni* is stoutly built, and has short legs and a stumpy tail; it is about the size of a rabbit. It lives in burrows in some of the mountainous regions, and is nocturnal; it feeds on grasses and the leaves of shrubs. It was once much sought after for the pot by the negroes, but is now legally protected. The two species of *Plagiodonta* from Hispaniola, are rat-like animals with medium-long almost hairless scaly tails. They are nocturnal and predominantly terrestrial, although according to Tate [469] they can climb in trees readily on occasion. They were formerly eaten by the Indians, but are now little regarded, and one species may still exist in some numbers. Several other genera of recently extinct hutias are known from their skeletal remains in caves and middens in various West Indian islands – indeed some of the living species such as *C. nana*, were first described from such remains and subsequently found to be still alive.

The single species of the genus *Myocastor*, the coypu, native to southern South America, has been widely introduced into Europe, Asia and North America for the sake of its fur which, beneath the coarse long guard hairs is soft and dense, and is the nutria of commerce. It is consequently well known, for it has everywhere escaped from fur farms and flourished in its new environment, and in some places has become a destructive pest. It looks much like an enormous musk rat with cylindrical tail; it is generally about the size of a rabbit but sometimes grows much bigger. The hind feet are webbed, and the animal is aquatic in its habits – it digs burrows in the banks of watercourses and lakes, but also makes surface nests among reeds and dense vegetation on the shore. It is vegetarian, and rapidly cuts up water plants and attacks agricultural crops in nearby fields with its huge orange-coloured incisor teeth. The coypu is valued not only for its fur but also for its flesh, which is very palatable, and has appeared on the market under the name of 'swamp beaver'. The flesh of many of the South American hystricomorphs is or has been esteemed as food – the hutias suffered much from human predation even before the conquest – and although some writers have recorded that the flesh of other species is indifferent other writers have praised it. Perhaps the disregard by the gauchos of the plains viscacha as recorded by Darwin was due to the abundance of beef. Pearson [370] says that the back and hind quarters of the mountain viscacha are as good as rabbit meat but that the abdominal muscles have 'an unpleasant, rat-viscera taste' – though few of his readers are likely to appreciate the simile from personal experience. He adds that some Peruvians believe that eating viscacha meat turns the hair grey, but that his party ate viscachas of both sexes almost daily for four months and 'did not tire of them, and noticed no acceleration of greying'.

Most of these hystricomorphs show an unusual arrangement of the mammary glands, which are placed on the flanks so that the nipples are set in peculiar positions, high up on the sides. The nipples are often reduced in number; in the Patagonian hare and the mountain viscacha, for example, there is a single nipple on the side behind the armpit, and the young often sit up to suckle from the mother as she sits on her haunches. In the hutias, there are two pairs of nipples set high up on the sides, and in the coypu the six nipples on each side are set so high up that the young can suck when the mother is swimming. These phenomena are perhaps correlated with the precocious state of the young at birth; the young are not sheltered in nests but run about and start eating some solid food very shortly after their birth.

FAMILY OCTODONTIDAE

The family Octodontidae contains a number of rat-like hystricomorphs found from southern Peru through Chile and Argentina; they live in many different habitats from the lowlands to high in the mountains. All are rather stoutly built, short legged animals with comparatively large heads and rounded moderately prominent ears. The three species of *Octodon* and the single ones of *Octomys* and of *Octodontomys* have long tails with terminal tufts of hair, particularly well developed in the last. Apart from local South American names these animals have no vernacular designations in European languages. Waterhouse [498] records that Darwin found *Octodon degus* by hundreds in the hedgerows and thickets in the central parts of Chile, where they make burrows close together, leading one into another. They feed by day in a fearless manner, and are very destructive to fields of young corn; when disturbed, they all run together towards their burrows, in the same manner as rabbits do in England when feeding outside a covert. When running they carry their tails elevated; and often they may be seen sitting on their haunches, like a squirrel.

They also store food for the winter, but do not become dormant.

The rock rat *Aconaemys fuscus*, the only species of the genus, with short, hairy, but untufted tail is about the size of a common rat. It is, according to Osgood [359], now reduced to a few small colonies and is on its way to extinction; over a hundred years ago Bridges, who discovered the animal, told Waterhouse [498] that in a valley five to seven thousand feet high on the east side of the Andes in some places 'the ground is completely undermined by the burrows of these little animals', and whilst riding over them the horses of his party frequently plunged into the ground almost up to the hock. The rats are nocturnal, and are active beneath the snow that covers their habitat for at least four months of the year. The single species of *Spalacopus* is even more subterranean; it is a stocky short legged rat-like animal, with small ears and black glossy fur. The incisor teeth are large and permanently exposed like those of gophers and mole rats. This animal feeds on roots, bulbs and tubers and makes large stores of them in its burrows. Bridges told Waterhouse [498] that the poorer natives probed the burrows to 'rob them of their store, which they ate'. He also remarked that 'the work of this little animal would surprise a person unacquainted with its habits; I have frequently seen a considerable surface of ground completely undermined by its burrows. It generally selects the slopes of hills and mountains where bulbs are found'. He added that it is nocturnal, since it seldom if ever makes its appearance during the day. It much resembles the members of the next family.

FAMILY CTENOMYIDAE

The family Ctenomyidae contains about two dozen species of a single genus, *Ctenomys*, the tuco-tucos. Tuco-tucos are medium rat-sized rodents with heavy bodies, large heads, short tails, small eyes and ears, and huge exposed incisors; they superficially much resemble the pocket gophers, except that they have no pockets or cheek pouches. The feet are rather large, bear stout claws, and are edged with stiff bristles which are used in grooming the fur. Tuco-tucos feed on roots and tubers, and spend nearly the whole of their lives underground. Their burrows lie fairly near the surface and are very extensive; local colonies completely undermine and denude the soil, and they throw up the earth in piles resembling mole-heaps. The nest chamber, and also chambers for the storage of food, are placed somewhat deeper. Tuco-tucos make a peculiar sound when under-ground – the name tuco-tuco is an attempted imitation of it. Darwin described it as a short, but not rough, nasal grunt, monotonously repeated about four times in quick succession. The young, like those of many other hystricomorphs, are precocious; Pearson [372] found that a litter consists of two to five young according to species. Tuco-tucos are widely distributed throughout southern South America, from about 15° S to Tierra del Fuego, wherever suitable soils occur, from the lowlands to great heights in the mountains; Cabrera and Yepes [65] record that one species burrows in the banks of streams and is semi-aquatic. In southern stock-raising areas their numbers have been much reduced.

FAMILY ABROCOMIDAE

The two species of *Abrocoma* or chinchilla-rats which form the family Abrocomidae are not well known; they live in the high mountains from southern Peru to northwest Chile and Argentina. They are rat-like, but have large eyes and ears, and soft fur resembling, but inferior to, that of the chinchilla. The animals are vegetarian, and live in crevices among rocks or in burrows beneath them. Cabrera and Yepes [65] describe the Argentine species as timid animals of the rocks, though able to climb in bushes where they occur in the high arid regions, and that they are colonial, and make their burrows in proximity in stony places among broken rocks.

FAMILY ECHIMYIDAE

The Echimyidae are the last family of hystricomorphs found exclusively in South America; they are the spiny rats, and are a comparatively large family of fourteen genera with over forty species in all. The family is found throughout the northern part of South America from Nicaragua to the tropic; it contains in addition several genera known only from bones in caves or middens in Puerto Rico and some other West Indian islands, representing animals that became extinct at about the time of the conquest. Of the fourteen genera only two contain more than three species, and six contain only one. Many species are rare in museum collections and little, if anything, is known of their lives, although a surprisingly large number have been known to the systematists for 150 years. *Echimys* and *Proechimys*, each with about a dozen species, thus contain about half the members of the family between them, and are the common spiny rats of northern South America. They are medium to large rats with tails about as long as the body and either nearly naked or covered with short hair, and with spines and bristles mixed with the fur. Some are terrestrial, others arboreal; all are vegetarian, and a few species, like common rats, live commensally in human dwellings.

Most members of the family have a peculiarly brittle tail, which readily fractures near the base if seized, much like the easily autotomised tails of some lizards, a phenomenon that is thought to have a protective value. Enders [133] found *P. semispinosus* to be very common on Barro Colorado Island, and caught one that had lost its tail, and several in which the tail had been partly broken at the base but had healed, one having a definite crook in the tail as a consequence. The animals were feeding on grass, but accepted bananas in captivity and in eating them, 'the skin at the stem was taken between the teeth, then with a turning and raising of the head the skin was stripped off as far as could be done without moving the body. At times the fore foot was placed on the banana to hold it on the floor while it was being peeled'. These animals were terrestrial, but the species of the more stiff-spined rats of the genus *Echimys* which Tate [468] found in the region of dry dwarfed forest in northern Venezuela were arboreal; he noted that they 'ran actively along the limbs and through the branches of trees in a rather squirrel-like manner, and in all probability it will be found that they live in holes high above the ground'. Several genera, such as *Isothrix* and *Dactylomys* have no spines among the fur, and the latter, with a few other genera, has very long toes arranged so that they are almost syndactylous; when climbing the grasp of a twig is between second and third toes of the hind foot, and the third and fourth toes of the fore. Tate [468] records *Dactylomys* from canebrakes on the islands of rivers in eastern Peru, where they could be found by their peculiar 'kuk-kuk kuk-kuk' calls. He also notes that *Isothrix* lives in holes in large trees along the borders of rivers. *Isothrix* has a rather well furred tail, and one species *I. pictus* differs from the other members of the family which are generally brown, black or reddish, in having a conspicuous pattern of black or dark brown and white.

FAMILY THRYONOMYIDAE

We now come to two families of hystricomorphs that are found only in Africa. The first, the Thryonomyidae, contains a single genus *Thryonomys* with six species – the cane rats or ground hogs found over most of Africa south of the Sahara. They are heavily built, thick set animals with blunt rounded muzzle, small rounded ears, moderately long tail, and harsh bristly fur. They are large rodents and weigh some eight to twelve pounds, occasionally reaching about twenty pounds. In some places they live in burrows, sometimes of their own digging, sometimes those of aardvarks or porcupines, but in reed swamps and among dense vegetation they lie up above ground. They are not gregarious, but in the swamps where they are plentiful many may be found fairly close to each other. The food is entirely vegetable and consists particularly of the roots and shoots of reeds and water grass; in cultivated areas they are very destructive to crops, particularly to sugar cane. They also bark trees which, according to Roberts [406], are often stripped from the ground to as high as the rats can reach by standing on their hind feet. They dive and swim well 'often disporting themselves in the water . . . especially at night'.

Shortridge [442] found cane rats breeding in southwest Africa between June and August and producing litters of two to four, but usually three, young. The nipples of the female, as in so many hystricomorphs, are set high on the side so that she suckles her young without lying on her side. The young, too, are precocious, and are born well haired and with eyes open. The tail of cane rats appears to be brittle like those of the echimyids, and specimens are often found with truncated or missing tails. These animals are much relished as food by the natives, although Shortridge says that the meat is rather flavourless and should be cooked with plenty of salt. Cane rats are hunted with dogs and assagais, and are driven out of their cover by firing the reeds, when they 'make lightning rushes from one unburnt reed patch to another, doubling and scampering about like small pigs'. Shortridge adds that to prepare cane rats or porcupines for the pot, the best method is to pluck them like fowls because the thin skin is tightly bound down to the panniculus carnosus muscle but the spines come away readily. 'The flattened spines of cane rats resemble the spines on the underside of porcupines in being loosely attached, and growing in unusually manifest tufts of three to five or six out of common root pores'.

FAMILY PETROMYIDAE

The family Petromyidae contains but a single species, *Petromus typicus*, the dassie rat of southwest Africa. 'Dassie' is the vernacular name in South Africa for the hyrax, and at a distance the dassie rat resembles the dassie so closely that Shortridge [442] says that when collecting in southwest Africa on more than one occasion new born dassies were shot in mistake for dassie rats. The dassie rat is about the size of a common rat, but the head is flattened and squirrel-like when seen from above. The tail is fairly long, and the tail hairs 'stand out as in a bottle brush'. Shortridge found that their tails snap off easily and at least a quarter of the specimens he trapped had lost the whole or part of the tail. The fur is extremely soft and silky although, owing to the absence of underfur, the hairs stand out separately in life and have a slightly wiry appearance. Dassie rats are plentiful in their habitat in hills covered with loose piled-up boulders between which there are plenty of crevices to give them shelter. They are generally diurnal, coming out of their dens in early morning and late afternoon, but also by moonlight; they are easily overlooked even where plentiful, for as Shortridge records

> they are protectively coloured and move warily like flat lizards between crevices and along the tops of boulders, never away from rocky hill slopes. As indicated by their short legs and squat build they are rock runners not rock jumpers. . . . The ribs of *Petromus* are flexible and the animal can be pressed almost flat without injury. The compressed and limp body, which touches the ground in life, enables the animal to squeeze into extremely narrow horizontal crevices.

They also like to lie out sunning themselves near their dens during the heat of the day. Their food consists mostly of green vegetable matter, with berries and seeds; Shortridge notes that the flesh has a 'pleasant aromatic smell perhaps due to a diet of mountain shrubs and herbs'. When startled they often squat motionless, or edge backwards out of sight. The litter is small, numbering not more than one or two, and the young are well furred at birth; the nipples of the females, like those of the cane rats, lie high up on the sides.

The following two families are included at the end of the hystricomorphs by Simpson [445] under the query '? incertae sedis' because their affinities are not understood, and they are pushed in here as there is no niche into which they fit.

FAMILY BATHYERGIDAE

The Bathyergidae contain five genera with about twenty species between them – the mole rats of Africa south of the Sahara but extending into the Sudan, Abyssinia and Somaliland. Superficially they resemble burrowing mole rats of the distantly related geomyid, rhizomyid and other families, but owing to their enormous exposed incisors and truncated snouts they are perhaps even more hideous to human eyes in facial expression. Darwin's comment about a snake with a hideous and fierce facial expression is apt: 'I imagine this repulsive aspect originates from the features being placed in positions, with respect to each other, somewhat proportional to those of the human race; and thus we obtain a scale of hideousness'.

About fifteen species of *Cryptomys*, three of *Heliophobius*, two of

Bathyergus and one of *Georychus* are all very similar superficially, and are separated mainly by characters of their anatomy and cheek teeth; they range in size from that of a large mouse to that of a very large rat. They are closely adapted to burrowing; their bodies are short and cylindrical, the head large, the eyes small, the external ear minute or absent, the limbs and tail are short. The outer sides of the fore and hind feet and both sides of the tail are fringed with stiff bristles; the colour of the short soft fur is usually grey or brown, but some species carry patches of lighter colour especially on the head. Bathyergids spend practically their whole life underground in extensive burrows, where they feed on the roots and bulbs of plants.

Although many anecdotal observations have been recorded about mole rats, the life history of only one species, *Cryptomys hottentotus*, has been studied in detail, by Genelly [164] – that of other species is probably in general similar. Genelly, who takes into account the previous results of other zoologists, worked in southern Rhodesia in gently rolling country, at an altitude of 4,860 feet, covered with tall grasses and large termite hills. The mole rats dig intricate burrow systems particularly below the large termite hills where they are safe from flooding during the wet season. Among the twisting and winding passages, spiralling to different levels, enlarged chambers are used as store rooms for food, and to hold the nest. When the earth is in suitable condition for digging, burrows extend for many yards away from the system of home burrows. The mole rat excavates its tunnel with the incisor teeth – the ceilings of most rooms examined by Genelly were engraved with incisor marks. When the soil is loosened the mole rat scrapes it back under its body with the cupped fore feet, and pushes it back with the hind. When a pile of earth as large as the animal has accumulated the mole rat goes backwards along the burrow, pushing the plug of earth with its hind feet and tail – the bristly hairs on feet and tail help to sweep the earth along if it is dry and friable and does not move as a coherent plug. On reaching a passage leading to the surface the mole rat goes through an almost acrobatic performance by standing on its fore feet to push the plug of earth up to the surface with the hind and to prevent it falling down on to its back. The final stage of the routine consists of vigorous vibration of the hind feet – about twenty-two to twenty-four vibrations a second for rather over a second – which consolidates the plug just as a builder vibrates wet concrete for the same purpose. The plug, pushed out of the burrow, at first stands up vertically from the centre of the mole-hill, but disintegrates or topples over as it dries; the entrance to the burrow, however, always remains plugged. In the dry season, when the surface earth is baked too hard for burrowing, the mole rats continue work at lower levels, and dispose of their excavated earth by using it to plug old workings.

Mole rats eat the roots, bulbs or tubers of many different plants – the locally abundant species is generally the most important – and the Rhodesian colonies fed almost exclusively on the bulbous bases of black-seed grass. The food is stored in chambers, or in sections of tunnel, of the home burrow system, and is eaten when the foraging of the animals is restricted during the wet season. One store examined by Genelly occupied five and a half feet of tunnel, and weighed nearly three and a half pounds; another weighed nearly six pounds. By observations on captive animals he calculated that each mole rat ate at least twelve grams of food daily – a quarter of its own body weight. Each home burrow system is occupied by a colony of about a dozen mole rats, and the different neighbouring systems appear to have little intercommunication. Genelly points out that the comparative isolation of the colonies leads to great variation in coat colour and the degree of development of a white spot on the head; the varieties are the result of the slow rate at which the gene pool of the species is stirred. Breeding apparently takes place shortly after the rainy season begins, and the one or two young are nursed in a globular nest made of plant fibres. Mole rats are fairly immune to predation since they always keep their burrow systems plugged, and seldom emerge onto the surface except when forced out by floods. In some places the natives use them for food.

Finally, the genus *Heterocephalus* completes the bathyergid family. *Heterocephalus*, the naked mole rat, has been mentioned in Volume I, and some of its adaptations to a fossorial existence discussed. It is a small mole rat, much resembling the other genera of the family, except that it has no fur and appears at first sight to be completely hairless. It lives in the arid parts of northern east Africa and spends its whole life underground; its burrow systems seem to be rather deeper below the surface than those of the other mole rats. It feeds on roots and bulbs, and some insect remains have been found in the stomachs, though whether these represented prey deliberately taken or things accidentally eaten together with the vegetable matter is unknown. It differs from the other mole rats in its method of digging, for it does not push out a plug of excavated earth but digs with its teeth, pushes back the earth with the fore feet, and then throws it out violently with the hind feet after pushing it towards the burrow entrance, where the spoil makes a large mole heap.

FAMILY CTENODACTYLIDAE

The Ctenodactylidae form the second family with no true home, which is included at the end of the hystricomorphs merely for convenience. The ctenodactyls are the gundis – eight species are known, portioned among the four genera *Ctenodactylus*, *Felovia*, *Massoutiera* and *Pectinator*. Gundis are rat-sized rodents with compact bodies, and short limbs and tail, which live in north Africa from Senegal through Morocco and Egypt to Somaliland. The two inner digits of the hind foot bear comblike brushes of bristles used in grooming the fur. Gundis look much like domesticated guinea pigs but little has been recorded about their lives beyond the observations of Kock and Schomber [271]. The animals are diurnal, but seek shelter during the heat of the day among the rocks and cliffs where they live. They live in small, probably family, groups, and give a warning high pitched whistle when alarmed. They run about on the rocks with the body held close to the ground and forage for plant material particularly in the evening. Gundis are alleged to sham dead like opossums when frightened, and to freeze motionless if danger threatens. The litter consists of three or four, and the young are well furred, have their eyes open, and are able to run at birth. The ears of the gundi are rather small and rounded, and are provided with a beautiful fringe of hairs along the upper margin, that is believed to help in keeping out sand and dust. In this feature they resemble some of the other desert rodents such as the jerboas.

Whales and Dolphins

ORDER CETACEA

THE MEMBERS of the order Cetacea, containing the whales, dolphins and porpoises, have departed farther from the typical mammalian patterns of structure and physiology than those of any other, in their high adaptation to aquatic life. They nevertheless retain all the basic mammalian characters, although many of them are greatly modified. The hairless, fish-shaped body cased in a layer of blubber, the fore limbs modified as paddles or 'flippers', the external absence of hind limbs, and the horizontal flukes of the tail, combine to produce a form far different from that of the land mammals. The anatomical and physiological adaptations to deep diving, specialised feeding, and echolocation that are so important to the aquatic way of life have been discussed in Volume I, and need not be repeated here – in this volume we examine in more detail the biology, as far as it is known, of the diverse kinds of animals that make up the order.

The living whales are divided into two suborders – some taxonomists think they should be separate orders – the Mysticeti or whalebone whales, and the Odontoceti or toothed whales. They differ not only in structure, food, and the use of echolocation, but in the chemical composition of the oils in their blubber and body fats. The structure of the placenta differs in the two suborders; placental structure is notoriously no guide to phylogeny, but if segregated on this character the whalebone whales would be cows and the toothed whales horses! The extinct suborder of Archaeoceti or zeuglodont whales is equally remote from the other two – the origin of all three in unknown, but they probably diverged from the early mammals as far back as the Cretaceous.

SUBORDER MYSTICETI

The whalebone whales form the smallest suborder, for there are only twelve species, partitioned among six genera in three families. They take their name from their salient character, the series of baleen or whalebone plates in the mouth, in the position occupied by the upper teeth in other mammals. The frayed-out inner edges of the baleen plates or blades together form a filter bed for straining from the sea water the planktonic animals which form the food of these whales. In correlation with the large size of the baleen apparatus the fore part of the skull is elongated as a rostrum for its attachment. As in all whales the nostrils open far back on the upper surface of the head, but in the mysticetes they are double and not fused into a single blow hole as in the toothed whales.

Some cetologists have tried to estimate the age of mysticetes by counting the minute ridges on the surface of the baleen, but better results have come from counting the laminations of the ear-plugs. The plug lies in the auditory meatus, which does not open to the outside, and is an accumulation of dead horny cells shed by the lining of the tube – rapid deposition during the season of feeding alternates with a slower rate, and produces rings similar to the growth rings of trees. In odontocetes growth rings in the teeth, which can be seen if a tooth is cut across, may be able to provide information about the age of the animals in some species.

Although various sounds made by mysticetes have been recorded by underwater listening apparatus, it is uncertain whether these whales make use of sonar. It might be expected that they would do so from the general similarity of their way of life to that of odontocetes. Their anatomy, however, gives no suggestion that they do; the shape of the skull, forming the foundation of the baleen apparatus, is quite unlike the dished reflector shape of the fore part of the skull of the odontocetes and, without the dish to hold it, there is nothing resembling the melon.

FAMILY BALAENIDAE

The family Balaenidae contains the right whales, the most highly modified of all the whalebone whales, and *Balaena mysticetus*, the only member of its genus, is the extreme example. This, the Greenland right whale or bow-head, was formerly numerous in Arctic seas but was reduced to near extermination through intense hunting by the whaling industry from the seventeenth to the late nineteenth centuries. It was first exploited in the waters round Spitzbergen, then known as 'Greenland' and believed to be an eastern extension of that country – it takes its other name from the high-arched bow-shaped rostrum. In the early days the whalers harpooned and killed the whales near the coast and towed the carcases to the shore to remove the baleen, strip the blubber, and boil out the oil. When whales became less plentiful they were pursued among the ice and these operations were carried out at sea, though the arctic whalers of the last century

P

usually did not boil out the oil on board but brought it home minced in casks for rendering on their return. The bowhead is an enormous plankton filter – it grows to about sixty feet long, and one third of its length is taken up by its head, most of which consists of mouth and its strainer. The baleen blades are narrow, and because of the bow-shape of the rostrum, those at the middle of the series are the longest and measure some twelve feet in length – even fifteen feet has been claimed but not authenticated. The lips of the lower jaw rise high on each side like a huge scoop to enclose the long baleen blades. The bowhead feeds largely on copepods, small planktonic crustacea, and pteropods or sea-butterflies, small planktonic molluscs collectively known as 'brit'.

Figure 12 Right whale (*Eubalaena*), drawn from a dead specimen; the mouth is held open by the swollen tongue distended by the gases of decomposition.

The bowhead has no dorsal fin, has tail flukes up to twenty feet across, and large rounded flippers set close behind the eyes which lie close to the angle of the jaws; the blowholes open on a raised mound far back on the top of the head. The general colour is black, but the chin and lower jaw are commonly white, and there is a variable amount of white splashing on the underside, sometimes enough to produce an almost piebald appearance. This species, like the other right whales, carries such a thickness of blubber that it floats when dead, and for this reason it was the quarry of the old-time whalers who were unable to cope with other species that sink. The bowhead inhabits all the circumpolar seas and the sea of Okhotsk. A few are killed annually by the Eskimo, but the species is protected from commercial hunting; it is now so scarce that most of our information about it is derived from two of the famous whaling captains who were also naturalists, Scoresby [431] and Scammon [416].

The closely related genus *Eubalaena* contains the right whales; three species are commonly recognised, the Atlantic, Pacific and southern right whales. These are certainly isolated populations, but it is doubtful whether they are truly specifically different, and it might be better to regard them as races of a single species. The right whale closely resembles the bowhead but has a rather smaller head and less highly arched rostrum; the baleen blades are shorter and fewer – eight to nine feet long, and about 250 in number as against 350. The colour is black with a variable amount of white patches on the under surface. The most peculiar character, however, is the structure called the bonnet, an area of cornified and irregularly roughened skin on the front of the rostrum, and some similar smaller patches on each side of the lower jaw. These excrescences appear at first sight to be pathological growths, but they are not, for they are present in the unborn foetus. The crevices of the bonnet are inhabited by a parasitic crustacean, the whale louse, a highly modified amphipod, *Cyamus*, which clings by hooked appendages and feeds on the epidermis of the whale. Another and smaller kind of *Cyamus* lives on the bowhead about the head and in the folds under the flipper, but it is not plentiful; that of the right whale, on the other hand, is often present in great numbers especially in the bonnet and other excrescences. It has been suggested that the bonnet and other growths represent some sort of adaptation to heavy parasitising by the lice – a sort of dane-geld to provide a special nest for them and buy off the irritation of a more widespread infestation. Be that as it may, the use, if any, of the bonnet to the whale is quite unknown.

The right whale inhabits the temperate and sub-polar waters of the oceans and does not, like the bowhead, penetrate into the polar ice. In southern oceans right whales come close into coastal waters and sheltered bays to give birth to their young – a single calf which closely follows its mother for some months. There appears to be some annual migration of the northern races, which go into the arctic during the summer and retire to warmer waters during the winter; a similar migration has not been recorded of the southern race. The planktonic food of the right whale varies greatly according to what may be locally abundant. The right whale of the Atlantic is the species that gave rise to the commercial whaling industry. Nearly a thousand years ago the fishermen of the Basque coasts of Spain and France were hunting the then plentiful right whales in their coastal waters; they invented the technique of harpooning and lancing the whale and towing the carcase to the shore for dismemberment and 'working up'. For centuries thay had a monopoly of whaling, and when the English Muscovy Company sent a whaling expedition to Spitzbergen in 1611 six Basque harpooners were taken to teach the English sailors the methods of whaling; that voyage was the beginning of the northern whale fishery, and thus the direct ancestor of the great modern industry which is now declining to its end.

The pygmy right whale *Caperea marginata* differs from the other right whales not only in its small size, which does not exceed twenty to twenty-one feet in length, but in possessing a comparatively large hook shaped dorsal fin set well back towards the tail. It is rather less tubby in proportions than the others, but it has unique internal features. There are seventeen pairs of ribs – more than in any other whale – which are very broad and flattened so that they form a close cage for the contained viscera. The skin is black with some greyish flank-streaks in some specimens, and the baleen is black on the outer margin of each blade but for the rest is white, as are the frayed inner fringes and the lining of the mouth and tongue. Nothing is known of its habits, but it has been suggested that the rib structure enables it to remain on or near the bottom for comparatively long periods. Less than three dozen pygmy right whales have ever been seen, and nearly all of those have been washed ashore dead, most of them in Tasmania and New Zealand, some on the southern coast of Australia and the Cape of Good Hope, and one or two on the east and west coasts of southern South America. Davies and Guiler [101], who are among those few fortunate cetologists who have had an opportunity of examining the pygmy right whale in the flesh, point out that the species has never been sighted at sea, although biologists working in the Southern Ocean have looked for it, but that it is probably present in Tasmanian waters for most of the year. They conclude that this little known whale is probably in the habit of lying on the bottom in moderate depths, and that 'its small size and peculiarly white mouth may have some significance in such an environment'.

FAMILY ESCHRICHTIIDAE

The family Eschrichtiidae contains only one species, *Eschrichtius gibbosus,* the grey whale of the north Pacific Ocean. The grey whale grows to about fifty feet in length, and is greyish in colour with an irregular white mottling. It has no dorsal fin, but a low hump or ridge in the place where a fin might be expected, followed by eight to ten low bumps along the ridge of the tail stock. The rostrum is broader and less bowed than in the right whales, and the baleen plates are short and narrow. A major distinction from the right whales is the presence of a pair of parallel grooves in the skin of the throat – sometimes increased to three or four, the function of which is not apparent. Grey whales are migratory; they spend the summer in the coastal waters of the Bering and Okhotsk seas and go south in the winter as far as Lower California and southern Korea. The eastern and western herds appear not to mix in the north by passing round the southern tip of Kamchatka. The California herd was nearly exterminated by excessive hunting, but has increased enormously since it was given complete protection in 1938, so that now the whales migrating close to the shore are one of the major tourist attractions in California – about a million visitors annually come to see them from various places on the coast from Cape Blanco, Oregon to Point Loma above San Diego Bay. The herd was estimated to number about 3,000 whales in 1953, and about 9,000 in 1969, an increase of about 10 per cent a year. The natural history of the grey whale has been admirably described by Gilmore [165] who says that 'it works itself more and more completely into the recreational and esthetic values of the great urban community that is southern California – in a manner that no other whale has done elsewhere in the world'. Grey whales feed during the four summer months, when they are in the Bering sea and beyond the Bering Straits, on planktonic euphausians, and also, it is said, on bottom-living amphipods which they get by stirring up the silt with their snouts. They become very fat, and then fast for eight months during which they travel some 12,000 miles or more – 6,000 to 7,000 miles there and back – to give birth to their young and breed in the lagoons and bays of Lower California. They stay in winter quarters for about two months, and then start north again for the summer feeding grounds. Although some thousands of grey whales migrate annually, they are not social animals and seldom form schools of more than a few individuals – even on the breeding grounds they are equally solitary.

From the middle of the last century an extensive whaling industry harrassed the grey whales both during the breeding season and in the Arctic, and in thirty or forty years nearly exterminated the species; among the whalers the grey whale had a reputation as a dangerous animal that often attacked small whale boats when harpooned. Scammon [416] says the whalers called them 'devil fish' for this reason, and also 'hard head' because they 'root the boats when coming in contact with them, in the same manner that hogs upset their empty troughs'. He also says they were called 'mussel diggers' in allusion to the habit of stirring up the bottom to obtain food. The grey whale formerly inhabited the Atlantic, for its remains have been found in sub-recent deposits on the Atlantic coasts of Europe and North America. It is a pity that the laws of priority have made obsolete the very appropriate and widely used former scientific name of the grey whale, *Rhachianectes glaucus*.

229

FAMILY BALAENOPTERIDAE

The family Balaenopteridae contains the five species of rorqual and the humpback whale. The rorquals are long slender baleen whales, some of great size, with a small hook-shaped dorsal fin set far back towards the tail, and a large number of parallel grooves or folds in the skin running from the chin over the throat to the chest. During the summer they migrate to high latitudes to feed on the abundant plankton – in the southern hemisphere almost entirely on one species of crustacean, *Euphausia superba* – and during the winter they retreat to the warmer waters of temperate and even tropical seas, where the young are born, and where they fast for several months. The rorquals of the northern hemisphere are in the Arctic when those of the southern hemisphere migrate north, and when they come south those of the southern hemisphere have returned to the Antarctic, so the two populations do not normally mix. Rorquals are usually found in small parties of up to half a dozen or so, but when feeding on concentrated shoals of 'krill' – euphausian crustacea – great numbers may come together. Forty-five years ago, on the whaling grounds off South Georgia I have seen them so abundant that literally hundreds, and perhaps thousands, of blows were to be seen in whichever direction one looked – even whaling captains of long experience were astonished – but such sights will not be seen again by anyone now living, for we have killed the whales so fast that their numbers will be long in recovering. The large rorquals reach sexual maturity at an age of six to ten years, and the females give birth once every two years; gestation lasts nearly a year, and lactation six months or more – the normal life span is probably not often more than thirty years.

The rorquals are the main quarry of the modern whaling industry, although other kinds are also taken. It was not until the Norwegian Svend Foyn invented the steam whale-catcher with whale cannon and large harpoon with explosive head, and the accumulator for giving spring to the whale line, that the rorquals became accessible to whalers. Rorquals sink when dead, but his ship with a heavy line was able to winch the carcase to the surface and make it float by injecting compressed air. Until the 1920s the dead whales were brought to a shore station for working-up, or to a floating factory-ship anchored in some sheltered harbour. Since then pelagic whaling has replaced shore-station whaling in the Antarctic and elsewhere, and huge factory ships hoist dead whales on deck up a special slipway, and work them upon the high seas. Over fifty years ago it was evident that the stocks of whales were being over-fished; scientific researches carried out over many years on the biology of whales provided the knowledge needed to regulate the industry so that an annual crop could be harvested without depleting the stocks. An International Whaling Commission to safeguard the industry was, however, unable to make scientific advice prevail over commercial greed, and consequently the world population of whales has been, and continues to be, catastrophically reduced, and the whaling industry is but a shadow of its former prosperous size.

The genus *Balaenoptera*, containing the five species of rorqual, includes the largest mammal that has ever existed, *B. musculus*, the blue whale, so named from the bluish-grey mottled skin. The blue whale has been recorded as reaching a length of slightly over 100 feet and a weight of about 130 tons. The young at birth measures over twenty-five feet in length and weighs several tons; it is the extreme example of rapid growth among the mammals for the fertilised egg, smaller than a pin-head, develops into this gigantic whale-calf in less than eleven months. The blue whale was particularly valued by the whalers because of its large size – fifty years ago there were some 30,000 of the species in the southern hemisphere, today there are probably less than 1,000. Not all blue whales reach so great a size; smaller specimens which used to appear on the southern whaling grounds towards the end of the season and were known to the Norwegian whalers as 'maibjørn' – May-bears – have been promoted to be a sub-species under the name of 'pygmy blue whale' by the Japanese whaling interests, anxious to find prey not prohibited and protected by international regulations.

The fin whale, *B. physalus*, is smaller than the blue whale, reaching a length of about eighty-five feet, and differs in colour – it is black above and white below, and both baleen and throat are asymmetrically coloured. In northern seas it feeds not only on planktonic crustacea but also takes small species of shoaling fish. This species is now the main prey of the whaling industry which has reduced the southern population from about 200,000 to less than 40,000. The Sei whale, *B. borealis*, much resembling a small fin whale but with finer shredded baleen and comparatively longer dorsal fin, becomes more important to whalers as fin whales decline in numbers, and the even smaller minke whale, or piked whale, *B. acutorostrata*, which in the northern hemisphere at least generally has a white armlet on the flipper, is being exploited not only for its blubber but for its meat which is much used in canned pet foods as well as for human consumption. Bryde's whale, *B. edeni*, closely resembling the Sei whale, does not migrate annually to polar regions but lives in the warmer waters of the equatorial belt. The humpback whale *Megaptera novaeangliae*, the only other member of the family, is a short tubby whale reaching a length of about fifty feet;

it has fewer and larger throat grooves, a small fin on a raised hump with smaller bumps between it and the tail, and very long narrow flippers. The edges of the flippers and the tail flukes are irregularly scalloped, and the snout, lower jaws, chin and fore-edges of the flippers are adorned with raised knobs. The colour ranges from all-black to black above and white below – no two specimens are alike. The humpback herds of the southern oceans mixed little, those that wintered off the coasts of the continents returned yearly to their own regions of the Antarctic. Humpbacks often came into coastal waters and were easy prey for the whalers who reduced their numbers early in the rise of the modern industry so that they are now commercially extinct.

SUBORDER ODONTOCETI

The Odontoceti are much more numerous in species than the Mysticeti – there are twelve families with thirty-three genera and some seventy-four species between them. The teeth are all alike – there is no differentiation into incisors, canines and molars – and are generally conical and often slightly curved. Although one species has more teeth than any other mammal, in others the dentition is reduced and some species are functionally toothless. In all toothed whales the forehead is rounded, and in many species the rounding is so pronounced that it is called the melon – *melon* in French means bowler hat, but the term applied to whales is not, as might be thought, a slang term of the modern whale-scientist, but is a whalers' technical name which first appeared in print in 1887 in the report on United States fisheries by Goode and others [172]. The melon consists largely of fat and oil and lies above the upper jaw and in front of the cranium which rises behind it; Norris [350] has shown that it probably acts as an acoustic lens and improves the resolution of the sonar emission. In a similar way the oil-filled lower jaw helps the transmission of the returning echo to the ear. Porpoise jaw oil, made from blackfish jaws was once greatly prized as a lubricant for small machinery. In the odontocetes the nostrils are joined to form a single blow-hole, and the passages leading from the hole to the larynx are often complicated with diverticula and valve-like plugs. The skull itself and its nasal passages are asymmetrical, often to an extreme degree; the asymmetry may have a function in echo-location by resolving ambiguities in orientation through producing an asymmetrical polar diagram. The epiglottis is tubular, and is inserted into the posterior nares and held there by pharyngeal and palatal muscles forming a sphincter, so that the channel for the breath is separated from the mouth and gullet. Squids form an important part of the food of toothed whales, as also do fish, but some species eat other things. The stomach in most species is complex, and consists of a number of separate compartments.

FAMILY ZIPHIIDAE

The family Ziphiidae contains five genera and eighteen species of beaked whales – not a very good name because although all the ziphiids are beaked, so are many other cetaceans. They are small to medium-sized whales of fairly slender build, with peculiar dentition, and two grooves diverging backwards on the throat below the jaw. The dorsal fin lies aft of the mid point of the back, and there is no notch at the centre separating the two halves of the tail flukes. These whales are especially squid eaters. Of the two species of *Hyperoodon*, the bottle-nosed whales, which grow to about thirty feet long, the northern one is hunted commercially, and formerly was the basis of an extensive fishery in the north Atlantic – the oil produced was known as arctic sperm oil. The bottle-nosed whales are functionally toothless; apart from some vestigial teeth embedded in the gums there is a single tooth on each side near the tip of the lower jaws, and it erupts through the gum only in old males – it is peculiar that a whale feeding on such active and slippery prey as squids should have no teeth. The melon is very large and prominent in the bottle-nosed whales, and also in the giant bottle-nosed whales of the allied genus *Berardius* in which there are a southern and a north Pacific species, the latter reaching a length of over forty feet – it is hunted commercially in Japanese waters. In *Berardius* there are two teeth on each side at the tip of the lower jaw; they are compressed laterally, and those of the front pair are larger than those of the second, but they cut the gum only in old males. The bottle-nosed whales are strongly social, and are generally found in close schools of a dozen or so – the north Atlantic species makes regular seasonal migrations northwards in the summer.

In all the other beaked whales the melon is much less conspicuous, so that the forehead and the beak are not so sharply separated. The goose-beaked or Cuvier's beaked whale, *Ziphius cavirostris*, the only species of its genus, reaches a length of about twenty-five feet. At the tip of the lower jaw there is one pair of teeth which cut the gum in adult males but do not erupt in females. This species has been recorded from practically all the seas of the world, but little is known of its life as it is seldom seen although, according to Omura and his collaborators [354], it is sometimes included in the commercial catch on Japanese and Norwegian coasts.

The genus *Mesoplodon* contains twelve species – more than any other in the family – but most of them are very rare and are known from only a few stranded specimens, probably because they are oceanic species that do not normally come close to land. The best known is Sowerby's whale *M. bidens*, an Atlantic species that has been cast ashore on both sides of the ocean on inhabited coasts where discovery is more likely than elsewhere. None of the members of the genus reaches any great size and generally they are between fifteen and twenty feet long. The most interesting point about these whales is the extraordinary series, well illustrated by Fraser [154], in the development of the single pair of teeth in the lower jaw shown by the various species. In True's beaked whale, *M. mirus*, of the north Atlantic the teeth are at the tip of the jaw and cut the gum in adult males. In the New Zealand beaked whale, *M. hectori*, they lie a little behind the tip, and in *M. europaeus*, a very rare species of the north Atlantic, they lie further back still. In Sowerby's whale and two little known southern species they lie about one third of the length of the jaw from the tip, as they do in the following three species. In *M. densirostris*, known from both Atlantic and Pacific, the teeth in the males are very large and laterally flattened and the part of the jaw forming the socket is correspondingly enlarged. *M. stejnegeri* from the Pacific also has large flattened teeth, but the jaw holding them is not unusually swollen. The series culminates in Layard's, or the strap-toothed whale, *M. layardi*, of southern seas, in which the teeth grow upwards and backwards out of the mouth somewhat like the tusks of a hog, and in old males form a loop over the upper jaw so that the mouth can be opened only to a limited extent. The teeth in all the species are male sexual characters – it is not known whether they are used in fighting or courtship, but the bodies of the animals are commonly marked with scratches presumably made by these teeth. Someday perhaps a fortunate biologist will find in what parts of the oceans the different species have their head-quarters, and will be able to study their lives by descending among them in an underwater research vehicle.

The final member of the family, *Tasmacetus shepherdi*, the only species of the genus, discovered on the coast of New Zealand as recently as 1937 and known from one skeleton and part of another, differs from all the other beaked ziphiids in having, in addition to a pair of large teeth at the tip of the lower jaw, a series of nineteen teeth in the upper jaw and twenty-six in the lower – a total of ninety, and a strong contrast to its almost toothless relatives.

FAMILY MONODONTIDAE

The family Monodontidae contains only two species, the narwhal and the white whale, comparatively small species which average about fifteen feet in length though some examples reach two or three feet more. They are arctic animals, and are found particularly at the edge of the arctic ice – consequently they move north in summer and south in winter as the ice breaks up or pushes south. Neither species possesses a dorsal fin, nor is there a beak below the rounded melon except for a slight projection of the upper jaw in the white whale. The young are pigmented at birth but the colour becomes lighter as maturity is reached.

The narwhal, *Monodon monoceros*, like the ziphiids, shows a remarkable sexual dimorphism in the development of the teeth, as the males are provided with the well-known unicorn tusk. The enlarged tooth, unlike that of the ziphiids, belongs to the upper instead of the lower jaw. Both sexes are toothless except for a pair of tusks pointing forwards in the upper jaw. In the females they are a few inches long and remain in their sockets without cutting the gum, as does that of the right side in males. The left tooth in males, however, grows out as a forwardly directed tusk up to eight feet or even more in length, ornamented with left-handed spiral grooves. Occasionally the right tusk grows out instead of the left, and very rarely both reach their full size; but whether the tusks are those of the left, or right, or both sides, the spiral pattern is always sinistral, and not opposite on opposite sides as might be expected. Very rarely, too, a tusk is grown by a female, probably because of some imbalance in the secretion of the sex hormones. What part the tusk plays in the life of the narwhal is quite unknown; all the obvious suggestions about its use – ice breaking, food digging, fighting and so on – prove invalid on consideration. The tusk, which consists of ivory, has been sought after commercially in the past, but its use is restricted because the pulp cavity extends far along its length so that there is a hole in a cross section of the tusk. These otherwise toothless animals feed on cephalopods, and to a lesser extent on fish and crustacea. They are gregarious, and commonly swim in small schools, but larger concentrations are seen during migration. The young narwhal is brownish, but as maturity is reached the skin becomes greyish white with darker grey spots.

In the white whale, or beluga, *Delphinapterus leucas*, the juvenile colouration disappears more completely, so that the adults are pure white, and are thus perhaps the most beautiful of all the whales. In shape the white whale is much like the narwhal except that it has the suggestion of a beak below the melon, and a very slight constriction behind the head suggestive of a neck; there are about nine conical teeth on each side of each jaw. White whales in captivity are gentle and docile; they will swim to a visitor offering a fish, and poke the fore part of the body out of the water to allow him to stroke the smooth skin – it is then noticeable that the blubber is much softer than in many whales. Observation of captive white whales has also shown that they can alter the shape of the melon considerably, presumably by muscular action, so that it is made more or less prominent. This action is almost certainly something to do with altering the focus or direction of the sonar emission; it is visible probably because of the softness of the blubber, and something similar may well occur internally in other species in which the harder blubber prevents visible external distortion. White whales are highly gregarious, and the usual small schools often join to make very large herds when the arctic ice breaks up. They come into inshore waters more than the narwhal, and enter the large arctic rivers in summer, often travelling many hundred miles upstream – indeed, a stray white whale found its way into the Rhine in 1966, and penetrated above Bonn before turning back and finally reaching the sea at Rotterdam. In the estuary of the St Lawrence there was for many years a fishery for white whales which were caught in a sort of enormous fish weir. They are hunted commercially in other parts of the arctic, and also form an important resource for eskimo subsistence fishermen and others. White whales eat a wide variety of food, but are particularly fish eaters. Odontocetes make many sorts of audible sounds in addition to their ultrasonic sonar, some of which at least, are certainly used in communication. These sounds are not generally heard by human listeners on, for instance, the deck of a ship because they are lost in the background noise of wind and wave. Some of the calls of the white whale, however, can be so heard, probably because the animals are then close to the surface and the hearer. The sound usually heard is a musical trill so characteristic that for over a century the animals were nicknamed 'sea canaries' by the old-time whalers.

FAMILY PHYSETERIDAE

The Physeteridae are a small family containing only two species; one of them, the sperm whale, *Physeter catodon*, is the largest odontocete. The sperm whale is perhaps the most widely familiar whale, owing to the great quantity of romantic literature about the old whaling industry which for a time made the sperm whale its main quarry. In its own way the sperm whale is as bizarre a creature as the bowhead among the mysticetes; its enormous barrel shaped head is about one third of the total length, the eye lies close above the angle of the long narrow underslung lower jaw in front of the broad rounded flipper. The dorsal fin is so broad and low that it resembles a hump, the first of a series of smaller ones diminishing in size towards the tail. The blowhole, single as in all odontocetes, opens at the apex of the front of the head, asymmetrically on the left side, and is of an elongated S-shape like the holes in the belly of a fiddle. In the skull the upper jaw roofs the mouth and sweeps back and upwards to the asymmetrical bones of the front of the cranium; the huge barrel of the head thus supported consists of soft tissues without bones. This part, the spermaceti organ, was known to whalers as the 'case', and consists of a mass of interlaced tendinous fibres filled with spermaceti; it is in fact no more than an enormously developed melon, and is probably an important part of the sonar system essential in the great depths to which sperm whales dive. The right nostril, though not used for breathing, is modified into a series of air sacs which, by providing an air-liquid interface, probably help in directing the sonar beam; they may also provide the source of the ultrasonic pulses. A mass of fatty fibrous tissue separating the spermaceti organ from the upper surface of the maxilla was known to the whalers as the 'junk'. The very narrow lower jaw carries about twenty-five large conical teeth; when the mouth is shut they fit into corresponding depressions in the gum of the upper jaw which has no functional teeth, though there is generally a number of rudimentary ones unerupted and lying unsocketed in the gums. The blubber of the sperm whale is very thick and hard – it may be over a foot thick in places – and the surface of the body bears irregular low longitudinal ridges, very unlike the smooth body surface of most whales.

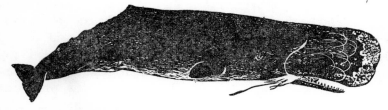

Figure 13 Sperm whale (*Physeter*).

The sexes of the sperm whale differ greatly in size; the males reach a length of about sixty feet, but the females little more than half that. The species is essentially an inhabitant of warm seas – it is found throughout the oceans of the world in all tropical and temperate waters, and concentrates on various feeding 'grounds' at different times of the year. Sperm whales are gregarious; schools may consist of females with young males and a few adult bulls, or of young adult males – the old bulls when not with the females tend to be solitary, and these lone bulls often migrate to high latitudes where they have been caught in some numbers. Sperm whales feed mainly on squid but also take fish; although they eat giant squid, and their skins are often covered with marks of giant squids' suckers and claws, the bulk of the food appears to be made up of smaller examples. They often hunt these creatures at great depths, for they are known to go to depths of several hundred fathoms, and they can remain underwater for up to an hour. The sperm whale was much hunted by the old sailing ship hand-harpoon whalers, especially in the first half of the nineteenth century, and now in the middle of the twentieth it has become important to the modern industry, particularly off the west coast of South America, owing to the growing scarcity of the rorquals. Sperm whale oil differs from that of the rorquals, and is not so suitable for making soap and margarine. Spermaceti is a clear liquid when in the case – it looks much like petrol – but it sets to a white wax on cooling; it was formerly believed to be the sperm of the male whale. Ambergris, thought to be the product of some pathological process in the intestine of sperm whales, is a waxy substance formerly of great price, used as a fixative in perfumery. Sperm whale teeth, and the hard bone of the lower jaw, used to be carved into souvenirs, known as scrimshaw, by the sailors in the old whale ships. Open boat sperm whaling with hand harpoons still survives in the Azores and Madeira.

The pygmy sperm whale, *Kogia breviceps*, is indeed a dwarf as it rarely exceeds twelve feet in length; there is probably only one species, though some workers such as Yamada [527], who has handled specimens in the flesh, and Handley [202] suggest there may be two. The general form of the body resembles that of a porpoise; there is a hooked dorsal fin, and the flippers are pointed. Although there is no huge melon as in the sperm whale, the front of the head is bluntly rounded, the blowhole is asymmetrical, the left nostril is the one used in breathing, and the head contains a spermaceti organ. Teeth are present only in the lower jaw, which carries eight to sixteen pairs of very sharply pointed teeth fitting into sockets in the gum of the upper. The small underhung jaw, and the eye on the side of the head far removed from the angle of the gape, give the pygmy sperm whale an appearance widely different from that of other odontocetes, and a remarkable superficial resemblance to a shark. The food of the species consists mainly of cephalopods, although some crustacea are also taken. The life of the pygmy sperm whale is little known; most of the specimens that have been examined were stranded ones, but a few are taken in the commercial catch made in Japanese waters. The stranded specimens show that the species occurs in all the oceans of the world.

FAMILY PLATANISTIDAE

The Platanistidae are a family of fresh-water and estuarine dolphins, found in South America and Asia; there are four genera, each with one species. All have small eyes – one is blind – in all the melon is prominent, and the jaws form a long narrow beak and carry numerous slender pointed teeth. *Platanista gangetica*, the susu, is confined to the river systems of the Indus, Ganges and Brahmaputra; it reaches a length of about eight feet, is dull black or grey in colour, and has a low ridge-like dorsal fin. The beak is long and very narrow, and the mouth contains a total of over a hundred teeth. The eyes are very small and have no lenses so that the animal is totally blind, a character unique among the cetacea. The melon is prominent and is backed by unusual bony plates in the skull; the susu must depend entirely on its sonar for finding its way about, though it is said to find its food, consisting of bottom living fishes and crustacea, by probing the mud with its beak.

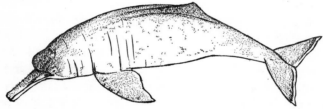

Figure 14 Susu (*Platanista*).

Inia geoffrensis, the bouto of the Amazon and Orinoco river systems, grows to about seven feet in length. The long dorsal fin is low and ridge-like, the flippers are long and pointed, the melon large, and the beak directed slightly downwards. The eyes are small but their vision appears to be good, though no doubt the sonar system is important when the animals are in muddy water. The food consists of fishes. During floods boutos leave the rivers and wander far among the forest. They vary considerably in colour from blackish above and lighter below when young, to an almost uniform whitish-pink in adults. The upper surface of the beak bears a number of scattered bristly hairs which are probably tactile in function; no other odontocete is similarly provided. The bouto is the subject of much superstition and Eskelund [138] says that the natives are unwilling to kill them, much more to eat their flesh. According to legend the males are said to be very fond of young girls and disguise themselves and come to dances at carnival time. 'It is not considered bad for an unmarried girl to become pregnant, for people believe that quite likely the bouto is responsible'. Legend apart, Layne [295] who studied river dolphins far up the Amazon found that they were rather sluggish in movement but could hustle when the occasion demanded. He also reported that they are inquisitive and will approach to look at a person drifting quietly in a canoe. He adds, 'This inquisitiveness was clearly evidenced on an occasion when two members of the party, swimming in a cove of one of the lakes, were investigated by a pair of boutos that were seen approaching from some distance away. The animals were unmistakeably interested in the swimmers and remained close by for a number of minutes. . . . Small *Inias* were much tamer than the larger adults'.

Lipotes vexillifer, the white flag dolphin, is known only from Tung Ting lake, 600 miles up the Yangtze Kiang river in China where it is, or was, numerous, though very few specimens have been sent to Europe or America. It was first described by Miller [324] in 1918, and a frozen specimen sent to London in 1922 was dissected by Hinton and Pycraft [231]. The species much resembles the susu externally, but the ridge-like dorsal fin is larger and gives the animal its name because it is conspicuous when the animal comes up to breathe. The long narrow beak contains about 130 teeth, the melon is prominent, and the eye small though functional. The blowhole is asymmetrically placed on the left, and there are extensive asymmetrical air sinuses connected with it; they are probably part of the sonar system. The asymmetry extends beyond the head to much of the skeleton so that Hinton [230] remarked that the whole body seems to have been twisted round inside its jacket. He added that in front view the face is so twisted that '*Lipotes* is by far the ugliest of all the cetaceans'. C.M.Hoy, who collected the first specimen in 1916 told Miller who described it in 1918 that 'the sudden appearance of a school of these whitish dolphins close to a small boat is very startling. . . . They are often seen working up the mud in their search for fish. The one I killed had about two quarts of catfish in its stomach'. In winter when the lake is low they are seen in great numbers usually in bunches of three or four, sometimes as many as ten or twelve. When the water level rises in summer the dolphins disperse and are said to breed in the small clear tributary rivers.

The La Plata dolphin, *Pontoporia blainvillei*, is the only member of the family not confined to fresh water; it lives in the estuary of the River Plate and in winter migrates along the coasts of Argentine and Brazil between latitudes 30° and 45° S. It rarely grows to as much as six feet in length, has a long beak and prominent melon, and its dorsal fin is more triangular and less ridge-like than that of other members of the family. The jaws carry up to sixty or more teeth on each side above and below making a total of nearly

250, the largest number of teeth found in any mammal. The dolphins are said to feed mainly on fish and the sharp-pointed slender teeth are noted by Kellog [263] as being 'marvellously adapted for catching and holding soft-bodied prey' – but what of the toothless ziphiids which feed on equally slippery and soft-bodied prey?

FAMILY STENIDAE

The family Stenidae contains three genera; its members live in the temperate and tropical waters of all seas, and in addition some species ascend estuaries and rivers, and some are inhabitants of freshwater only. They are dolphins of moderate size, few exceeding or reaching seven feet in length, with well developed but not extremely long beaks, rounded melon, and triangular dorsal fin. The genus *Steno* contains a single species, *S. bredanensis*, the rough-toothed dolphin, in which the teeth instead of being smooth are furrowed. Specimens have been stranded on the tropical and temperate coasts of all the oceans, but nothing is known of its biology. Several species of the genus *Sousa* live in coastal waters of India, southeast Asia and southern China. They are all pale in colour, the darkest being grey rather than black, *S. borneensis* is white dappled with grey, and *S. chinensis*, which lives in the Yangtze river as well as Chinese coastal waters, is milky white. *S. teuszi* is common in the river estuaries of West Africa, but is not, as once thought, herbivorous – it feeds on fish. Of the two species of *Sotalia* one, *S. fluviatilis*, the bufeo, is found only in the Amazon and Orinoco river systems. It is a small species, and varies greatly in colour from almost black to almost white; it is very distinct from the river dolphin *Inia*, the bouto, which lives in the same rivers with it. The other, *S. guianensis*, is a marine species of the coasts and estuaries of north-east South America; it is common in the bay of Rio de Janeiro where its presence is welcomed because dolphins are believed to drive sharks away.

FAMILY PHOCOENIDAE

The family Phocoenidae contains the true porpoises, which have no beak extending beyond the melon. In Europe the name porpoise refers to *Phocoena phocaena*, the common porpoise; in America, where *P. phocaena* is called the harbor porpoise, the name porpoise is also applied to the beaked delphinids called dolphins in Europe, and even to the white whale – the resulting confusion is unfortunate. The common porpoise, like the other members of the family, lives in coastal waters and often ascends the estuaries of large rivers. It is a small species never exceeding and rarely reaching six feet in length, black above and white below, the two colours

R

fading into each other through grey. The dorsal fin is triangular and, as in all members of the genus, the teeth are not conical but spade-shaped; the food consists almost entirely of fishes. Rae [387] found that porpoises in the North Sea feed mainly on herrings and small whiting, and he quotes other authors' reports that in the Baltic sea herring, sprats and small cod under twelve inches long are eaten, and that off eastern Canada herrings are the principal food, though other fishes and squids are also taken. The species occurs on both sides of the north Atlantic and the east side of the Pacific. Porpoises were formerly esteemed as food in Europe, and in the middle ages they were in demand for consumption particularly during Lent because they were considered to be fishes. They were also valued for the oil extracted from them. Large fisheries existed on several coasts, and one still survives in Denmark where porpoises are netted during the winter as they migrate through the Little Belt from the Baltic to the warmer water of the North Sea (Plate 16).

Of the three other species, all from the southern parts of the oceans, the spectacled porpoise, *P. dioptrica*, has a white ring round the eye, and the black of the back is sharply divided from the white of the underside, and *P. spinipinnis* is entirely black. The two species of *Phocoenoides*, *P. dalli* and *P. truei*, are handsomely marked black and white porpoises of the north Pacific, from the arctic to Japan and California; in both species the teeth are very small. Both species are found in Japanese waters where some are taken in the large commercial catch of dolphins. Wilke, Taniwaki and Kuroda [506] found that, although Dall's porpoise is the more northerly, their ranges overlap by 200 miles, but that they always school separately, and that True's porpoise ate mainly lantern fish, whereas Dall's porpoise fed principally on squids. On the other hand Cowan [85] found that the stomachs of four Dall's porpoises from Queen Charlotte Sound, British Columbia, were full of herrings. Cowan reported that Dall's porpoise is common in the many channels between the islands flanking the coast of Vancouver Island and in Alaskan waters. Yocom [528] also saw the species in Californian coastal waters, but it does not go far beyond the continental shelf for he saw none during many weeks observation between four and eight hundred miles from the coast. Benson [34] suggests on ecological grounds that Dall's and True's porpoises may be but one species.

The finless black porpoise, *Neomeris phocaenoides*, is the only species in its genus; its colour is uniformly dark grey, it has no dorsal fin, and its melon is more rounded and protruding than in the other porpoises. Romer [408] points out that the name of this porpoise is misleading because it is not black when alive; two specimens he examined 'were of a steel-grey colour in life, becoming entirely black some hours after death'. It is a small species that does not exceed five feet in length. It inhabits the Indian Ocean, the seas of the East Indies and the coasts of east Asia to Japan. It is a coastal and estuarine species, and ascends the Yangtze river to the Tung Ting lake where it is found with *Lipotes*, and beyond to 1,000 miles from the sea. Blandford [44] records that it does not herd into schools and is slow in its movements and 'is, on the whole, a sluggish little porpoise'.

FAMILY DELPHINIDAE

The Delphinidae are a large family with fourteen genera and over thirty species; they are typically the beaked true dolphins, though in some genera the beak is small or even absent. The size ranges greatly from that of several species about six feet long to that of the killer whale some thirty feet long. It is members of this family in particular that have been brought into captivity, and trained to give amusing displays in the huge aquaria built in several countries during the last twenty years – the bottle-nosed dolphin *Tursiops truncatus* was the first species so exploited, and is still the commonest performer. A few members of other families have also been similarly subjugated – specimens of *Inia* have been flown from far up the Amazon to an aquarium in Florida, and the white whale has often been exhibited. It must not be thought, however, that the public exhibition of cetaceans is anything new, for over a hundred years ago trained white whales and bottle-nosed dolphins were giving performances in Barnum's 'museum' in New York, and captive common porpoises lived for several months in England at the Brighton aquarium. In 1861 and 1862 Barnum brought white whales from the St Lawrence river to New York and one, a ten foot male weighing 700 pounds, lived there for two years. Wyman [525] reported that

it became so tame that it would allow itself to be harnessed to a car in which it drew a young lady round the tank. It learned to recognise its keeper, would allow itself to be handled, and at the proper time would come and put its head out of the water to receive the harness, and take food from the hand. . . . It was, however, less tractable than a bottle-nosed dolphin which was for a time its companion in the tank.

In 1877 and 1878 several white whales were brought to England and exhibited at the former Westminster aquarium, in a tank forty feet long, twenty feet wide, and six feet deep, and in shows at Manchester and Blackpool. These whales travelled from the St Lawrence to England in boxes packed with wet seaweed, and broke their journey at New York where they lived in the aquarium at Coney Island while awaiting their ocean passage. None of them survived for long after their arrival – if antibiotics had then been available they might have recovered from the pneumonia that killed them. The study of captive cetaceans in modern aquariums has led to the discovery of their echolocation and much else about their biology (Plate 15).

SUBFAMILY ORCINAE

The subfamily Orcinae contains six genera which differ from those of the other subfamilies in having no beak, or the merest trace of one shown by a groove separating the upper lip from the melon. The pilot whales or blackfish – two species of *Globiocephala* – are examples of the last arrangement, and take their generic name from the large melon that bulges out above the jaws which project forward slightly below it. Blackfish grow to a maximum of nearly thirty feet in length; the dorsal fin is not high but its base is long, and the flippers are narrow, long and pointed. Blackfish are found throughout the oceans of the world, and although several species have been named it is doubtful whether there is more than one widely distributed species, *G. melaena*. The colour is black with a light patch beneath the chin, and often some other white markings on the under side, except in the animals of the north Pacific which generally lack the white markings and have consequently been named *G. scammoni* – a distinct race if not a distinct species. The blackfish of the warmer waters of the Atlantic and Indian oceans, *G. macrorhyncha*, are entirely black with no markings and have shorter flippers and fewer teeth than *G. melaena*. Blackfish are highly gregarious, and often occur in very large schools; this habit of keeping close together gives the chance for fishing them commercially in some parts of the world. In the Faroes, and formerly in Shetland and Orkney, when a school approaches the land a fleet of small boats gently drives them into a bay or fjord and, when the animals are safely enclosed, panics them so that they strand on the beach. Thereupon the whalers slaughter the animals for their oil and meat. The techniques of such 'grinds' as they are called in Faroe, are of great antiquity, and the methods of dividing the spoils among the participants follow ancient rules.

Blackfish are hunted commercially in many parts of the northern hemisphere, including Japanese waters, both for their oil and for their meat which is used for human consumption as well as for making canned pet food and for feeding ranched fox and mink. Blackfish are migratory, at least in the north Atlantic, for the largest catches of the Faroe hunts are made during the summer months. The blackfish of the north Pacific is one of the cetaceans that have been brought into captivity and trained to perform tricks for the amusement of onlookers and themselves. The Irawadi dolphin, *Orcaella brevirostris*, resembles the blackfish in the shape of its head, with a large melon bulging above the upper jaw. It is a small species, about seven feet long, with a rather small dorsal fin and broad flippers, grey in colour, slightly lighter beneath. It is found in the seas of south-east Asia, and in the Irawadi river as far as 900 miles from the sea. *Feresa attenuata*, a species so little known that it has no vernacular name, is rather similar in shape, but has a comparatively larger dorsal fin and less bulbous melon. A few specimens are known from the Atlantic and Pacific, but in 1963 some Japanese whalers found a school of fourteen and drove them gently thirty miles to their port and netted them. Nishiwaki [248] says 'the porpoises moved slowly and were very tractable'. They were transferred to the Ito aquarium, but all except one refused to feed and died within a week. The one with an appetite ate sardines, squid, sauries and horse mackerel, but it too died, of pneumonia, after three weeks, so that little beyond their details of their anatomy could be recorded.

Members of the genera *Orcinus* and *Pseudorca* each containing a single species, have no constricted beak and no protruding melon – the profile of the head runs in a smooth curve from the tip of the snout and merges into the back as in the common porpoise. The well-known killer whale, *Orcinus orca*, is the largest of the dolphins; adult males reach a length of thirty feet whereas females are only about half as long. In adult males, too, the flipper and the dorsal fin are very large so that the difference between the sexes is great; both sexes are handsomely marked in a pattern of black and white. There are up to a dozen large pointed teeth in both sides of each jaw which give the mouth a formidable appearance. Killers are generally found in small schools, but as many as forty have been recorded swimming together. They are found throughout the oceans of the world from the polar seas to the tropics. They feed on fish, squids, sea birds, seals, porpoises and dolphins – they are the only cetaceans that prey on warm-blooded vertebrates – and have a wide reputation for fierce aggressiveness, which some writers tend to exaggerate. Slijper [449] gives a dramatic diagram of a killer with thirteen porpoises and fourteen seals below it which he says were found in its stomach, but if he had consulted the original account of this finding, published by Eschricht [137] in 1862, he would have seen that 'most of them were half digested, or already fallen to pieces, some only remaining in the shape of loose parts of the skeleton'. A beef-eating man might eat a dozen bullocks – but not at one meal.

Killer whales are often said to kill and feed on the larger whalebone whales; the story goes that they attack the large mysticete by biting its lips and tearing out its tongue. This yarn has been copied from book to book for over two centuries, and has been used by natural history illustrators as a subject for dramatic drawings. The story seems to have started from the letter sent by Dudley [119] to the Royal Society published in 1725, describing whaling on the coast of New England. He says that killers

set upon a young whale, and will bait him like so many bull-dogs; some will lay hold of his tail to keep him from threshing, while others lay hold of his head, and bite and thresh him, till the poor creature, being thus heated, lolls out his tongue, and then some of the killers catch hold of his lips, and if possible of his tongue; and after they have killed him they chiefly feed upon the tongue and head, but when he begins to putrify, they leave him.

He also says they 'often kill the young ones, for they will not venture upon an old one, unless much wounded'. Scammon [416] 150 years later saw killers attacking a grey whale and her calf in a lagoon of Lower California; the calf was killed, sank to the bottom, and was eaten by the killers but the parent, though bitten, 'made her escape, leaving a track of gory water behind'.

Copiers of this story have overlooked the point that it is the young whale, not the adult, that is killed and eaten. Both Dudley and Scammon tell that killers commonly feed on dead whales that have been killed by whalers, and modern whalers of this century confirm this observation. Indeed, a story is told of a killer actually helping whalers; one, known as Old Tom, of Twofold Bay, about thirty miles north of Cape Howe on the boundary between New South Wales and Victoria, was said to have helped drive right and humpback whales into the bay where, after the whalers had killed them, Old Tom was allowed to eat the lips and tongue of the dead whale.

On the other hand Hancock [201] saw killer whales attacking a minke whale in a bay off the coast of Vancouver Island, and he thought they killed it by drowning it; he saw no blood in the water, but a thin film of oil spread over the waters of the bay. The killers, however, did not eat much of the minke whale, for the body was recovered and towed ashore. Only the dorsal fin, the tongue, and the flesh of the lower jaw were gone, but the skin had been neatly peeled off the surface of the body leaving the blubber intact. It is difficult to understand how the killers skinned the minke whale, and it is surprising that they did not eat the blubber or flesh. Morejohn

[336] saw killers harassing grey whales off the coast of California, but they broke off the attack, and apparently killed none, when the grey whales with their calves swam close inshore into shallow water.

Killers, like other whales, sometimes poke their heads more or less vertically out of the water in order to look at things, such as a boat, on the surface – they, as the others, also do this when among broken ice on coming up to blow because the area of open water is restricted. It is not proven, however, that they try to upset ice floes to throw the seals sleeping upon them into the water, and the account of their similarly trying to obtain a human victim seems to be based on a mistaken interpretation of what was seen. Nevertheless, killers are certainly voracious carnivores which eat, among other things, seals and the smaller cetaceans up to the size of white whales, and the calves of mysticetes – and yet, when in recent years adult killers have been brought into captivity they have entirely belied their reputation as ferocious creatures. Several killers have been caught and exhibited on the Pacific coast of America, and they have turned out to be docile, tractable, and able to learn circus tricks. It is truly astonishing to see the trainer swimming in the water with one of these monsters, riding upon its back and thrusting his arm into its frightening mouth as he feeds it with large fish. The killer could easily bite him in two and swallow the bits – but no tragedy has yet happened. Can it be that the killer knows it must not misbehave? There is, however, reason to think that some of these captive killers, though of large size, are not sexually mature, and that sexually adult killers show behaviour more in keeping with their reputation. Bull killers in the north Pacific have recently been found to be feeding on seals, sea-lions, porpoises, minke whale, as well as large fishes, and apparently co-operating in harassing and killing their prey [399].

The false killer whale, *Pseudorca crassidens*, is smaller than the killer; it never reaches twenty feet in length and is generally about twelve to fifteen feet long. It is all black, has narrow pointed flippers, hook-shaped dorsal fin, a slightly underhung lower jaw, and about ten large teeth on each side of both jaws. When Owen [362] described this species in 1846 from a subfossil skeleton found in the great fen of Lincolnshire near Stamford in 1843, it was thought to be extinct, until Reinhardt [400] confirmed Owen's speculation that it might still be living. He reported a large shoal of living ones in Kiel bay in 1861 of which one was captured, and several more specimens stranded on the shores of Denmark in 1862. Since then it has been found in all the oceans of the world; it is gregarious and often goes in large schools. When exceptionally a school comes into shallow water the animals sometimes panic and get stranded in large numbers, even apparently jostling each other to come ashore, and returning to the beach if put into the water by people trying to help them.

False killers are fish-eaters, and seem to prefer large species which they do not swallow whole but tear to pieces, discarding the head and skin – they have even become a nuisance to 'sport fishermen' in the Gulf of Mexico by taking large game fish off their lines before they can reel them in. When a school of whales finds a good shoal of fish and starts feeding the animals are noisy and give out piercing harsh whistles which can be heard several hundred yards away. Two or three false killers have been kept in captivity where they have proved to be very gentle; almost at once they took food from their keeper's hand, and they quickly made friends with cetaceans of other species confined in the same tanks. They were easily trained; indeed, they learnt many tricks without training, merely by imitating what their companions did. All kinds of foreign objects dropped or thrown into dolphin tanks by visitors are a perpetual menace to the lives of the dolphins, which pick up and swallow things; false killers seem particularly tempted by them. Brown and the Caldwells [55] reported that the female false killer in their charge was 'peculiarly prone to playing with objects of this kind. Fortunately, however, when observed indulging in this habit she will usually swim to the feeding platform when summoned and permit removal of the object from her mouth. Combs, pens, plastic toys, coins and a variety of objects have been recovered in this manner'. Like the true killer, the false killer is voracious when chasing its prey, but unexpectedly docile in its behaviour towards man when in captivity. *Peponocephala electra*, very similar in shape to the false killer but only about seven feet in length, entirely black or very dark grey except for lighter areas below, and white lips, is a rare species that has been sighted in the tropical and temperate waters of all the oceans – Nishiwaki and Norris [349] have suggested the vernacular names Hawaiian porpoise or many-toothed blackfish for this species.

SUBFAMILY LISSODELPHINAE

Two species of right-whale dolphin make up the genus *Lissodelphis*; they are small species, six to eight feet long; the absence of a dorsal fin gives them their name. They have a well-defined, but not very long beak. *L. borealis* of the north Pacific is mainly black with a white area beneath; it forms a large part of the catch of the Japanese dolphin fishery in which many hundreds of right whale dolphins are killed annually – they are used for food, oil, leather and fertiliser. Wilke, Taniwaki and Kuroda [506] record that the Japanese hunt dolphins 'up to thirty miles offshore, from twenty to thirty ton, motor-driven, harpoon vessels with a platform projecting from the bow for shooting and harpooning. Most animals are killed by shotgun but are harpooned to forestall sinking'. *L. peroni* which lives in all the southern oceans between approximately 40° and 50° S, is brilliant shining white with a cloak of black, demarcated by a sharp line, thrown over the upper parts from the blowhole to the tail. Little is known of the life of this dolphin and indeed it has seldom been seen; I can never forget the sight of two of these beautiful little creatures, in their clear-cut quaker-pattern dress, revealed for a few brief moments in the mountain-side of a huge south Atlantic swell on the edge of the roaring forties – the image is as bright now as it was forty and more years ago.

SUBFAMILY CEPHALORHYNCHINAE

The genus *Cephalorhynchus* contains four species about which little is known. They are small porpoise-like dolphins with ill-defined beaks but with a dorsal fin; they are patterned with black and white. They are known from the waters of the southern oceans, off South Africa, New Zealand and South America. Commerson's dolphin, *C. commersoni*, is the species that has been most frequently seen; it lives round the coasts of southern South America, and in the Straits of Magellan. The head and flippers are black, as is the back from the front of the dorsal fin to the tail, there is a white patch under the chin and a black one surrounding the vent; the black and the white areas are sharply defined. Heaviside's dolphin, *C. heavisidei*, from the other side of the south Atlantic, with smaller white areas confined to the flanks and underside, has somehow gained the book-name of 'tonine' which in fact does not belong to it specifically; *tonina* in Spanish, *toninha* in Portuguese, and *tonijn* in Dutch are imprecise vernacular names applied to any porpoise or dolphin – Heaviside's dolphin is a tonine, but not the tonine.

SUBFAMILY DELPHININAE

The Delphininae or true dolphins, although often called porpoises in America, are the last subfamily of cetaceans; they comprise six genera with about twenty species between them. All, with one exception, have a beak, though in some it is not very well defined, and all have a dorsal fin. Risso's dolphin, *Grampus griseus*, the only species in its genus, is the beakless form in which the melon rises almost vertically from the snout. It grows to about a dozen feet or a little more in length, the dorsal fin is high and pointed, and the flippers are rather narrow. The colour is grey, darker on the flippers and flukes, shading to lighter below; the skin is generally covered with long white lines which are said to be made by the teeth of companion dolphins of the same species; though why Risso's dolphins should bite each other more than other species is not explained. It has also been suggested that the marks are made by the cephalopods on which Risso's dolphin is believed to feed exclusively but they do not resemble the cephalopod marks found on sperm whales. On the other hand the teeth of the dolphins could make the marks, for they number not more than about half a dozen, and often fewer, on each side at the end of the lower jaw – the upper jaws are toothless. The species has been recorded from all the oceans, and is found in small schools of up to a dozen or more. The famous dolphin named Pelorus Jack that for about twenty years up to 1913 accompanied ships through French Pass on their way between Wellington and Nelson in New Zealand, and was given official protection by an Order in Council in 1904, has generally been identified as a Risso's dolphin; recently, however, it has been suggested that he – or she – was a large bottle-nosed, *Tursiops truncatus*. The question is not likely to be settled, for he disappeared over half a century ago.

The best-known species of the genus *Delphinus* is *D. delphis*, the common dolphin, often called the saddle-backed dolphin in America, found in warm and temperate seas throughout the world. It is a slender dolphin with a well-defined narrow beak; it grows to about eight feet long. the colour is black above and white below with irregular bands of grey, yellow, brown and white running obliquely on the sides. A black ring surrounding the eye, from which a dark streak runs towards the snout, gives it a facial expression pleasing to human eyes, and is a feature commonly accentuated by the artists of ancient Mediterranean civilizations with whom the dolphin was a favourite subject in painting and sculptural ornament. Common dolphins feed on shoaling fish near the surface – their long beaks carry up to fifty pairs of small sharp teeth in both jaws. They are very gregarious and are often found in schools numbering many hundreds; in the Black Sea they are the object of an extensive fishery. Many other species have been described as belonging to the genus *Delphinus*, but most of them are probably no more than subspecific races of *D. delphis*. The genus and species *Lagenodelphis hosei* were established by Fraser [154] in 1956 as a result of examining the skeleton of a dolphin found on a beach in Sarawak about 1895; it has structural affinities with *Delphinus* and *Lagenorhynchus*, but nothing is known of it in life.

The genus *Lagenorhynchus* contains at least six species – there may be more, but some species are not well known and the synonymy is confusing. The dolphins of this genus have a rather poorly defined beak, a comparatively large dorsal fin, and a well-marked ridge or keel along both upper and under sides of the tail stock; they are marked with distinctive patterns of black and white. They grow to seven or eight feet in length, some rather less, and have from twenty-five to over thirty pairs of teeth in both jaws according to species. The white-sided dolphin, *L. acutus*, and the white-beaked dolphin *L. albirostris*, are both common in the north Atlantic, where they are often seen in very large schools – their common names give their distinguishing features. The rather similar striped dolphin, *L. obliquidens*, of the north Pacific, is also gregarious, and with the northern right-whale dolphin forms most of the catch in the Japanese dolphin fishery. *L. obscurus*, the dusky dolphin is common throughout the southern Atlantic and Pacific, where it is often seen in large schools; much of its light-coloured areas, like that on the rear of the flank in *L. acutus*, is creamy yellow in tint. *L. australis*, from the waters of southern South America, is similar, but the patches on the sides are greyish.

L. cruciger, the most striking of all the species, is black along the back, white on the sides and below, with black flippers and a black band on the sides extending upwards to meet the black of the back below the dorsal fin – the effect when seen from above is of a white dolphin with a black cross on its back; it is found in the southern oceans as far south as the antarctic pack ice. There is much uncertainty about the status of some of the described species; Bierman and Slijper [40] showed that the patterns of many *Lagenorhynchus* species are very variable; for example, the white-beaked dolphin is sometimes black beaked. Their rather wholesale lumping of species was disputed by Scheffer [420], for the striped dolphin *L. obliquidens* of the north Pacific is separated by thousands of miles from the allied species of the southern oceans.

It is surprising that so little is known of the biology and relationships of dolphins that are so numerous. Some of them, such as *L. albirostris* and *L. acutus*, are frequently stranded in some numbers on the coasts of Europe; Jonsgård and Nordli [261] record the capture of large schools of *L. acutus* from time to time in the fjords of Norway between 1745 and 1952. The fishermen took advantage of the approach of schools numbering up to a thousand to surround and kill them for their oil and other products. A shoal of fifty-two animals was driven into the harbour of Kalvåg on Norway's west coast in March 1952: 'this was a relatively easy task because they kept together and at once turned away when a boat drew near'. The school was surrounded with nets and the animals harpooned and clubbed; 'both meat, blubber and bones were taken care of and subsequently sold. The meat, which weighed altogether 3,162 kg, was used as fox food.

The weight of the blubber and bones was about 1,300 and 1,800 kg respectively'. Some of the species are probably migratory; the north Atlantic species eat herrings, mackerel and whiting as well as cephalopods.

Stenella is another genus, most of the species in which are rare and imperfectly known, with consequent nomenclatural confusion. The beak in this genus is long and narrow, the dorsal fin is fairly large and hooked, and the flippers slender; the small pointed teeth range in number from about three dozen to over fifty pairs in each jaw. These dolphins inhabit the temperate and tropical parts of all the oceans. Most of the species are dark above and light below with no sharp demarcation, and with dark longitudinal stripes or oblique bands on the sides, but some are more uniformly coloured and speckled with dark or lighter spots. There are probably at least seven distinct species; they usually keep away from coastal waters and feed, often in large schools, on oceanic fishes and cephalopods.

Two species, *S. longirostris* and *S. roseiventris*, the spinner dolphins of the Pacific, have a most remarkable habit; like many other dolphins they jump clear of the water – 'porpoising' as it is called, though the common porpoise seldom does this – but instead of plunging straight back, when they shoot out obliquely they spin on the long axis of the body before diving in nose first. They can complete two 'barrel-rolls' as they spin in the air clear of the water – a school of some hundreds of these dolphins leaping into the air and spinning before splashing back into the water is an extraordinary and exciting spectacle. Although this habit was long known to the tuna fishermen of the Pacific it was not known to biologists until reported in 1963 by Hester, Hunter and Whitney [227] who made a beautiful film of the spinning. They found that the dolphins 'often made several spinning jumps in sequence. Clockwise and counter-clockwise rotation appeared to be equally prevalent. The same animal sometimes rotated in a reversed direction on subsequent leaps'. At first these authors thought the spinning might be to detach remoras, but only some of the dolphins carried these peculiar sucker-fish; they also rejected courtship display as a motive because both mature and immature animals of both sexes indulged in spinning. There seems to be no reason for this acrobatic performance beyond the assumption that the dolphins enjoy doing it.

Finally there is the bottle-nose dolphin *Tursiops truncatus*, the best known and most popular of all the smaller cetacea. Although some dozen or so species of cetaceans have been exhibited in captivity, it is the bottle-nose dolphin that is the mainstay of the performing dolphin shows, one of the most amenable and tractable species – the striped dolphin *L. obliquidens* can be trained to magnificent displays of jumping, but *T.*

truncatus is probably one of the most versatile species. It has a short stout beak – hence its name – a hooked dorsal fin, and rather broad tapering flippers; the colour is black or slaty-grey on the back and tail, lighter below. Bottle-nosed dolphins are more stoutly built than the slender-bodied species of the last two genera mentioned, and reach a maximum length of about twelve feet. This dolphin is a coastal rather than an oceanic species, and is found usually in small schools of about half a dozen in all the seas of the world apart from the coldest. *T. truncatus* is very common along the Atlantic coast of America, and there was formerly a commercial fishery for the animals at Cape Hatteras, where they were caught in nets. It was no doubt the ready availability of the species that led to its being captured for exhibition in the pioneer marine aquariums, a fortunate choice as it turned out. Some members of the species seem to be specially ready to enter into friendly relationships with man; Pelorus Jack, already mentioned, may have been a *Tursiops* and not a *Grampus* as was generally assumed. More recently a young *Tursiops* became a tourist attraction at Opononi beach, New Zealand, where she associated with human bathers, playing ball with them and even allowing children to ride on her back – like Pelorus Jack she was the subject of a preservation Order in Council, but just as the Order was published she was stranded on the rocks and died. These examples of spontaneous friendliness give credibility to some of the stories of antiquity about friendship between men and dolphins, though in the world of Mediterranean civilization the species concerned was probably *Delphinus delphis*, an animal which is rather highly strung and difficult to keep in captivity.

Dolphins of many species help a comrade in trouble by pushing it to the surface so that it can breathe, as a mother dolphin does with her newborn young, and dolphins are said to have similarly helped drowning men. As recently as 1966 the press reported that a man was saved from drowning in the Gulf of Suez by a school of dolphins which 'carried him on their backs and protected him from sharks'. The authors of antiquity stated that some fishermen were able to call dolphins to them, and in modern times such happenings have been confirmed. Sir Arthur Grimble [186] tells that a head man in the Gilbert Islands was able to call large schools of dolphins to the shore, where the natives kill them for food. In 1966 Dawbin [105] saw canoes in the Solomon Islands going as far as ten miles out to sea in search of a species of *Stenella*; when they found the dolphins the men in the canoes reached over the side and clashed two heavy stones together underwater, and the noise drove the dolphins towards the shore. When near the beach the animals suddenly plunge vertically and bury part of their heads in the softish sandy mud bottom. Here they remain with tails oscillating above the surface ... hundreds of natives dash into the sea shouting and waving and ... by grabbing beak, fin or flipper can escort them passively without a struggle to the shore. Within minutes there may be hundreds lined up along the beach.

The natives eat the meat, and extract the dolphins' teeth, which are used as currency.

In captivity bottle-nose dolphins readily learn many tricks; they also play with foreign objects thrown into their pools, and even invent games of their own. It is interesting that they learn tricks not only from their trainers but by observing and imitating the actions of animals already trained. An astonishing example is recorded by Brown, Caldwell and Caldwell [55]; a *Tursiops* caught in the Atlantic was placed in a tank with a spinner *Stenella* from the Pacific – she could never have seen a spinner spinning, yet almost at once she copied her.

The spinning by the *Tursiops* was observed only minutes after she was placed for the first time in a show tank with the *Stennella*. The *Tursiops'* spinning leap was made almost immediately after the *Stenella* had made her spin upon a previously reinforced cue ... she was clearly learning to make the spinning motion without human instruction. Although not polished, the spin out of water consisted of almost $1\frac{1}{2}$ complete revolutions. ...

These authors add that although the striped dolphin *L. obliquidens* jumps far out of the sea in the wild, the high leaps of a team of captives in unison when cued by their trainer 'seem at least in large part to be a result of observational learning by new animals from old residents'. It is these natural capacities for being trained that have made dolphins give such extraordinary co-operation to biologists investigating their sonar and behaviour, subjects discussed in detail in Volume I (Plate 15).

A last thought on dolphins in captivity; if, as claimed, they are extraordinarily intelligent, why do they docilely submit to being forced to swim round and round in a tank and perform tricks, apparently with enjoyment, before hordes of gaping bipeds, instead of splashing about at liberty in the ocean? One would expect a sensitive intelligent creature would be more likely to pine and die.

Chapter 9

Beasts of Prey

IN SIMPSON'S classification of the mammals the carnivores contain two suborders, the Fissipedia and Pinnipedia, the land carnivores and the seals. Many systematists now prefer to treat them as separate orders, but here they are placed in separate chapters more for convenience than from a conviction that the two closely related groups should not be regarded as forming a single order. Although the fissiped Carnivora are typically meat-eaters and predators on other animals, not all of them are exclusively carnivorous, and some are solely herbivorous. There are about 250 living species, classified in seven families, and members of the order inhabit all the continents and adjoining islands throughout the world.

The jaw joint is transverse hinge, which facilitates biting and cutting, but does not permit a grinding action of the teeth. The canines are large and pointed, and the fourth upper premolar and first lower molar teeth are enlarged and used in shearing, especially in the cats – these teeth are known as the carnassials. In spite of the fact that one species has an unusually numerous dentition the number of premolars and molars tends to be reduced, and is smallest also in the cats. The anal scent glands are well developed in most of the carnivores, and are used in marking territory, social recognition, or for defence; some, such as the foxes (*Vulpes*), have in addition a scent gland on the upper surface of the tail. Nearly all the carnivores are clothed in soft fur beneath longer guard hairs, and some of them are the bearers of the most valuable furs in commerce. The gait of the carnivores ranges from plantigrade to digitigrade. The superfamily Canoidea, the dog-like carnivores, contains the wolves, foxes, bears, raccoons, badgers and weasels; the Feloidea contains the cat-like forms, the civets, mongooses, hyaenas and cats.

FAMILY CANIDAE

Most of the members of the family Canidae, which contains the dogs, wolves and foxes, use their voices in communication; their cries range from barks and howls, to whimpers and snarls. The canids are world-wide in their distribution and are particularly varied in South America – of the fifteen genera totalling about forty species in the family no less than six, totalling fifteen species, are peculiar to that continent. The genus *Canis* includes the wolves and jackals, and the domestic dog, probably descended from various races of the wolf. *C. lupus*, the wolf, was formerly widely distributed throughout the palearctic and nearctic regions but has been greatly reduced in numbers because it is a destructive predator on domestic animals. Nevertheless the wolf was reported in 1953 to be increasing in numbers in Scandinavia, Canada and Alaska, and it has since also increased in eastern Europe. The destruction of wolves to increase the numbers of deer for recreational killing has proved to be a mistaken policy in many parts of America; the deer population has exploded so much that the animals have destroyed their feeding range. Leopold and Darling [296] emphasise that 'deer irruptions all over the United States attest to the fallacy of closely regulating hunting and predation without reference to range limitations'. The wolf is a highly social animal, and lives in groups based on the family; a loose combination of family groups makes a pack – but the large packs of travellers' tales and dramatic illustrations are more legendary than real. Wolves maintain a social hierarchy within their groups, and the domestic dog, inheriting the social instinct, owes to this its ability to take its place in human society by treating its master as the dominant individual of its group – though some dog lovers seem to allow the relationship to be reversed.

In contrast to the reduced numbers of the wolf, those of the smaller coyote, *C. latrans*, of North America have increased, and its range has extended in spite of intense trapping and poisoning by man; the coyote is one of those species that can survive in settled country. Although coyotes occasionally prey upon animals as large as deer or bighorn sheep, such animals can generally defend themselves against attack. Wolf and O'Shea [523] saw a bighorn ewe chase away a prowling coyote, and a band of ten rams charge another; they add 'except for early in the encounter when they banded together and charged the coyote, the sheep displayed little concern over its close presence'. Coyotes are notorious for their ability to survive and recover from serious injuries. Wolves kill large animals such as deer and domestic live stock, but they also eat small animals of many kinds; coyotes prey largely on rabbits and smaller rodents; and the three species of jackal, found from central Asia to South Africa, prey on small animals and also eat carrion. All the members of the genus breed in burrows, often dug by themselves.

The genus *Vulpes* contains about ten species of foxes, found in North America, Europe, Asia and Africa. The red fox, *V. vulpes*, of Europe – and also of Asia and much of North America – is another carnivore that is able to flourish in settled countries in spite of human persecution. The silver fox, which has been extensively farmed for its fur, is a colour mutant of the North American red fox *V. fulva*, which may be conspecific with *V. vulpes*. The smaller Arctic fox, *Alopex lagopus*, the only member of its genus, has also been farmed for its fur which is white in winter and either brown or greyish blue in summer – the two colour phases are found together, but their proportions differ in different parts of the range of the species, which covers all the arctic lands. The fennec, *Fennecus zerda*, is a small fox inhabiting the deserts of north Africa and Arabia; its soft fur is cream or pale fawn in colour, the snout short and pointed, the eyes large and dark, and the ears enormous. This very pretty little fox feeds on small animals and insects; it is nocturnal and lives by day in burrows that it digs in the sand. Two species of *Urocyon*, the grey foxes, live in America, from southern Canada to northern South America; they are rather short-legged, and have the unusual habit among the family of climbing trees, although anatomically they show no special arboreal adaptations. The raccoon dog, *Nyctereutes procyonoides*, of eastern Asia and Japan, is also unusual in being the only canid that hibernates. It deserves its name, for it bears a remarkable resemblance in build and colouration to the raccoon of North America. Its fur is valued, and consequently it has been farmed for pelt production, and introduced into parts of Russia, Poland and Finland.

The six species of the genus *Dusicyon* are the South American foxes; they are rather long legged and resemble small coyotes – the species that formerly lived in the Falkland Islands was known as the Antarctic wolf. This species, *D. australis*, has been extinct for nearly a hundred years; it was killed for its fur and poisoned by settlers to protect their sheep. It showed no fear of man; Darwin relates that the gauchos 'frequently in the evening killed them, by holding out a piece of meat in one hand, and in the other a knife ready to stick them'. He also found a related species, *D. fulvipes*, in the island of Chiloe nearly as confiding, for one evening when two of the *Beagle*'s officers were taking a round of angles by theodolite he saw one sitting on a rock: 'He was so intently absorbed in watching the work of the officers, that I was able, by quietly walking up behind, to knock him on the head with my geological hammer'. He brought its skin home and had it stuffed.

Two genera, each with a single species, of dark-coloured short-legged dogs occur in tropical South America. The small-eared dog, *Atelocynus microtis*, and the bush dog, *Speothos venaticus*, are both rare, and little is known of their lives, but the bush dog is said by Cabrera and Yepes [65] to hunt in packs and to swim after capybaras and pacas when they seek refuge by taking to the water of lakes or streams. The very long-legged and graceful maned wolf, *Chrysocyon brachyurus*, contrasts strikingly with these small dogs. It is a large animal with pointed snout and big ears, clothed in long rather loose fur, golden red over most of the body and black on the legs. Maned wolves inhabit tropical and subtropical South America; they are timid animals and avoid man and his works, and consequently they do little damage to his domestic animals. They are solitary and nocturnal, and feed not only on all kinds of small animals but eat various fruits and other vegetable matter – Cabrera and Yepes [65] say that they are fond of figs and sugar cane. By night they utter a mournful cry. *Cerdocyon thous*, the crab-eating fox, is another tropical South American species, which inhabits open woodlands and grasslands. It is a small grey fox of nocturnal habits that feeds on small animals, insects, fruits and berries, crabs, and the eggs of tortoises.

Figure 15 South American maned wolf (*Chrysocyon*).

The dhole or wild dog, *Cuon alpinus*, of east Asia and India looks much like a medium-sized domestic dog with reddish fur. Dholes are gregarious and hunt in packs of up to twenty or more, mainly by day, following their prey more by scent than sight; they can kill animals as large as sambar deer. The hunting dog, *Lycaon pictus*, of most of Africa south of the Sahara is also a gregarious hunter by daylight. It is a large, long-legged rangy dog with a heavily built head and large ears. The hair is short and often sparse, and is mottled irregularly with black, yellow and white patches, so that no two individuals are alike – the heavy head and broad black muzzle give it some resemblance to a hyaena. Hunting dogs form packs of a dozen to twenty or more and lope after their intended victim in a long file, frequently changing the leader. They eat large ungulates as well as smaller animals, chasing their prey until it tires, and then biting at the flanks and hind-quarters, and often pulling out the entrails, until it drops, whereupon they quickly fall upon and devour it. Kruuk and Turner [283] found that most of the hunting is done by the first two dogs, and that the rest of the group take part only at the end of the chase by heading the prey when it starts dodging from side to side; only a few members of the pack do the actual catching. The dogs spend most of the day sleeping, generally in the shade, but an hour or so before dark 'one dog would get up, and after stretching and yawning, walk over to another dog and greet it'. Other dogs got up, and soon the pack was playing together, chasing and biting one another. Then the pack moved off, with one dog leading. When the dogs sighted prey, often as much as two kilometres away, they advanced at a slow trot and then a walk; when the prey saw the dogs it generally stood watching the approach for a while, and then suddenly fled. At this the dogs started running and kept up a steady 60–65 kilometres an hour for as much as 13 kilometres, though the chase was usually over in 6 to 8 kilometres. If the last members of the pack arrive late, and find all the prey has been eaten, the other dogs feed them by regurgitation; Kruuk and Turner saw many lame dogs fed in this way. The hunting habits and unprepossessing appearance of these wild dogs have given them a bad name, and people who are ruled by their emotions rather than their intelligence have thought them repulsive. Consequently some game wardens in Africa used to shoot hunting dogs whenever possible, not realising that by doing so they were destroying the ecological relationships of the fauna they were paid to protect (Plate 17).

The bat-eared fox, *Otocyon megalotis*, of eastern and southern Africa has enormous ears, and is light yellowish brown in colour so that it bears some resemblance to the fennec, though it is much larger. The teeth are rather small and weak, but as there are four premolars in each jaw, and up to five molars above and four below, the total number of teeth can be fifty, more than in any other placental mammal with heterodont dentition. Bat-eared foxes eat small vertebrates, fruits, and insects especially termites. They are common in dry open country and are mainly nocturnal, though often seen also by day; they live in pairs or small parties, and lie up among bushes or rocks but give birth to their young in a burrow. Their fur is valued for making into karrosses.

FAMILY URSIDAE

The family Ursidae contains five genera of bears with a total of seven species. Bears are medium to very large carnivores; they are heavily built and short-legged, are plantigrade and have large claws, and very short tails. The eyes and ears are small, and the coat usually long and shaggy, generally uniform in colour but with a light coloured patch on the chest in the young, which persists throughout life in some species. Bears are omnivorous, and in correlation their molar teeth are flattened, with low tubercles, and the carnassials are very weakly developed. Bears get fat in autumn and den-up for the winter in the cold parts of their range, but they do not become torpid or enter into true hibernation. The gestation period ranges up to nine months in length owing to the occurrence of delayed implantation in the early stages. The young at birth are extremely small compared with the adult – so small and naked that it was formerly thought that the mother had to 'lick them into shape'. They are born in the winter den, and nursed there for some weeks. Bears are generally rather slow and shambling in their gait, but they can be surprisingly swift and agile when necessary; the smaller species are good tree-climbers. They can stand upright and even walk a few steps on their hind legs; the femur has a strong superficial resemblance to that of habitually bipedal man, and has more than once provided a puzzle for the police. Bears are usually solitary, but may be seen in some numbers together at garbage dumps, or at rivers during a salmon run. Bears of one kind or another are found throughout Eurasia and North America; one species inhabits the Andes, and another until recently the Atlas mountains of north Africa.

The brown bear, *Ursus arctos*, called the grizzly in America, is the most widespread of all the bears, inhabiting Europe, Asia, North America and formerly the mountains of north Africa. It is a very variable species both in size, colouration, and habits, and the different forms have been classified under a host of specific names. Merriam [320] gave specific names to no less than eighty-six kinds that he thought he could distinguish in America alone, forms that are now not even regarded as subspecies – as Osgood [360] says, what he had done was to marshall his facts and label them without any effort to interpret them. The nomenclature of *U. arctos* was put straight by Couturier [84] and other systematists who showed that all the brown bears of the Old and New Worlds belong to a single species. The largest race inhabits Kodiak Island and the neighbouring islands off

Alaska; these giants are the largest living carnivores, and may reach a weight of over half a ton and a height of nearly twelve feet when standing upright. Like all of their kind they are omnivorous and not above catching mice or rooting for insects – brown bears in some parts of their range are almost exclusively vegetarian, and lives on leaves, berries and roots.

The American black bear, *U. americanus*, formerly placed in a separate genus *Euarctos*, is a smaller species that manages to survive even close to settled areas – it is less destructive to man's live stock and has consqeuently been less persecuted; black bears are a great tourist attraction in some American wildlife parks. The black bear, lighter in weight than the huge grizzlies, is a good tree-climber. It is polymorphic in colour and, though generally black, some individuals are brown, cinnamon, blue-grey or even yellowish-white; all the colour varieties were once known by specific names but, as different colours can occur in the same litter, it is obvious that there is but one species – although some variants such as the white black bears of the American north Pacific coast regions are found mostly in restricted areas.

The polar bear, *U. maritimus*, formerly placed in a separate genus *Thalarctos*, is the last species of *Ursus*. It inhabits the floes and floating ice-fields, and the shores of the arctic ocean, and seldom strays far from the sea; it feeds on seals, fish, beach carrion, arctic berries and leaves. The female dens up for the winter among ice and snow, but her body-heat raises the temperature of the igloo so much that it is a warm nursery for her naked and blind young. Polar bears have been so much hunted for sport – they have even been shot from aeroplanes – in the American arctic that their numbers have been seriously reduced; they are now given some protection, especially by the Russians, and winter dens are numerous near the settlement on Wrangel Island. The polar bear is the end of a food chain which starts with the plankton and is linked to the bear by fish and seals. The concentration of vitamin A increases along the chain, and the vertebrates store the excess of the vitamin in the liver. The concentration of vitamin A in the liver of the polar bear is so great that it is poisonous to man and many other animals if they eat it. Rodahl [407] found as much as 26,700 international units of vitamin A per gram in the liver of a polar bear from Greenland; rats were killed by eating 0·5–0·7 gm of the liver daily.

Selenarctos thibetanus, the Asiatic black bear, inhabits the mountains and forests of much of Asia from Afghanistan to Korea and Japan. It is a fairly large species, black in colour with a white patch on the chest. Its habits are similar to those of the brown and black bears, but it also makes sleeping nests in the open. Hirasaka [232] describes a basking couch 'found among deep snow on a Japanese mountain' as being built of branches and twigs of rhododendrons and cedar near a wintering den:

With those branches, which the bear must have broken off and carried by himself to the place (as was clearly seen by the marks of his claws and fangs) he made a comfortable cushion, elliptical in form (130 cm x 100 cm), its thickness being quite sufficient to support his body softly, when he lay down on it. It is believed that bears usually lie on such cushions to dry their fur during a few days preceding their retirement to the wintering places.

Helarctos malayanus, the sun bear, is the smallest of the bears; it inhabits tropical and subtropical southeast Asia. Its black coat, usually with a white chest patch, is short and comparatively sleek. In spite of its stocky build it is active, agile and a good tree climber; it too makes basking nests with branches, sometimes well above the ground.

The larger sloth bear, *Melursus ursinus*, of India and Ceylon, has very long shaggy fur, black in colour with a white crescent on the chest and light hair on the snout. The snout is long and narrow, the lips very mobile, and the claws of the fore feet long and powerful. This bear is omnivorous, but although it eats much vegetable matter it is particularly an insect eater. Tickell, quoted by Jerdon [256] says,

The power of suction in the bear as well as of propelling wind from its mouth is very great. It is by this means it is enabled to procure its common food of white ants and larvae with ease. On arriving at an ant-hill, the bear scrapes away with its fore-feet until he reaches the large combs at the bottom of the galleries. He then with violent puffs dissipates the dust and crumbled particles of the nest, and sucks out the inhabitants of the comb by such forcible inhalations as to be heard at two hundred yards distance or more. Large larvae are in this way sucked out from great depths under the soil. . . . The sucking of the paw accompanied by a drumming noise when at rest, and especially after meals is common to all bears, and during the heat of the day they may often be heard puffing and humming far down in caverns and fissures of rocks.

Finally, the spectacled bear, *Tremarctos ornatus*, of the northern part of the Andes, is a medium-sized bear with dark fur, and the ends of the white chest patch carried up on to the face where they encircle the eyes and give it a rather lugubrious expression of countenance. Little is known of its biology, but Tate [468] came across it in the rain forest of southwestern Ecuador about 4,000 feet up in the Andes. He found that the bears had broken down many trees two inches in diameter, and were feeding on the seeds of a palm 80 to 100 feet high. 'They evidently climbed the trees and brought down the whole fruit stalk, which looks something like that of the royal palm. . . . Other food was secured by breaking down palms, tearing open the green stalk and eating the unopened leaves in the interior'. He said also that 'they construct large nests in the tree tops, whence they sally forth daily to feed, and to which they return to sleep'. Although the spectacled bear eats much vegetable matter it is probably omnivorous, for in captivity it will readily eat meat.

The giant panda, *Ailuropoda*, has generally been considered to be a member of the family Procyonidae, but Davis [102] as a result of a detailed anatomical investigation, has recently concluded that it is more closely related to the bears, and should be placed by itself in a subfamily of the Ursidae. It is here treated conservatively as a member of the next family.

FAMILY PROCYONIDAE

The family Procyonidae contains the raccoons and related animals, including the red and giant pandas; it is not a large family as it contains only eight genera and less than twenty species. All its members, except the pandas of Asia, are found only in the Americas. Most of them are small to medium sized animals with long or moderately long tails; their gait is plantigrade or semiplantigrade. The carnassials are generally hardly developed, and the crowns of the molars are flattened and tuberculated. Most of the species are omnivorous, and are agile tree-climbers. They are comparatively silent animals, unlike the loud voiced canids and some others.

The raccoons, about half a dozen species of *Procyon*, are found from southern Canada to northern South America; their grey fur is dense, there is a dark mark on the face across the eyes, and the tail has black rings. They are good climbers and swimmers, but on the ground their gait is rather shambling, with the body carried in hunchback curve. They live in a wide variety of habitats, but prefer timbered or bushy country and the neighbourhood of water. The most familiar species are *P. lotor* of North America which extends into central America, and *P. cancrivorus* from further south – the two species are similar but the latter has shorter hair. Raccoons are omnivorous, and in captivity they habitually dunk their food in the water trough – this habit is so well known that Linnaeus referred to it by the specific name *lotor*, the washer, when he named the species. The habit, however, has never been seen in the wild – it occurs only in captivity. Lyall-Watson [306] showed that in the wild raccoons find their food either on land or in shallow water. When they find food on land they rub it under the very sensitively tactile palms of the hands to investigate it before eating it; in shallow water they find their food by dabbling with the hands. Lyall-Watson points out that

The dry-feeding activity pattern involves searching for the food, an investigatory rubbing of the food once it has been found, a manipulative grasping and the consummatory eating of the prey. When a raccoon is transferred from the wild into a captive situation, this dry-feeding activity is very little affected. . . . Wet-feeding does not translate from the wild into the captive situation easily as does dry-feeding. Dabbling, which is the search component of the wet-feeding activity, is a fixed motor action pattern for searching for food in water which would normally be rewarded in the wild, but which is starved in captivity due to the fact that the raccoon can never find anything in the water dish which in captivity corresponds to the river's edge. . . . By carrying food objects to the water, 'losing' them and then 'relocating' them during dabbling, both types of feeding activity are rewarded, and the raccoon is able to satisfy all the appetitive fixed action patterns of its normal behaviour.

The action is, therefore, neither for washing nor for moistening the food, as was formerly supposed.

Figure 16 Coati (*Nasua*).

The coati, *Nasua nasua*, is found throughout most of South America; it is very variable in colouration, and was formerly classified into about twenty species, but is now known to form only one. Two other species occur in the southern United States and in Central America. The coati (pronounced *co-ah-tee*) is a short-legged active creature, characterised by its long mobile snout truncated like that of a pig, its short ears and its long ringed tail; the colour ranges from red to black, and is lighter below. Coatis are diurnal as well as nocturnal, and generally inhabit woods and forests; they are agile tree-climbers and spend much time among the branches as well as on the ground. They are omnivorous and intensely active, and investigate every crevice for insects and other food with their sensitive piggy snouts – their liveliness and inquisitive habits make them endearing pets. Coatis are social animals and are commonly found in parties of up to twenty, but these bands consist solely of females and young males; the adult males are solitary and are often thought to be a separate species by people in South America. A smaller species from the northern part of the Andes is placed in a separate genus *Nasuella*.

The kinkajou, *Potos flavus*, the only species of its genus, lives in the forests from southern Mexico to central Brazil. It is a short limbed arboreal animal with small ears, and is the only New World carnivore with a prehensile tail; there is, indeed, only one other prehensile-tailed carnivore, the Asiatic binturong of the family Viverridae. The short dense fur of the kinkajou is light buffy brown, sometimes with a black streak along the middle of the back. Kinkajous are arboreal and seldom descend to the ground; they are almost entirely herbivorous and feed mainly on fruits. They are nocturnal and spend the day sleeping in a hollow; in captivity they give the impression of being slow and sluggish animals because they are generally seen only by day when they are asleep. This impression is far from the truth, for at night they are as active and agile as weasels. Kinkajous are well provided with skin scent glands; paired mandibular glands, a throat gland, and an abdominal gland. Poglayen-Neuwall [379] who studied the behaviour of about a dozen captive kinkajous and allowed them singly or in pairs to romp at large in his house for periods of up to three hours at night, found that all the glands are used for marking the surroundings. In addition, the mandibular and throat glands are used for sexual stimulation. The marking is not to define a territory; both sexes have the scent-marking behavioural pattern which is not connected with breeding. Marking is done within the home range to facilitate contact between the individuals of the rather loosely-knit social groups, and to make trails between the sleeping dens and feeding places. In the wild they evidently mark tree branches, for the captive ones always marked objects with rounded contours such as door-knobs, telephone receivers, table legs, Mrs Poglayen-Neuwall's occipital region, and her thigh.

The four species of olingo of the genus *Bassaricyon* resemble small kinkajous but their very long thickly furred tails are not prehensile, and the body build is more slender. Little is known of their lives, but they are arboreal, nocturnal and frugivorous, and live in small bands, sometimes associated with kinkajous. They inhabit the forests of central and southern South America. Of the two cacomistles or ring-tail cats forming the genus *Bassaricus* one, *B. astutus*, is found from Oregon to southern Mexico, the other, *B. sumichrasti*, thence through Central America. They are short legged animals, brownish buff in colour, with long bushy tails marked with dark rings. They are arboreal, but also live in rocky places, and are as agile among cliffs as among trees. They are omnivorous but much more carnivorous than the kinkajou and olingos. Although cacomistles are nocturnal they are easily tamed and even domesticated as house cats to catch mice.

The pandas are found only in some of the mountainous regions of Asia. The red or lesser panda, *Ailurus fulgens*, inhabits parts of western China and the slopes of the Himalayas. It is about the size of a large cat, and is clothed in thick chestnut-red fur. The muzzle, cheeks and insides of the rather large ears are white, and the bushy tail is slightly ringed. Lesser pandas live among rocks and in trees which they climb readily, but are said to forage mostly on the ground. Their food is mainly vegetable matter – fruits, roots, bamboo shoots, acorns and so on. They sleep by day, and are crepuscular, feeding morning and evening. Jerdon [256] misprints his quotation from Hodgson and makes him say 'they drink by lapping with the tongue, piss and spit like cats when angered, and now and then utter a short deep grunt like a young bear'. The giant panda, *Ailuropoda melanoleuca* is a large bear-like creature found in western China and neighbouring mountains of Tibet, where it inhabits the bamboo forests at an altitude of five to ten thousand feet. It is a heavily built animal with dense woolly coat, white or yellowish in colour, with black legs, a black band over the shoulders, black ears and eye patches. A false thumb on the fore foot, formed of a radial sesamoid bone, enables the paw to be used as a hand in grasping, by apposition of the digits. Bamboo forms the main part of the diet, but other plants and small rodents are also eaten occasionally. Giant pandas, as Sheldon [439] points out, live among unlimited supplies of their staple food and have no enemies except man; consequently they are sluggish, dull creatures – he calls the panda 'an extremely stupid beast'. Few specimens have been exhibited in captivity because of the difficulty of collecting them from the remote habitat where, however, giant pandas seem to be more plentiful than was formerly believed. The species has thus received much publicity but, except when young, giant pandas are attractive exhibits only because of their unusual colouration and the mystique that has been built up about their supposed rarity. Without the black eye-patch, which gives the face an expression attractive to some, the small piggy eyes would produce a less endearing appearance. The opinion that the giant panda would be more appropriately classified with the bears has already been mentioned.

FAMILY MUSTELIDAE

The family Mustelidae is a large one, and contains animals as diverse as the weasels, polecats, skunks, badgers and otters. Its members are classified in twenty-five genera with about seventy species; they are found in all the continents except Australia, Antarctica and Madagascar. Mustelids are small to medium sized carnivores with long bodies, short legs, and usually medium to long tails; they are plantigrade or digitigrade, and the carnassials are generally large. They have well developed anal scent glands, which are used for defence by some species. In many mustelids the period of gestation is lengthened by the occurrence of delayed implantation of the embryo. Some members of the family are much sought after fur-bearers.

In the genus *Mustela*, which is divided into several subgenera, the species inhabiting the palearctic region moult into a white pelage during winter in the northern parts of their range. The stoat *M. erminea* – called weasel in North America – with black-tipped tail, provides ermine when in the white coat. The weasels and stoats are extremely lithe and active animals; they are generally terrestrial, but are also agile climbers in trees – they are carnivorous and prey largely upon small rodents. The least weasel, *M. rixosa*, of North America, about four inches long, is the smallest carnivore. Members of the genus are found throughout Eurasia and North America, and extend into South America together with the subgenus *Grammogale* containing a single species from the Amazon region misnamed *M. (G.) africana*. The subgenus *Putorius* distinguished by a dark mask across the eyes, contains the polecats, *M. (P.) putorius*, the common polecat, extends in several subspecies through Europe, Asia and North Africa – it is the wild ancestor of the domesticated ferret. Another species *M. (P.) nigripes*, the black footed ferret, is confined to North America; it feeds almost exclusively on prairie-dogs, and finds both food and shelter in prairie-dog towns. It is surprising that a carnivore so closely linked with a single prey species which, apparently, has no defence against it, should not have brought the food species to extermination or near it. There is some unknown factor which prevents its numbers building up to the point where it endangers its own food supply. The black-footed ferret is so closely linked to the prairie-dog that it is now very rare, because prairie-dogs have been eliminated by man from large areas of their range, owing to the nuisance they cause to agricultural and pastoral farming. The subgenus *Lutreola* contains the minks, *M. (L.) vison* of North America, and *M. (L.) lutreola* of Europe and Asia. Their dense fur is highly valued in commerce, and they differ from the other members of the genus in having bushy tails. They are semi-aquatic in habit, and are generally found near water; their food includes fish as well as small land mammals and birds.

The polecat proverbially stinks – the secretion of the anal glands is highly offensive – but other genera, culminating with the skunks, are even worse. Their scent is so powerful and unpleasant that it is used as a defence, and other animals quickly learn not to molest them. In correlation with this method of defence the animals are coloured with conspicuous contrasting patterns that advertise their presence and warn other creatures to beware. The protection is so effective that the animals take no trouble to conceal themselves and flaunt their warning colours openly, as though knowing they are immune to attack. If they are disturbed they erect the tail in preparation for a shot from the glands, and at the same time spread the hairs of the tail and of the conspicuous black and white dorsal pattern in threat; this action is generally enough to ensure their safety without firing a shot. The genera thus protected include the marbled polecat, *Vormela peregusna*, found from southeast Europe to Mongolia, the zorilla or African striped polecat, *Ictonyx striatus*, found throughout most of Africa south of the Sahara, the north African banded weasel, *Poecilictis libyca*, the two species of striped skunks, *Mephitis*, of North and Central America, the two spotted skunks, *Spilogale*, of similar distribution, and the several species of hog-nosed skunks of the genus *Conepatus* from South America. Because of the confidence their defence gives them and their consequent lack of fear of man, skunks make good pets even when their stink glands have not been surgically removed to avoid accidents – the stink is so overpowering that it almost chokes the victim, for it temporarily inhibits breathing.

Members of several other genera have colour patterns that are conspicuous, but their stink defences are less pungent, though that of the African striped weasel, *Poecilogale albinucha*, an animal coloured black with white stripes along the back and a white tail, is said to be nauseating. The Patagonian weasel, *Lyncodon patagonicus*, and the two species of grison, *Galictis vittatus* and *G. cuja*, which between them cover most of South America, are light grey on the back, black below, and have a white band across the forehead extending onto the sides of the neck, a pattern of colour that appears to link with that of the ratel and the badgers; all these mustelids have well developed anal glands, though the secretion is not so awful as that of the skunks.

The ratel *Mellivora capensis*, is a medium sized heavily-built mustelid with strong claws especially on the fore feet. The top of the head and back are light grey, the rest of the body black. It inhabits most of Africa south of the Sudan, and much of southern Asia, and lives in many different habitats. It is both carnivorous and vegetarian, and takes its name from its partiality to honey, which it obtains by digging up wild bee nests. In tropical Africa it has a remarkable association with a small bird, the honey-guide *Indicator indicator*. When the honey guide finds a bees' nest it starts calling in a characteristic way if a ratel, or indeed any other mammal, including man, is nearby. On hearing these cries the ratel follows the bird, which keeps five or six yards ahead, and flies towards the nest; when the ratel reaches the nest it digs it up to eat the comb with its honey and grubs. The honey-guide shares the meal, and eats the comb as well as the contents – it is the only bird that is known to possess a digestive enzyme capable of breaking down beeswax.

The genus *Martes*, containing the martens, is more closely related to *Mustela*; the American marten and the two species of Eurasia look much like large bushy-tailed stoats. They are extremely agile and quick-moving animals, as acrobatic as squirrels when leaping among the branches of trees, as also are the yellow-throated martens of southeast Asia. The fisher, *M. pennanti*, of North America and the Russian sable, *M. zibellina*, have both been much reduced in numbers through being killed for their beautiful dark pelts. The tayra, *Eira barbara*, is a rather large species, longer legged and with a long tail, that lives in Central and South America. It is dark brown or black in colour with a lighter head. It is both terrestrial and arboreal, and as active as a marten among the branches. It is omnivorous and eats large amounts of fruit as well as smaller mammals. The wolverine or glutton, *Gulo gulo*, is a large heavily built mustelid with dense blackish brown fur marked with a lighter band on each flank; it is one of the more plantigrade of its family. It inhabits the northern parts of Eurasia and America, especially the coniferous forests. Wolverines are mainly terrestrial, but can climb; they are chiefly carnivorous and are strong enough to kill such large animals as deer – they are indeed very strong animals for their size, but they have a reputation for wanton ferocity that, according to Krott [276] is quite undeserved. Like the badger and many other mustelids the wolverine has a period of delayed implantation,

269

which Wright and Rausch [524] found to last in Alaska for about six months, from midsummer to the following January.

The badger, *Meles meles*, found from western Europe to China, is smaller and has shorter legs; its colour pattern of grey above, black below, with white stripes on the black head and neck, has been claimed as a warning pattern, but the facial pattern is not as conspicuous by night as might be expected. The badger is omnivorous but not generally destructive to mammals larger than nestling rabbits; it eats large quantities of earth-worms and other invertebrates and much vegetable matter such as bulbs and fruits. This species and the American badger, *Taxidea taxus*, both dig extensive burrow systems. They, and the east Asian hog-badger, *Arctonyx collaris*, a species with a less clear-cut 'warning' pattern, are well provided with anal glands, the secretions of which, however, are mild compared with those of the skunk-badgers, of two genera, *Mydaus* and *Suillotaxus*, each with a single species, of Malaya and some East Indian islands. The colour of the skunk badger is black or dark brown with a light or white stripe on the crown and back – not nearly so striking a warning pattern as might be expected on animals protected by such dreadful scent, said to be as bad as that of the skunks. The ferret badgers, three species of *Melogale*, inhabiting eastern Asia, Borneo and Java, also have an offensive though less disabling scent; the head and throat are light coloured, with a dark mask across the eyes extending on to the snout. It is peculiar that the stink secretion of the mustelids, like many other offensively smelling substances, though unendurable in concentration, is often sweet-smelling and agreeable in great dilution.

Some seventeen or more species of otter make up the mustelid subfamily Lutrinae; twelve of them belong to the genus *Lutra*. These are the familar otters of Europe, Asia, America and Africa, medium sized animals with long bodies and tails, short legs, small eyes and ears. They are highly aquatic and the digits are generally webbed and, as in some other aquatic mammals of widely different orders, the long vibrissae are set in a thick pad on each side of the snout. The underfur is particularly dense, and when submerged it carries a layer of trapped air that prevents the skin being wetted. Otters inhabit all kinds of freshwaters, and in some places the sea coast as well; they feed not only on fish but on many kinds of aquatic invertebrates and small land mammals. They are nocturnal and crepus-cular, and lie up by day among dense vegetation or in hollows in river banks and such places. Like most mustelids they are very active creatures and are extremely graceful in the water; some species indulge in play, even when adult, by sliding on their bellies down slippery mud or snow slopes into the water. The giant otter, *Pteronura brasiliensis*, of South America from the Guianas to Paraguay and Uruguay, reaches a length of over six feet including the tail. The digits are fully webbed and the tail is wide and strongly flattened from above downwards, with a crest or keel along each edge. The profile of the head differs from the weasel-like outline of the other otters, the forehead rising behind the snout in the shape called 'apple-headed' by dog fanciers. The otters of the genus *Aonyx* resemble those of *Lutra* in general, but their claws are reduced or absent. The toes are not webbed in the small-clawed otter of southeast Asia, or in the African clawless otter widespread throughout the continent, but webs are present in the African small-clawed otter, of the west coast.

The sea otter, *Enhydra lutra*, is by far the most highly adapted of the fissipede carnivores for aquatic existence – it spends nearly the whole of its life in the water and seldom comes ashore. The head merges imperceptibly into the cylindrical body, the limbs are very short, and the rather short tail is flattened from above down. The eyes are small, and the external ear is reduced and narrow so that it much resembles that of an otariid seal. The feet are highly modified; on those of the fore limbs the digits are joined together and recognisable only on the under surface, where grooves in the pads show their limits – the claws on the upper surface, curved in the young, are soon worn to stumps in the adult. The digits of the hind foot are joined by webs, and the claws are small; the fifth digit is the longest and the others progressively smaller so that the foot is much like the fore flipper of a seal. When the animal is on land the shape of the hind flipper and the hunched position taken up by the hind end of the body give a strong resemblance to the attitude of an otariid seal. The sea otter is a coastal animal and seldom goes far to sea; it was formerly found all round the shores of the north Pacific from Japan to California, with a gap between Vancouver and San Francisco. Sea otters were so intensely hunted for their very valuable pelts during most of the eighteenth and nineteenth centuries that they were nearly exterminated. For the last fifty years or more they have been given international protection, and have increased in numbers in some places – notably in California – though they have not yet re-occupied all their former haunts.

Except when in a hurry sea otters usually float at the surface on their backs; they sleep in the same position anchored among the floating strands of kelp beds. Over half of the food eaten by sea otters consists of sea urchins, and nearly a quarter of molluscs, the remainder beings crabs and fish. The food is collected from the bottom in depths of up to but not much over 100 feet, and is brought to the surface to be eaten. The sea otter then lies on its back and uses its chest as a dining table to hold the food while it is being consumed. The molar teeth used in breaking the shells of the food are wide and flattened but the animal also calls on extraneous aid for cracking shells. It brings a stone to the surface together with the food and, placing it on the chest, uses it as an anvil on which the food is battered with the fore paws. The sea otter is thus one of the few mammals that makes use of a tool but, stranger still, this habit has been seen to be regularly used only among the sea otters of the Californian coast, and never in those of the northern habitats. Sea otters have been kept in captivity in America on a few occasions but extensive experiments, described by Barabash-Nikiforov and his colleagues [20] have been made in Russia on breeding them in captivity, and in introducing the species into the west on the Murman coast. Young sea otters are born on land but soon accompany their mothers to the water, where the parents carry them on their chests when resting. The main natural enemy of the sea otter is the killer whale.

FAMILY VIVERRIDAE

The family Viverridae contains about seventy-five species of civets, genets, mongooses and their allies, apportioned among thirty-six genera; so many that it is convenient to consider them under the headings of the six sub-families into which it is divided. Viverrids are small to medium sized animals, generally with long tails, and often with rather short legs; many of them are marked with conspicuous spots or stripes. They are semi-plantigrade or digitigrade, and the claws are partly retractile in the digiti-grade species. Many of the viverrids have well developed scent glands, some of which produce secretions of agreeable odour much valued by man. The family is confined to the Old World where it is widely distributed, and forms the counterpart to the procyonids of the New.

The subfamily Viverrinae contains the genets, linsangs and civets. All except one species are spotted or blotched, and have ringed tails. They are omnivorous, and although terrestrial, most of them are agile climbers of trees. About five or six species of genet, *Genetta*, inhabit Africa, and one of them extends into southern Europe. They are long bodied slim animals with short legs, long tails and soft fur beautifully marked with rows of dark blotches. Their food consists mainly of small rodents and birds, but some vegetable matter is also eaten. They are said to be easily tamed when young and to make charming pets. The African linsang, *Poiana richardsoni* of West Africa is similar, but has rows of smaller spots; it is nocturnal and omnivorous but it is rare and little known. The oriental linsangs, two species of *Prionodon*, of southeast Asia, are similar in build, but are marked with larger blotches or bands. They, like the animals of the preceding genera, are nocturnal and omnivorous, and are good tree climbers. The African civet, *Viverra civetta*, is the largest species of the subfamily. It is rather heavily built and has longer legs than the genets and linsangs. The fur is fairly coarse, and is heavily spotted, but that of the oriental civets, three species of the same genus inhabiting India and southeast Asia, is finer, and the spots, though numerous, are smaller and confluent into bands round the neck. All members of the genus are nocturnal and terrestrial, live in forest and bush country, and seldom attempt to climb trees; in diet they are omnivorous. When alarmed the animals of this genus erect the hairs along the crest of the back and bush out the tail. The rasse, *Viverricula indica*, widespread in India and southeast Asia, is smaller but similar in diet and habits. The scent of the civets is secreted into a pouch below the anus, where it accumulates as a greasy secretion with a sweet musky smell. It is used in marking territory by the animals, and is prized for use in perfumery and traditional medicine by man. Animals of all the species are kept in captivity, and the secretion is removed with a spatula at short intervals – it is particularly prized in eastern medicine. *Osbornictis piscivora*, the Congo water civet, is the only member of the subfamily that is unspotted and has an unringed tail; it is chestnut brown and the tail is black. It is said to be semi-aquatic, but as it is very rare and few specimens have ever been seen this assumption may not be true.

The subfamily Paradoxurinae contains the palm civets and the binturong. The palm civets are small to medium sized animals with long tails; their markings are spots, stripes and tail rings, which are obscure and not strongly contrasted with the ground colour as they are in the civets and genets. All species are arboreal and omnivorous. One of the best known is the palm civet or toddy-cat *Paradoxurus hermaphroditus*, of India and southeast Asia, where it lives in wooded places and often near or even in the habitations of man – Jerdon [256] found a large colony of them living in the roof of his house. The degree and distinctness of the spotting varies much individually; Jerdon says that in some specimens the tail is twisted so that the tip has the lower surface uppermost, and that the generic name refers to this feature. The specific name refers to the glandular fold forming a rudimentary scent pouch which lies between the anus and scrotum in the males and superficially resembles a vulva. The toddy-cat is nocturnal; it feeds on small mammals, birds and insects, but also eats fruit, and is supposed to like toddy, the juice of the palm, a tree in which it is often found. Jerdon reported that it is often tamed and

T

becomes quite domestic, and even affectionate in its manners. One . . . went about at large, and late every night used to work itself under the pillow of its owner, roll itself up into a ball, with its tail coiled round its body, and sleep till a late hour of the day. . . . Their skill in climbing is very great and they used to ascend and descend my house at one of the corners of the building in a most surprising manner.

The African palm civet, *Nandina binotata*, inhabiting central and southern Africa, is equally readily tamed and can be semi-domesticated as a house cat to keep down rats and mice. The colour of the fur is grey or brownish with obscure darker spots and tail rings; the two cream spots often present on the shoulders give it its specific name. It is nocturnal and omnivorous, as are the small-toothed palm civet, *Arctogalidia trivirgata*, and the masked palm civet, *Paguma larvata*, both found in Burma and widely in southeast Asia. The former has a very long tail, and has three obscure longitudinal stripes on its tawny fur; the latter is greyish with no spots on the body or rings on the tail, and has a light patch on the head and snout from which it takes its name. The rare Celebes palm civet, *Macrogalidia musschenbroeki*, a faintly spotted species, probably has similar habits.

The binturong, *Arctitis binturong*, lives in southeast Asia from Burma to Malaya and the East Indies; it is much larger than the palm civets, the head and body reaching a length of nearly a yard and the tail nearly as much again. The fur is long, coarse and dishevelled in appearance, and is black in colour with a white edge on the ears. The binturong is nearly plantigrade, and its tail is prehensile; it is the only Old World carnivore thus provided. It is a rather slow-moving animal and although it is arboreal it creeps among the branches in a careful and deliberate manner. Like the other members of the subfamily, it is nocturnal and omnivorous; like them too it possesses anal scent glands – the secretion has a peculiar odour, neither pleasant nor particularly repulsive.

The subfamily Hemigalinae contains five genera of rare or little known viverrids from Madagascar and southeast Asia. Three of them, *Fossa*, *Hemigalus* and *Chrotogale* are like rather slender viverrines, one, *Eupleres*, resembles a mongoose, but *Cynogale* is unlike other members of the family. *Fossa fossa*, the Malagasy civet, resembles a slender, long-legged genet with diffuse spots which tend to coalesce into lines. It is found only in Madagascar, as are the small-toothed mongooses, *Eupleres*, two species of which are known. They have long narrow heads, pointed snouts, and short legs. The dense soft fur is greyish brown, and that on the tail is long and bushy. The jaws are slender and the teeth very small; the animals are believed to be mainly insectivorous. There are two species of *Hemigalus*,

H. derbyanus, the banded palm civet from Malaya and the East Indies, and *H. hosei* from Borneo. They are long-bodied animals with elongated pointed snouts. The first is light in ground colour conspicuously marked with dark cross bands; the second is brownish above lighter below and unbanded. Both are believed to be arboreal and omnivorous, and to prey extensively on insects and other invertebrates, but little definite is known about them. The same must be said of *Chrotogale owstoni*, coloured much like *Hemigalus*, but with some spots in addition to the bands; it has been found in Indo-China. The otter civet, *Cynogale bennetti* of Indo-China, Malaya and some of the East Indies, differs strikingly from all other viverrids. As its name shows, it is somewhat like a small otter in build, but it has a rather short tail. The ears are small, but the toes are only slightly webbed. The coat is greyish brown in colour and the underfur, like that of the otters, is very dense; the vibrissae are long and set in pads on the sides of the snout, like those of otters and some other aquatic mammals. The otter civet lives near rivers and marshes, and is semi-aquatic, for although it catches fish, crabs and other aquatic creatures, it can also climb trees with agility in search of fruits and small mammals and birds. It has occasionally been seen in captivity, but little is known of its biology.

All the members of the small subfamily Galidiinae are found only in Madagascar; some are rare, and nothing of importance has appeared about their biology. *Galidia elegans*, the ring-tailed mongoose, is brown with darker rings on the tail; the two species of *Galidiactis*, the Madagascar striped mongooses, have dark longitudinal stripes on a lighter ground colour, and long bushy tails; the two species of *Salanoia* are much like mongooses – the colour of their fur is brown with darker or lighter spots. The two species of *Mungotictis*, the narrow-striped mongooses, are greyish brown with narrow darker longitudinal stripes. It is regrettable that nothing of interest can be added about the life of these animals.

The subfamily Herpestinae, containing the mongooses and their relatives is a comparatively large one, of thirteen genera. Its members are long, narrow bodied animals with short legs, generally long tails, and pointed snouts. The fur is brown or grey, speckled with a pepper-and-salt effect through individual hairs bearing alternating light and dark rings. The subfamily is essentially African, and most of its species are found only there They are terrestrial and many are social and diurnal – all are mainly or exclusively carnivorous.

The mierkats, three not closely related species placed in separate genera, inhabit open country in southern Africa. They are small herpestines with eyes comparatively rather larger than those of the mongooses, giving them a bright, foxy facial expression. The slender-tailed mierkat, *Suricata suricatta*, and the bushy-tailed mierkat, *Cynictis penicillata*, are burrowing species which live in colonies and dig extensive warrens with many entrances, and the burrows communicating below ground. Their habitats overlap, although the bushy-tailed mierkat often lives in less arid and more scrub-covered regions; their warrens also often overlap so that the two species may live in one warren. Stranger still, the warrens are also often shared with the ground squirrel *Xerus*; the three species are of about the same size, and all are in the habit of sitting upright at the burrow mouths. The mierkats are diurnal and sit on their haunches when thus basking in the sun, particularly in the early morning, but they stand up on the toes of the hind feet with the fore legs dangling over the belly when they are alerted and on the lookout. Both species are largely insectivorous and scratch about digging up grubs, termites, spiders and other invertebrates; they also take mice and other small vertebrates, *Cynictis* more so than *Suricata*. *Suricata* is easily tamed and makes a good pet. Selous' mierkat, *Paracynictis selousi*, of southern Africa is grey with black feet and a white tip to the tail. It also is a burrower, but is not known to live in communal warrens.

Figure 17 Mierkat (*Suricata*).

The genus *Herpestes* contains about a dozen species of mongoose and is, apart from *Ichneumia*, the only one found outside Africa. One species, *H. ichneumon*, the Egyptian mongoose, inhabiting the whole of Africa south to the Cape, extends into Europe and is found in southern Spain and

Portugal; other species inhabit southern Asia from Arabia through India and Ceylon to Malaya and the East Indies. Mongooses live in many habitats, but most species prefer bushy or wooded country, where they lie up in burrows or natural hollows by day, for they are mainly nocturnal although sometimes active by day as well. Although they are not gregarious as mierkats some species occur in small parties of up to half a dozen or sometimes more. They prey upon all kinds of small vertebrates and sometimes kill and eat poisonous snakes. They are not, as is often supposed, immune to snake venom, but avoid being bitten by their agility and quickness. Mongooses were introduced into some of the West Indian islands and into Hawaii in the hope that they would kill rats devastating the sugar cane fields; they did so, but also preyed upon species of native vertebrates with disastrous effects on the indigenous faunas. The small *Helogale parvula*, the dwarf mongoose, of eastern and southern Africa, is diurnal and feeds on insects and small vertebrates; Shortridge [442] says 'they hunt in compact troops of from six to a dozen on the lower slopes of open or scrub-covered hills, on rocky ridges, or in flat sandy country, keeping all the time in close formation like small packs of beagles'. They are small enough to squeeze down the burrows of rodents or into the crevices of termite hills. Several other species of *Helogale* inhabit various parts of Africa. The banded mongoose, *Mungos mungo*, widely distributed through Africa south of the Sahara, is distinguished by the dark cross bands on the back and rump. It is diurnal and social, and lives in small parties like *H. parvula* – Shortridge reports that an entire colony sometimes takes refuge in a warren or rock shelter with only one or two entrances. He adds, 'when suddenly alarmed in the neighbourhood of a warren, all the individuals rush together and bolt down the nearest entrance one after the other with a noisy scuffle'. They are largely insectivorous, and when feeding they scratch about among dead leaves all the time keeping up a low chattering. On two occasions a pack of these mongooses trotted through Shortridge's camp and 'investigated pots and pans and whatever else was lying about'. Although very active, they do not climb trees.

About four species of cusimanse of the genus *Crossarchus* live in western and Central Africa. Their fur is rather long and coarse, and some species are marked with dark cross bands. They are gregarious and diurnal and seek their food in parties of up to two dozen or so, usually in open grassy country. They feed on insects, and other invertebrates, and on small vertebrates, berries and fruit. They share an interesting food-preparing habit with some other genera of mongooses; when they find a large snail or a bird's egg too large to be cracked between the jaws, they take the object between the fore paws and hurl it violently back between the hind legs against a stone or onto the ground, repeating the throwing until the shell is broken. The savannah mongoose, *Dologale dybowski*, of Central Africa, resembles *Crossarchus* but has a longer tail and shorter fur. Little is known of its habits, and less of those of *Liberiictis kuhni* of Liberia which has never been seen in the flesh by Europeans, and is known only from a few skulls in museums.

The species of the four remaining genera are larger than those we have discussed, and have proportionately longer legs. *Atilax paludinosus*, the marsh mongoose, inhabits most of Africa south of the Sahara, and lives among the dense vegetation bordering rivers, lakes and tidal estuaries. It is nocturnal and not gregarious; the dark brown fur is long and rather coarse. The marsh mongoose dives and swims well, and is nearly as aquatic as an otter, an animal it much resembles when in the water. Although it swims so well its feet are not webbed; Pocock [378] pointed out that its feet resemble those of the raccoon, and that it may have similar feeding habits, and find its prey by feeling for it in mud and under stones. It feeds on frogs, crabs, fish and invertebrates, and also catches swamp rodents, birds and reptiles. The white-tailed mongoose, *Ichneumia albicauda*, grey with dark legs and long-haired white tail, inhabits most of Africa and southern Arabia – it is the only mongoose apart from those in the genus *Herpestes* found outside Africa. It is nocturnal and solitary in habit, and lives in places where there is plenty of thick cover; it preys upon small vertebrates including snakes, insects, and also eats some vegetable matter. Unlike the other mongooses it is arboreal as well as terrestrial, and climbs trees in pursuit of tree-hyraxes which form much of its diet in some regions. Although it is a common animal its biology is not well known. Four or five species of black-legged mongooses of the genus *Bdeogale* live in East and West Africa south to Angola and Mozambique. They are grey above, with long but not bushy tails, and with black legs and feet. Their fur is close, dense and short, especially on the head; this gives them a sleek, well-groomed appearance very different from that of other mongooses – they look as though they have just had a hair-cut. They feed on small vertebrates including snakes, and on insects and other invertebrates. Little is recorded about their habits, or about those of the related genus *Rhynchogale* which contains but one species, *R. melleri*, of eastern Africa.

The remaining subfamily of viverrids the Cryptoproctinae, contains a single genus with one species, *Cryptoprocta ferox*, the fossa of Madagascar – the vernacular name fossa is unfortunately the same as the scientific name of the Madagascar civet, an animal belonging to another subfamily. The fossa is about five feet long overall, over half the length being that of the long cylindrical tail. The face is comparatively short – the skull has a close superficial resemblance to that of a felid – the ears are short and round, and the feet are semi-plantigrade. The close dense fur is red-brown to nearly black. The fossa, the largest carnivore in Madagascar, is nocturnal and is said to include lemurs among its prey; it is certainly destructive to poultry and small livestock, but very little beyond highly coloured and probably fabulous native stories of its ferocity and savagery has been recorded of its habits, although it is said not to be a rare animal.

FAMILY HYAENIDAE

The family Hyaenidae is a small one, and contains the three species of hyaena and the aardwolf. Hyaenas are fairly large animals about the size of large dogs, thickset with stout bodies, blunt muzzles and powerful jaw and neck muscles. The back slopes downwards from the high shoulders to the hind quarters, the legs are thick, and the gait is digitigrade. The teeth are large and the carnassials well developed – when worn to sharp cutting edges the teeth in the dried skull of a spotted hyaena will shear a piece of thin paper as neatly as sharp scissors. The hyaenas possess scent glands that open into a depression or small pouch under the tail. The striped hyaena, *Hyaena hyaena*, inhabits most of India and southern Asia west to the shore of the Mediterranean, Arabia, and northern Africa south to the Serengeti Plains. Its shaggy light grey coat is marked with vertical stripes and blotches which alter direction on the limbs to become bars on the legs. The ears are large and pointed, and a dark mark across the eyes gives the animal a scowling expression. The hair of the neck is long and forms a mane extending along the ridge of the back. Striped hyaenas live in open rather than bushy country; they are nocturnal and lie up by day in dens among rocks or in other natural hollows, or sometimes in burrows of their own digging. They feed upon any smaller animal they can snap up, but are particularly scavengers of the remains of prey killed by larger carnivores; their powerful jaws and teeth enable them to shear and break up large bones – presumably their digestive juices are able to remove most of the organic matter from the mineral parts of bones, for their faeces dry to characteristic hard white chalky nodules. The brown hyaena, *H. brunnea*, inhabits southern Africa south of the Zambezi, north of which it becomes scarce; it has been exterminated in the southern part of South Africa, and in the more closely settled areas. In build it resembles the striped hyaena, but is dark brown in colour with diffuse lighter clouding – the long mane hairs are grey, and the legs grey with dark bars. In habits and food it

279

resembles the striped hyaena, except that near the coast it frequents the sea beach to feed upon marine refuse thrown up by the waves – hence its Afrikaans name of strandwolf. Both species of hyaena have a variety of calls, but they are on the whole not noisy animals; both are nocturnal and rather shy.

The spotted hyaena, *Crocuta crocuta*, differs considerably; it is placed in a separate genus on details of its dentition, but the structure of the female genitalia is equally distinctive. It is a larger and much less shaggy animal than the other hyaenas, the ears are shorter and more rounded, and the coarse short coat is spotted. The colour varies greatly between individuals – the ground colour is anything from sandy to deep red-brown sprinkled with large and small spots of dark brown to black. The hair on the neck forms a sparse mane which, like that of all species, is erected when the animal is excited; the tail carries a black tuft. The wide range of individual variation in colour and cranial characters has misled museum naturalists into naming a large number of species or subspecies as 'geographical races'. The German systematist P.Matschie named many; like Merriam, lost in the forest of bear-nomenclature, he was unable to recognise the absurdity of his results. Schwarz [429] tells how he several times described two species or subspecies of mammal from a single specimen taken on a water-shed: 'the right side of the animal was believed to present the characters of one, and the left side those of the other of the two neighbouring and hybridising forms'. In 1935 I collected a series of spotted hyaenas in a comparatively small area of Tanganyika; it contained all the named subspecies, and enough material to describe as many more [315].

Spotted hyaenas inhabit all of Africa south of the Sahara, generally preferring the more open country to forest; they have been exterminated in the more civilised and southern parts of their range. They are nocturnal, and den up in natural holes among rocks and elsewhere, in the burrows of aardvarks, often enlarged by themselves, or in patches of bush. Spotted hyaenas are loosely gregarious, and in some places form 'clans' which gang up in search of food. They not only eat the remains of any dead animals, and carrion of all sorts, but catch living prey, especially the young of the ungulates – they also sometimes pull down the adults. Although they habitually clean up the remains of the lion's kills, the role is sometimes reversed for Kruuk [282] has found that in parts of Tanzania lions feed on the prey killed by hyaenas. Hyaenas kill domestic stock, especially sheep and goats, occasionally kill children or mutilate a sleeping adult, scavenge native villages, and among some tribes, are the recognised disposers of the dead, who are 'given to the hyaenas'. The spotted hyaena is much noisier than the other species; when it starts prowling in search of food at dusk it repeatedly utters a loud 'woof' starting as a gruff rumble and finishing on a high shrill note. It is, however, when hyaenas find some carrion and settle to feed that the characteristic laughing cry is heard – it is a chuckling sound much resembling human laughter, and though loud and noisy, is nothing compared with that produced under sexual stimulation. A number of hyaenas, among which there was a female on heat, once spent several hours rushing about my camp in great excitement [314]. 'The noise they made was indescribably hideous. Shrill shrieks and yells, accompanied by deep emetic gurgling and groans, made a background for wild peals of maniacal laughter. . . . The hair-raising din is indescribable, and is truly horrible to human ears'.

The external genitalia of the female spotted hyaena are peculiar, for there is no vulva, and the large clitoris closely resembles the penis of the male; in addition a pouch-like swelling simulates the scrotum of the male. At oestrus and parturition the aperture of the urinogenital canal enlarges, and the tissues surrounding it and the canal become hypertrophied and loose, so that copulation and parturition can occur. After the first parturition there is some return towards the former condition. During lactation the mammary glands and their teats are much enlarged, and the teats do not return to their original small size afterwards. It is no wonder, therefore, that an apparently male animal with large mammary development should have given rise to endless legends that the hyaena is hermaphrodite, or that it can change its sex at will – such yarns have been current for thousands of years and were recorded by Aristotle, Aelian, Pliny and others. The corresponding genitalia of the striped and brown hyaenas, on the other hand, do not differ from the pattern generally found in the mammals [314]. The one or two cubs of the spotted hyaena are clothed in dark brown hair at birth, and acquire the spotted pattern gradually as they grow up.

The aardwolf, *Proteles cristatus*, looks much like a small, sparsely marked, striped hyaena with slender legs. It lives in southern and eastern Africa, becoming rarer northwards to the Sudan and Abyssinia; it inhabits open and bushy country but not forest. The teeth of the aardwolf, except for the canines, are very small and weak; the cheek teeth are reduced in number and are no more than small conical points. The dentition is correlated with the animal's diet, which consists almost entirely of insects, and mainly of termites. Aaardwolves are nocturnal and live by day in burrows, often those dug by aardvarks, sometimes dug by themselves, or enlarged from those of smaller animals. Farmers in South Africa have thought that aardwolves attack young stock; this has been generally denied on the grounds that they are entirely insectivorous though Bothma [49] found that the stomach of one from the Transvaal contained a few insects, the remains of a rat and a tortoise, some green grass, and a large quantity of carrion. The nocturnal and burrowing habits of the aardwolf often lead to its presence being overlooked even where it is common, and it is not, as sometimes supposed, a very rare animal.

FAMILY FELIDAE

The last family of the carnivores is the Felidae, containing four genera and nearly forty species of cat, large and small. Members of the family are found throughout all the continents except Australia. Felids are medium to large sized animals, nearly all clothed in soft fur; they are short-faced, and the eyes look forwards so that they have good binocular vision. All are digitigrade, and all except one have claws retractile into fleshy sheaths. The canine teeth are large, but the cheek teeth are much reduced in number though the carnassials are well developed; the tongue is covered with curved horny papillae with the points directed backwards. Felids are typically predators on other mammals, and most of them are nocturnal.

The majority of the felids are classified in the genus *Felis*, a genus in which there is so wide a range in size, build and particularly in colour pattern, that it was formerly split into numerous genera, which are now regarded as not more than subgenera or are entirely discarded. A similar range of variation is found within some species, such as *F. silvestris* the wild cat of Europe, Africa, and much of Asia, in which several forms once thought to be separate species are now included. Haltenorth [195] has sorted these out into no less than forty named subspecies. The colour pattern of the wild cat ranges from the familiar tabby through lightly and heavily spotted and streaked, to almost unmarked self colour. *F. silvestris*, particularly the north African subspecies *F. s. libyca*, is thought to be the ancestor of the domestic cat; but no doubt there has been much mixing of the various subspecies from time to time in different parts of the world. *F. chaus*, the jungle cat of Asia and North Africa, and the smaller sandy-coloured *F. margarita* of north Africa and southwest Asia, are species closely related to *F. silvestris*. *F. nigripes*, the black-footed cat of southern Africa, is the smallest species of the family; it is a prettily spotted desert species, now rather scarce. The leopard cat *F. bengalensis* from much of southern and eastern Asia is strongly spotted; the fishing cat *F. viverrina*, of southern Asia and the East Indies, feeds much on fish. Several other Asian species, such as the marbled cat *F. marmorata*, and the golden cat *F. temmincki*, are beautifully marked, the latter with much rich chestnut.

The lynx, *F. lynx*, of the northern parts of both Old and New Worlds, the bobcat *F. rufa*, of North America, and the caracal, *F. caracal*, of Africa and southwest Asia, are long-legged, short-tailed cats with conspicuous tufts at the tips of the ears, and in the first, two tufts on the cheeks as well. The lynx and bobcat have dense slightly spotted fur; the caracal has shorter fur, sandy or reddish in colour, without spots. The serval, *F. serval*, of Africa south of the Sahara, is another long-legged cat, but of slim build, and yellow colour with black spots. The South American cats include some very handsomely marked species with black spots or rosettes on a yellow ground colour, such as the ocelot *F. pardalis*, the smaller margay, *F. weidi*, the tiger cat, *F. tigrina*, and several others. The jaguarondi, *F. eyra*, found from Texas to Argentina, is a short-legged small-headed cat, with long tail and short rounded ears, dark brown, red or buff in colour entirely without spots – the colour phases are so distinct that they were formerly thought to be different species. The pampas cat, *F. colocolo*, the Andean cat *F. jacobita*, and Geoffroy's cat *F. geoffroyi*, are striped and spotted in patterns more closely resembling the tabby. The puma *F. concolor*, found or formerly found throughout most of North and South America, is the largest member of the genus – in colour and build it superficially resembles a small maneless lion.

The five big cats form the genus *Leo* – until recently *Panthera*. The tiger, leopard and jaguar are forest animals, but the lion lives on dry plains and in bush country, and has a pelage of short hair unlike the fur coats of the others. The tiger and leopard, however, are not confined to tropical jungles, but are found far to the north among the snows of Manchuria and Mongolia. The snow leopard, *L. uncia*, of the Himalayas and Tibet, is more fully adapted to a cold climate; its fur is longer and paler and more diffusely spotted than that of the common leopard. Although the male lion *L. leo* with its shaggy mane on head, neck, chest and belly looks so utterly different from the striped tiger with its ruff round the face, the two animals are surprisingly alike in their anatomy. Indeed, once the hide has been removed it is almost impossible to distinguish the carcase of a lion from that of a tiger. The dried skulls can be separated by a slight difference in the proportions of the nasal and maxillary bones, and in the profile of the lower jaw which makes the skulls balance differently when placed on a flat surface. Yet in life the two species differ widely in appearance, voice and habits. The lion, unlike the other big cats which are solitary except when a female is bringing up her cubs, shows some social behaviour, and often occurs in loosely knit groups generally based upon the family. Lion cubs are spotted, but the spots disappear as the cubs grow up; nevertheless many adult and subadult lions retain some faint spotting, particularly on the limbs. The lion, now almost confined to Africa, was in historic times widely spread also in southeast Europe and western Asia – a small population still survives in Kathiawar.

Kruuk and Turner [283] point out that lions do not co-operate in hunting and that any appearance of one driving prey towards another is merely accidental. Lions stalk their prey and catch it with a final short rush; they grab the prey with one paw on the back and the other on flank or chest, drag it down, and kill by biting the throat and holding on until it dies of suffocation, never by neck breaking. Sometimes stalking is replaced by lying in ambush, but if the prey is not caught in the first rush the lion seldom pursues more than fifty to one hundred yards. Lions prey mainly on medium-sized animals, but can take large ones and often take small ones. They also commonly eat carrion, and scavenge after hyaenas; they locate carrion by watching vultures and sometimes thus find it at a distance of over a kilometre. Kruuk and Turner estimated that the 700 lions living in the 8,000 square miles of their study area on the Serengeti plains, killed 7,800 prey animals yearly, three quarters of them wildbeest and zebra, and consumed over a million and a quarter kilograms of meat.

The tiger is confined to Asia, and is much less abundant than it was a century ago. Schaller [417], in his comprehensive study of wild tigers, says that in India, owing to the drastic decline in the country's wildlife in recent years, 'the tigers in most forests subsist at present partly to wholly on cattle and buffalo; in fact, it is likely that these two domestic animals are the chief source of food of the country's tiger population'. In spite of the depredations of man-eater tigers, which have sometimes caused a local panic, Schaller remarks that 'although man is the most easily obtainable source of food throughout the tiger's range, he is for unknown reasons rarely eaten'. Is it possible that during thousands of years man has selected a strain of tigers that are not usually man-eaters, by hunting down and killing those that are, just as he does today? As Schaller says, 'the tiger's reputation for savagery appears to be largely based on its potential to do harm. . . . A person on foot rarely sees a tiger because the animal perceives him first and avoids encounter by hiding or sneaking away'. Apart from man-eaters, the legend of the ferocity of tigers appears to have been created by sportsmen anxious to boast of their prowess and to inflate their social status.

The leopard, *L. pardus*, often called panther in India, is very widely distributed through Asia and Africa though its numbers, like those of all the big cats, are much reduced in many places. It is the most variable of the genus in size; the degree of spotting, too, ranges widely so that different species were formerly named. A black mutant is well known and appears to occur more frequently in regions of dense humid forest. In East Africa Kruuk and Turner [283] found that the leopard hunts chiefly by stalking or from an ambush; while stalking the body is often held low or even touching the ground, the leopard 'proceeding with almost snake-like movements, eyes and ears fixed on the spot where the prey is'. The final rush is very short, often only a single bound, and the prey is grabbed by the fore-paws; both animals roll over on the ground while the leopard bites at the throat. Leopards appear to have individual preferences for different kinds of prey – Kruuk and Turner heard of one that selected jackals. When leopards make a kill too large to be eaten at one meal they cache the remains in the fork of a tree, secure from scavengers, and live on it for several days. Lions have been known to rob a leopard's larder, and hyaenas have several times been seen to chase a leopard away from its kill on the ground. The leopard is more of an opportunist than the lion, and takes a much greater variety of prey species. The jaguar, *L. onca*, is the South American equivalent of the leopard – although it is there known as 'tigre' – and similarly has a fairly common black mutant. It is rather more heavily built than the leopard, but is almost equally agile as a tree-climber. All these big cats are nocturnal, and prey upon any other animals that they can overpower. The clouded leopard, *Neofelis nebulosa*, although placed in a separate genus is generally regarded as one of the big-cat group – it inhabits southeast Asia and the East Indies. Its pattern of large blotches with darker borders makes its pelage one of the most beautiful among the felines.

Finally, the cheetah, *Acinonyx jubata*, the long-legged hunting leopard, differs from the other cats in not having retractile claws. Its range formerly extended from South Africa through western Asia to India, but it has been exterminated over much of this vast area, and is now found mainly in East Africa. There has been a widely held belief that the cheetah is decreasing in numbers in East Africa, and Graham [182] undertook a survey on behalf of the East African Wildlife Society to find if there is any foundation for it. He reported that cheetah are found over some 406,000 square miles of East Africa wherever the habitat is suitable, and that 'nothing was found during the survey to suggest that the animal is declining or that population densities are lower now than they were, except possibly in the Narok district of Kenya'. It lives on open plains and catches its prey, which consists of the smaller antelopes, by running it down in a fast sprint – the cheetah is the swiftest mammal over a short distance. Cheetahs can be readily tamed and were once much used in India for hunting game for sport – they have even been tried for track-racing, but were not a success; for when they saw that they were unable to catch the electric hare they lost interest and would not run. A beautiful mutant with the spots confluent into longitudinal stripes and loops has been found several times in Rhodesia and erroneously described as a separate species, the 'king cheetah'. Eaton [128] has recently observed a most interesting relationship between cheetahs and jackals in Kenya. It is an isolated example, and not a regular habit, but it has implications that are important. The two species co-operated in hunting;

one adult jackal distracted the prey by running in the herd [of Grant's gazelles] and barking. While the herd's attention was centred on the jackal, the adult female cheetah got within killing range before being seen, and attacked. The jackals scavenged the remains. Mutualistic hunting by these two species lends credibility to the theory that the Canidae were domesticated by a hunting relationship by man.

He goes on to suggest that the presence of scavenging canines might have led to co-operative hunting between them and man, and adds that it is equally possible that it was man who scavenged the canid and thereby established the symbiosis.

Schaller [418] found that in the Serengeti National Park nearly ninety per cent of the prey of the cheetah consists of Thompson's gazelle, though in the Kruger National Park the main prey is the impala; the prey taken is the species most readily available. Over half the gazelles taken were subadult, and there was a strong selection for fawns; large fawns and adults escaped from about half the pursuits observed. The cheetah spends most of the day lying under the shade of a bush or in thick grass 'seemingly waiting for prey to wander into the vicinity'. It hunts at any time of day, but mostly in the early morning or late afternoon. Schaller found that cheetahs stalk up to about 100 yards from their prey, although Kruuk and Turner [283] who worked in the same area considered that stalking is of little importance in their hunting. Schaller says that sometimes the cheetah walks towards a herd of gazelle with no attempt at concealment; the gazelle watch the cheetah until it is about seventy yards away and then stampede whereupon the cheetah, having selected its victim, makes a rush. At other times the cheetah bounds towards an unsuspecting herd, and follows a selected individual when the herd flees. When the cheetah catches up with its prey it slaps a hindleg or the flank and bowls it over, and then grabs it by the throat and kills it by strangling it – a process that may take ten minutes. Twice females have been seen to carry a young fawn alive to her cubs and release it before them so that they could chase it. On another occasion three cubs were seen playing with a fawn:

one at a time, a total of eighteen times, the cubs swatted the fawn and bowled it over, yet it continued its attempts to escape. The mother cheetah suddenly rushed up, bit the fawn in the neck, but then released it. Again the fawn tried to run away but, after being knocked over once more, it merely crouched. The three cubs surrounded it and one grabbed its throat.

The cheetah eats all but the bones, skin and intestines, and drinks the blood that drains into the body cavity. Eight per cent of the kills were taken away from the cheetahs by lions, and four per cent by hyaenas.

Ewer [141] points out that the apparent teaching of young to hunt by the mother, which is common among many species of carnivores, is really the 'creation by the parent of a situation in which the responses of the young automatically lead to their learning'. There is no need to suppose that the parent necessarily understands what she is doing; a suricate reared in isolation showed the typical feeding behaviour as soon as the young of her first litter were the right age to start catching prey, but in the artificial conditions of captivity her efforts sometimes actually hindered the feeding of the young, and 'she quite clearly did not comprehend the function of her own behaviour'.

All the big cats are well on the way to extinction; most of them bear valuable fur, all are destructive pests on man's flocks and herds, the larger ones are favourite quarry for the sportsman, some are frequently man-killers, and the widespread availability of modern firearms makes it easy for savage as well as civilised man to kill them for pleasure or profit (Plate 18).

Chapter 10

Seals. Sea Cows

PLACING THE seals and sea cows in the same chapter is merely a matter of convenience, and in no way implies any close relationship between these animals, which belong to very different orders; their only common characteristic is that they are all aquatic mammals.

The order, or suborder, Pinnipedia contains three families, the Otariidae, the otaries or eared seals, the Odobenidae, the walrus, and the Phocidae the true or earless seals. The pinnipeds are distinguished by the fusiform shape of the body, the webbed feet modified as flippers with the first or second digit of the fore flipper generally the longest, and the first and fifth of the hind approximately equal and longer than the other three. The tail is very short, and a layer of blubber covers the body beneath the skin. Although seals are aquatic, with rare exceptions they must haul out on to land or floating ice to give birth to their young and nurse them during lactation. A single young at birth is the rule, and delayed implantation occurs in those species in which the reproductive physiology has been investigated. This phenomenon is apparently correlated with the annual congregations for breeding – on arrival the females give birth and are shortly after re-impregnated, thus having an apparent gestation period of nearly a year though the active period of development of the foetus is much less. The main natural enemy of seals is the killer whale, and for some species the polar bear, but the large herds that leave the water at the breeding season are very vulnerable to predation by man, and have been excessively exploited for their skins for fur or leather, for their blubber-oil, and to a lesser extent for their meat, bones, and other products. The history of commercial sealing from the sixteenth to the twentieth century has been one of short sighted extermination, but now wiser policies are gaining ground and the future for most species and for their rational commercial use is brighter – but the change has come only just before it would have been too late.

The interesting adaptations for deep diving and prolonged apnoea in the circulatory and respiratory physiology have been discussed in Volume I, as have the territory holding habits of fur seals, elephant seals and others. Many species habitually swallow stones – the stone-filled stomachs of fur seals were known to the sealers of the last century as the seals' 'ballast bag' – but the reason for the habit, and what use the stones may be in seals' economy have not been discovered. The eyes of seals have an unusually flat cornea which enables an image to be brought to sharp focus when the eye is immersed in water. In air an eye with a flat cornea cannot focus sharply; when out of the water, therefore, the iris of a seal's eye is strongly contracted and functions as a pin-hole lens, thus giving a sharp image though with some loss of illumination. When a seal dives under water the iris opens widely to admit all possible light through the comparatively large pupil. Seals utter many sounds underwater, clicks, and high pitched whistles, and some of them are used in echolocating – in one species of otariid, the Californian sea lion, however, Schusterman [428] found that underwater echolocation was not highly critical and that sight under water was also needed for discrimination. Nevertheless, he observed that the amount of sonar clicking was inversely proportional to the degree of visibility.

The pinnipeds are essentially marine, though they often frequent estuaries and enter rivers, and some races of two species are landlocked in large freshwater lakes. Some species have regular and long seasonal migrations; others are comparatively sedentary. Although seals are generally thought of as animals of polar and temperate seas a few are confined to the tropics and even live on the equator. The Otariidae and Odobenidae are fairly closely related, but both are widely separated from the Phocidae. The first two families and the third have evolved in parallel, perhaps from a remote common ancestor, but more probably from separate stocks. Some zoologists suggest that the phocids originated from the stock that gave rise to the otters, and the otaries either from that ancestral to the dogs or, more probably, the bears; however that may be, the polyphyletic origin of the pinnipeds is generally agreed. Limitations of space prevent more than a mention here of some of the many interesting matters known about the biology of seals and their populations; the recent works of Scheffer [421] and of King [268] comprehensively review the taxonomy and biology of the members of the Pinnipedia.

FAMILY OTARIIDAE

The Otariids are so named from their possession of external ears, small though they are, but a much more fundamental difference from the

U

phocids lies in their method of locomotion both on land and in the water, which is linked with their bodily architecture, especially that of the limbs. The long fore flippers are the main agents for propulsion when swimming – their surface area is increased by a comparatively wide axillary fold at the base, and by cartilaginous extensions of the digits at the free ends. When moving on land the otaries lift the body clear of the ground, supporting it on the fore limbs the flippers of which are turned outwards, and the hind limbs the flippers of which are turned forward; when hurried they can reach a surprising speed at a lumbering gallop. The toes of the hind flippers also are extended by cartilaginous rods; as a consequence the nails which lie at the ends of the terminal phalanges are remote from the edge, as are those of the fore flippers. The cartilagious extensions of the hind flipper can be folded back, and the nails of the three inner digits can then be used for scratching the body; the nails of the outer digits of the hind flipper, and all those of the fore flipper are rudimentary. In the otaries sexual dimorphism is much more pronounced than in most of the seals. The adult bulls are much larger and several times heavier than the females, and show an enormous and apparently disproportionate development of the neck – a bull Steller's sea lion may weigh as much as a ton whereas an adult female is little more than a quarter of that weight. The otaries are social during the breeding season and assemble in rookeries, often of a large size, where the bulls guard a territory on which their harem of cows is crowded. Outside the breeding season the otaries are pelagic, seeking their food of squids, fish and crustacea; some species do not go far from their home beaches, but others travel great distances, and at least one, the northern fur-seal, makes long migrations (Plate 19).

The seven genera of otariids fall naturally into two groups, the sea lions and the fur seals, differing in that the latter possess a dense undercoat of rich soft fur under the guard hairs, whereas in the former the undercoat is negligible. The five species of sea lion are all placed in separate genera – one wonders if so much splitting is really necessary – and the seven of fur seals in only two. Otaries inhabit the southern oceans between the tropic and the antarctic circle, but are found north of the equator only in the Pacific.

Of the sea lions, the northern Steller's sea lion, *Eumetopias jubatus*, breeds on the islands of the north Pacific, and ranges from California to Japan, but the Californian sea lion, *Zalophus californianus*, is more restricted to the waters of the Californian coast, with isolated populations on the Galapagos Islands and off Japan – the latter is the species commonly exhibited doing tricks in circuses. Of the southern species the southern sea lion, *Otaria byronia*, is widely distributed on the coasts and islands of

South America, the Australian sea lion, *Neophoca cinerea*, on the southern coasts and islands of Australia, but Hooker's sea lion, *Phocarctos hookeri*, is confined to the subantarctic islands of New Zealand. The sea lions take their name from the shaggy mane of long hairs on the neck and shoulders of the bulls; the short head and rounded muzzle adds to the resemblance. The sea lions carry a thick coating of blubber, which has led to the commercial exploitation of all species for the sake of the oil that can be extracted and which has similar uses to whale oil; the skins too were much used for making leather. The commercial killing of sea lions started in the seventeenth century and was at its height in the eighteenth and early nineteenth – it is by comparison very limited at the present day. The toll taken by the sealers, though heavy, was not so devastating to the populations as was the slaughter of the more valuable fur seals.

The fur seals, one northern species, the Pribilof fur seal, *Callorhinus ursinus*, and seven southern species of the genus *Arctocephalus*, differ from the sea lions not only by their fur, but by the much more sharply pointed snout. Wherever they lived they were hunted almost to extermination by sealers in the past although, because of their gregarious and polygamous habits, it is easy for man to take an annual crop of the preferred younger males without damaging the stock. There was, however, one exception; the sea lions and fur seals on the Lobos Islands at the mouth of the River Plate were before 1825, as Weddell [502] tells us, farmed out by the Governor of Montevideo to sealers on whom he imposed restrictions such that the taking of a valuable annual crop did not injure the breeding stocks. Weddell reckoned that if the South Shetlands herds had been rationally cropped in a similar way they would have 'been spared to render annually 100,000 furs for many years to come'. Nearly 100 years were to pass before sensible management was given to another species, the Pribilof fur seal. The herd breeding on the Pribilof Islands, drastically reduced in the nineteenth century, made a spectacular recovery from less than half a million to over one and a half millions in about twenty-five years. The Pribilof seal was threatened with extermination not only by the slaughter on its breeding grounds but also by pelagic sealing in the north Pacific during the winter dispersal of the herds towards the south. Pelagic sealing was stopped by international agreement, and the nations concerned receive part of the profits of the crop from the American owned Pribilof Islands in compensation for refraining.

The South African fur seal, *Arctocephalus pusillus*, a coastal species of southwest and southern South Africa, was similarly overfished in early days, but the population made some recovery in the intervals when sealing was reduced owing to lack of quarry. The large population of this species is now managed by government. The once numerous populations of other southern species have not yet completed their recovery. The South American fur seal, *A. australis*, was killed extensively on the coasts of South America, the Galapagos and the Falkland Islands during the eighteenth century, as were the species *A. doriferus* and *forsteri* from southern Australia where their ranges overlap, and New Zealand. The widespread species, *A. tropicalis*, which is never found in the tropics but inhabits the shores and waters of the subantarctic islands from the South Shetlands east to Kergulen and Tristan and Gough Islands, was practically exterminated at South Georgia by the early nineteenth century although immense herds lived there before. In 1819 the South Shetlands were accidentally discovered, and the enormous numbers of fur seals of the same species there immediately attracted attention; in two seasons of slaughter they were exterminated – at the cost of a number of shipwrecks among the uncharted and dangerous islands.

Fifty years ago, in the 1920s, one or two fur seals were occasionally reported as sighted on remote parts of the inhospitable shores of South Georgia, but the species was rare and showed no sign of recovering its numbers. In 1933 a small breeding colony totalling about sixty seals was discovered on an offlying island, and by 1962 Bonner [47] found it had built up to about 40,000, and small colonies were appearing on the mainland. Small numbers have also been seen in the South Orkneys and Shetlands and elsewhere, so it is possible that the huge populations that inhabited these regions before the great exterminations may now build up again. Throughout the nineteenth century recovery was impossible because occasional raids by sealers wiped out any stocks that had managed to make even slight increases. A similar history applies to the fur seals of all the southern oceans, and even to the very rare Guadalupe fur seal, *A. philippi*, a tropical species from the Lower Californian island from which it takes its name, though it has not yet returned to the Juan Fernandez group. The fur seals feed on fish, squid and crustacea – in some southern waters *A. tropicalis* feeds almost exclusively on the shrimp *Euphausia*, the krill that forms the food of the whalebone whales. The annual crop harvested on the Pribilof islands is about 70,000 pelts, which find a ready market, and it may well be asked, where do they all go, for how often does one see a person wearing a sealskin coat? In the early days of sealing immense quantities were sent to China, but there, according to Bonner, they were probably not used as furs – the fur was clipped off for making felt and the skins probably discarded.

FAMILY ODOBENIDAE

The family Odobenidae contains a single species, *Odobenus rosmarus*, the walrus, circumpolar in the shallower waters of the arctic seas, but divided into two subspecies, the Atlantic and the Pacific races. In general body shape and in the ability to turn the hind flippers forwards when moving on land or ice the walrus resembles the otaries, and probably evolved from the same basic stock – it is certainly more closely related to them than the phocid seals. Walruses are large animals, for adult bulls can reach a weight of a ton and a quarter and females about half that. They are characterised by their thick wrinkled skin which becomes practically hairless in old animals, the large canine teeth which form the projecting tusks of both sexes, and the large mystacial pads carrying vibrissae so stout that they are more like quills than hairs. Walruses congregate in large herds often of many hundreds on sea beaches or ice floes, and seek their food on the bottom in depths up to about forty fathoms. Their food consists mainly of bivalve molluscs, which they are said to expose by stirring up the mud or gravel with their tusks; they certainly use their moustache and lips in manipulating their food, and for holding clams so that they can suck out the contents. In addition to the tusks there are four teeth on each side of both jaws; they are conical when erupted but soon become worn down to a flat surface. The wear might be caused through cracking clam shells to allow the meat to be sucked out, after which the shells are spat out, but there seems to be no observation on this point and the wear is unexplained. The food is swallowed without being chewed – a walrus stomach filled with slightly digested clams is, or was, regarded as one of the greatest delicacies of the arctic by the Eskimo. Although the usual food consists of molluscs with some other invertebrates certain walruses are reported sometimes to eat seals, and indeed seal blubber has been found in their stomachs; but it is not clear how a walrus can seize and tear a seal to pieces with its small flattened teeth and its gape obstructed by the tusks.

The tusks are not used solely for submarine gardening but are also brought into play when a walrus clambers out of the water on to ice or rocks, and in squabbling among the bulls. Lamont [284] wrote,

From the animal's unwieldy appearance and the position of his tusks one is apt to fancy that the latter can only be used in a stroke downwards but on the contrary they can turn their necks with great facility and quickness and can strike either upwards, downwards, or sideways with equal dexterity. I frequently observed them fighting with great ferocity on the ice and the skins of the old bulls which are light coloured and nearly devoid of hair, are often covered with scars and wounds received from these encounters.

The walrus is polygamous but the bulls do not hold harems of females as do the otaries; the young, born in the spring, suckle for at least a year and often for much longer, in great contrast to the short lactation periods of some phocid seals. There is a general seasonal movement, not amounting to a migration, as the herds follow the ice north in spring and south in the fall, but not all of the population takes part.

It has long been known to arctic hunters that the walruses can make a ringing noise resembling the sound of distant church bells, in addition to the roars and bellows that might be expected, but it is only recently that zoologists have learned of the matter. Fay [142] made careful dissections, and found that the pharynx is extremely elastic and is expandable as a pair of large pouches, which extend far back under the skin on each side and may reach the hind end of the thorax. They are often asymmetrical and are developed only in adults, always in bulls and sometimes in cows, and can be inflated with air at will. Fay found that the pouches are used for increasing buoyancy both when sleeping and swimming; one sleeping animal was shot and killed instantly as it slept 'whereupon the pouches deflated and the carcass sank immediately, demonstrating the high specific gravity of the animal and the importance of the inflated pouches in maintaining his position at the surface'. Fay also recorded that the pouches might be connected with the production of the bell note, and this was confirmed by Schevill, Watkins and Ray [424] who made recordings of the sound from a tame captive walrus, and concluded that, 'the pouches are a secondary sexual characteristic, used both for sound production and flotation during courtship and coition, as seen in our captive. Flotation during rest would be useful as well'. They also confirmed that the pouches act as resonators for the bell; Ray took part in an Eskimo walrus hunt, when an adult was shot dead with the one pouch inflated; after the pouch was exposed by removing the skin over it, 'when it was struck with the flat of a knife blade, a bell-like tone almost identical to that recorded was produced'.

The hunting of walruses has always provided an important contribution to the livelihood of people living in the arctic: oil, meat, bones, skin, tusks and many other things. The commercial exploitation of walruses by people from elsewhere began in the sixteenth century with the voyages of Europeans to hunt 'sea horses' at Spitzbergen. Thereafter the killing of walruses was combined with whaling and sealing wherever ships could penetrate into the northern Atlantic and Pacific; the populations were overcropped, the animals became scarce, and were exterminated at the more accessible places such as the Pribilof Islands and Spitzbergen. The commercial hunters sought the walrus for its ivory, its oil and its hide – the skin from the neck of a bull, nearly three inches thick is still valued for making discs for buffing or polishing metals. It is astonishing to read of an English yachtsman [284] who in 1876 was able to sail to Novaya Zemlya with no political let or hindrance, and hunt walruses in order to pay the expenses of fitting out his yacht for a pleasure cruise in arctic waters. The reduced stocks of walruses are now protected in parts of the Russian and American arctic, and are killed only by resident natives to whom they are a necessary part of their living.

FAMILY PHOCIDAE

Thirteen genera with a total of eighteen species of true or earless seals make up the family Phocidae; their fore flippers are comparatively small, and their hind flippers cannot be turned forwards under the body for locomotion on land or ice. When out of the water most species progress by a looping movement, alternately hitching the body along with the fore flippers and pushing with the pelvic region. The fore flippers are pressed to the sides of the body when swimming, and propulsion is effected by the action of the hind flippers, which sweep from side to side. The plantar surfaces are used alternately; each flipper is extended by separating the digits on the power stroke, and is relaxed and partly folded on the return. Although they do not have an external ear pinna like that of the otaries, some species are able to erect a rudimentary ear cartilage slightly above the surface of the head when listening out of the water. The young are active as soon as born and grow rapidly during the comparatively short lactation. The milk teeth in most species are rudimentary and are resorbed or shed before or soon after birth; in many the permanent teeth have cut the gums before birth. Seals are easily killed by a blow on the snout because the interorbital region of the skull is narrow and easily broken with a club. The bulls of large species with more massive skulls were traditionally attacked with lances, but now a rifle is preferred.

The pelage of the seals consists of stiff hair, smooth when stroked with the lay but harsh against it – hence the use of sealskins in ski-ing – and has only sparse undercoat hairs, except in bull elephant seals, which are nearly naked. Most species are spotted, some are uniform, and a few strikingly patterned, but the colour and appearance of the coat varies greatly. A newly moulted animal is very different from its appearance when the coat is old and faded, and the difference in the coat when wet and when dry is great. If the coat dries when a seal is out of the water a wet patch remains round and below the eyes because seals have no naso-lachrymal ducts to lead the excess tears from the eyes into the nose – the duct is present in most mammals – and the abundant tears overflow on to the face. It may be that a physiological function is served by this perpetual weeping. Most seals are entirely marine – how therefore do they manage to do without drinking fresh water? Although they will drink fresh water readily when they have the opportunity, even when ashore on sea beaches or rocks their opportunities must be limited, and when they are away at sea for months on end there are none. They must therefore obtain the water needed in their metabolism from their food such as fish, squids, crustacea and so on, or manufacture it from their blubber, unless they can drink sea water and separate the salt from it. The kidney of mammals in general is not capable of selectively removing salt from the blood, and it has not been shown that the kidney of seals differs on this point. Although the organ is lobulated and has an unusually large venous drainage, it is not similar to the complex and compound kidney of whales. Is there, then, any other way of dealing with unwanted salt? Many kinds of sea birds – gulls, petrels, auks, penguins and others – drink sea water, and consequently have to get rid of the excess salt; they possess a large gland above the eye, the lateral nasal or supra-orbital gland, whose duct runs directly into the nose cavity. It is the lateral nasal gland that excretes the unwanted salt and preserves the electrolyte balance in the body tissues. If these birds are fed with salt, or are injected with salt intra-venously, the gland at once starts to work, and a stream of drops of salt solution runs from the tip of the beak; the drops are flicked away by a characteristic head-shake. Marine turtles, too, excrete salt in their tears. It is possible that seals drink sea water, and preserve their electrolyte balance by excreting the salt in their copious tears – even human tears are salt, as everyone knows – and it may well be that those of seals are even saltier. The matter is worth testing experimentally, as this suggestion appears not to have been made hitherto.

At birth seals are clothed in a coat very different from that of the adult; it is soft and woolly and either uniform white, light grey, or dark brown in colour. It is lost by moult usually in a few weeks after birth, though it may, as in the common seal *Phoca vitulina*, be shed before birth; the coat that succeeds it consists of hair resembling that of the adult but is softer and sometimes differs in colour from that of the adult. Scheffer [421] suggests that the species with white coats are, or were formerly, those that bred in polar regions, and those with dark coats have always bred on sea beaches and not floating ice; the colour thus serves to conceal the young seals from predators. The white colour of arctic animals is believed to have only a procryptic function because it gives no advantage in temperature regulation, a point already discussed in Volume I. The suggestion thus appears to be plausible, but one may ask what are the predators on young seals; in the arctic there is the polar bear, and sometimes the arctic fox, but white coats give no protection from them, and elsewhere there is no habitual predator on young seals, white or black, except man. The correlation between light coats for breeding on ice, and dark ones on land certainly holds – all the otaries, too, have dark natal coats – and no doubt the colour makes the animals less conspicuous, but it is difficult to find any protective advantage in the colour, in the absence of potential heavy predation. The grey seal, *Halichoerus grypus*, has a white natal coat which renders the young very conspicuous where they are born on land as on British coasts, but not on ice as in the Baltic Sea – but the only predator is man, who protects them in Great Britain, but pays a bounty on those killed in the Baltic. A dark natal coat can be a disadvantage; the dark pups of the elephant seal, *Mirounga leonina*, are sometimes born on snow compacted into ice, and if a few days of sunshine occur the ice melts below the pups and they sink below the surface as does a lump of coal or any dark body placed on it. They are then unable to climb out of their narrow cells, where they starve to death.

Many, but not all of the phocids, congregate in large numbers to give birth to their pups, either on certain beaches or on floating ice generally in specific geographical areas; after breeding a few species show definite migrations, but more generally there is a wide dispersal over the feeding range, and often some wandering by odd individuals, which turn up in unexpected places. In some species that haul out for the duration of the moult there is another gathering later in the year for this annual process. The animals generally do not feed during their haul-outs, and some species such as the elephant seal then fast for several months. The seals form three geographical groups – those of the arctic and north temperate, of the tropical, and of the antarctic and south temperate regions.

In the arctic group the bearded seal, *Erignathus barbatus*, is circumpolar; it is a large species and, like the walrus, is a bottom feeder – its name is derived from its very long and abundant mystacial whiskers, which are probably used in feeding on invertebrates. They have the unusual property of curling into a fuzzy bush at the tips when dry, but straightening out into a splendid fan when wet (Plate 20). The bearded seal breeds on the ice, but is not gregarious. The ringed seal, or floe rat, *Pusa hispida*, is also circumpolar; it lives as well in the Baltic Sea and, on the Pacific side, reaches as far as Japan. It is one of the smallest seals, and the darker spots on its grey pelt are bordered with light rings; it feeds on the larger planktonic crustacea and small bottom fish. Like some antarctic seals it winters under the ice – the water is much warmer than the air – and keeps breathing holes open. The young are born in dens under the snow or among hummocky ice. In the spring the rutting males give off a strong foetid smell; indeed all seals have a characteristic smell, and as no glands are known that could produce it, it is perhaps a by-product of their food – the musky, ambergris-like odour of some fur seals might be correlated with a diet of squid. The ringed seals that live in the freshwater lakes connected with the Gulf of Finland are regarded as subspecies, as are those of the Baltic, but the seals of Lake Baikal, *P. sibirica*, and of the Caspian Sea, *P. caspica*, are land-locked species probably descended from ringed seals that were cut off from the arctic in the Miocene.

Figure 18 Harp seal (*Pagophilus*).

The harp seal, *Pagophilus groenlandicus*, and the hooded seal, *Cystophora cristata*, inhabit the Atlantic side of the Arctic. The adult male harp seal is white with black face and an irregular black band running from the shoulders along each side and ending above the tail; the females are duller, and young males are grey with irregular spotting. Harp seals are migratory and assemble in great numbers on the ice in spring for breeding. The pups are clothed in a thick woolly white coat, and are the subject of an extensive fishery at the three main breeding areas, in the White sea, on the western ice east of Greenland, and off Newfoundland and in the Gulf of St Lawrence. Adult and immature harp seals also are taken for their blubber and skins. Accounts of the fishery have disturbed humanitarians, who have recently been re-assured that it is not carried on with needless brutality. Sivertsen [446], who made a long study of the biology of the harp seal, concluded that the seals eat little food when on their breeding grounds, and that pelagic and demersal crustacea are seasonally equally as important as fish in the diet. The hooded or bladder-nosed seal reaches a large size, old bulls being as much as ten feet long; it is not a gregarious species but comes on to the ice in family groups which lie nearer the ice-edge than the harp seals. The young shed the natal coat before birth, and the second coat, in which they are born, is soft and blue grey above, light below; blueblack skins are commercially valuable.

In the male there is an extraordinary enlargement of the nose cavity which extends up on to the head between the eyes; this hood can be inflated with air at will and blown up like a balloon. Olds [353] found that the anatomical features of the hood are present also in females, though they are not known to inflate it. The hood is thus a secondary sexual character, but it is also inflated when the bulls are attacked by man. Olds found that it holds about six quarts of water and that its valvular append-ages make it air-tight; the walls consist of fibrous and elastic tissue with many muscle fibres so that it is a 'self sealing tank. . . . The muscle fibres close on any number of bullet holes. . . . I have seen them make no move or other evidence of annoyance from bullets going through the inflated hood'. The nostrils are shut with strong sphincters when the hood is inflated, 'aided by very redundant mucous membrane'. Even more strange are the bladders that the seal can extrude from the nostrils. Olds writes, 'Sealers will tell you that if you get him mad enough "he will blow gert bladders out of his nose". This is so. The redundant mucous membrane is extruded as fiery red paired "bladders" six to seven inches long and five to six inches in diameter'. On the other hand a captive seal in a zoo blew a bladder only out of the left nostril; the bladder in this animal seemed to consist of the nasal septum itself. The use of the hood and bladders to the seal is not plain; it does protect the seals from the sealers' usual technique of attack, for it is impossible to kill them with clubs – this may explain their aggressive reaction to sealers, very different from the timidity of other species – but the hood was certainly not evolved in response to such predation. It probably has some meaning in the sex life of the seal, perhaps during aggressive encounters between males, but nothing of this is known with certainty. The hooded seal has generally been classified as related to the elephant seal, which also possesses an inflatable nose, but King [268] has recently shown that they should be separated.

The ribbon seal, *Histriophoca fasciata*, is the Pacific arctic counterpart of the harp seal, and similarly breeds on the ice. The colour pattern of adult bulls is almost the negative of that of the harp seal – black, with a broad white ribbon round the neck, another round the body above the tail, and a ribbon forming a large ring round each fore flipper.

The two species inhabiting the waters of northern temperate latitudes, the common seal *Phoca vitulina* and the grey seal *Halichoerus grypus*, both

299

extend into the Arctic; the first, known as the harbor seal in America, inhabits the Atlantic and the Pacific, the second only the Atlantic. Both are seals of coastal waters, and feed largely on fish. The common seal is classified into five subspecies, four based on the geographical separation of the populations in the eastern and western Atlantic and Pacific, and the fifth on a landlocked population in the Seal Lakes on the east side of Hudson Bay. The young of all but the western Pacific subspecies shed the natal white coat completely or partly before birth; they are generally born between tide marks, and can swim from the first – occasionally they are born in the water. The subspecies of the western Pacific, however, breeds on the ice, and the natal coat of the young is not shed until two or three weeks after birth; this correlation has interesting evolutionary implications. In the larger grey seal, which tends to inhabit more rocky coasts than the estuaries and sandbanks favoured by the common seal, the young are less precocious and retain their white coats for two or three weeks whether they are born on the ice of the Baltic, the Murmansk coast, the beaches and grassy slopes of the offlying islands of Great Britain, or of the Gulf of St Lawrence. The white-coats generally remain ashore until the first moult, but they can swim if they get into the water even from their first days. During the breeding season the grey seal shows a well marked social structure based on the holding of territory containing a number of cows by the dominant bulls. Isolated populations of the species inhabit the western and eastern Atlantic and the Baltic, but are not classified as subspecies.

The three species of *Monachus*, the monk seals, are warm water seals. Two species are confined to the tropics, and the third *M. monachus* inhabits the Mediterranean and Black seas, the Atlantic coast of Africa, and the island groups south to Cape Blanco. The monk seals are large animals and reach a length of over nine feet; their pelage is brown or grey with little spotting, and the natal coat of the young is black. None of them is abundant owing to past hunting by man for oil, hides and meat. *M. monachus* is the least rare as it has the widest area of distribution; *M. tropicalis* of the islands and shores of the Caribbean is thought to be on the verge of extinction though it may yet recover from the few possibly still living if they are unmolested. *M. schauinslandi*, of the northwestern islands of Hawaii numbers between one and two thousand but is increasing under protection, although Wirtz [512] found an annual production of only fifteen pups per 100 adults at Kure Atoll. It is peculiar, as pointed out by King [267] that the monk seals show no anatomical or physiological adaptation for life in tropical seas and carry the usual thickness of blubber, though the 'less woolly and more silky nature of the pup's coat may be adaptive'.

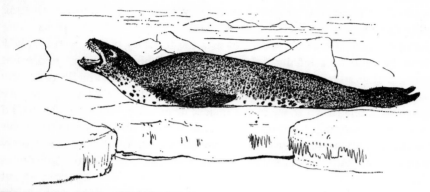

Figure 19 Leopard seal (*Hydrurga*).

The four species of the Antarctic inhabit progressively higher latitudes. The leopard seal, *Hydrurga leptonyx,* a large but slender-bodied seal with a beautiful pelage, silver grey above and white below with light and dark spots, is an inhabitant of the outer edge of the pack ice and the shores of the subantarctic islands – it feeds on fish and birds, especially penguins as they come and go from their rookeries. The crab-eater seal, *Lobodon carcinophagus,* with a dark coat which fades to nearly white before the moult, lives in great numbers on the drifting pack ice. It feeds mainly on krill – the planktonic shrimp *Euphausia* – which it sieves out from the water through the complicated cusps of the cheek teeth. Its peculiar locomotion on the ice has been mentioned in Volume 1. The Weddell seal, *Leptonychotes weddelli,* the most southerly species, lives on the fast ice, and spends much time below it, especially in winter when it breathes through air holes which it keeps open with its teeth, and through tide cracks, or by using air pockets under the ice. It feeds mainly on fish, and is known to be able to dive to depths of over 1,000 feet; Lindsey [299] writing in 1937 before echolocation had been discovered in mammals was unable to suggest how it can catch fish in the submarine darkness of the polar night.

Weddell seals make several audible underwater sounds, and Watkins and Schevill [499] by playing recordings of the sounds underwater were able to observe the seals' reactions. One characteristic call, a trill, appears to be given when a seal comes up to breathe, and conveys the message 'this hole is now occupied' as a warning to others to keep away. This behaviour was further studied by Ray [394] who spent many hours scuba-diving under the ice with the seals; he found that just as shorebreeding seals hold territories on the land, Weddell seals hold and defend underwater territories. The trill is probably made only by adult males in establishing and maintaining territories near natural tide- or pressure-cracks. The dominant male allows subordinates into the territory but he restricts their activities, and occasionally fights with them if his trill warnings are not heeded. Ray adds, 'Females also claim territories, individually or jointly with other females, for themselves and their pups. These territories are less well defined than those of the males; nevertheless, the females defend them with a variety of sounds'. The females lead their pups into the water for short swims when they are a week or so old, but by the time the woolly coat is shed the pups can go 'several dozens of metres and to depths of at least ten metres'. The young are born on the ice and Ray has pointed out their astonishing ability to survive the shock of birth. Birth occurs very quickly, and the pup, which has enjoyed an even temperature of about 37° C for many months, suddenly finds itself expelled from this warm comfort, soaking wet, into a temperature of many degrees below zero. The mother cannot lick her offspring dry as do most mammals, and the amniotic fluid almost instantly freezes, not into an armour of ice, but into powdery crystals which are quickly shaken off as the natal coat fluffs up to keep the pup warm as it shelters from the wind under the lee of its mother. The Ross seal, *Ommatophoca rossi*, inhabits the heavy pack and is so inaccessible that it is seldom seen; it is a fairly large animal but remarkable not for the length of nearly ten feet that it reaches, but for the great thickness of blubber on the neck and throat which makes the short face and jaws appear disproportionately small.

Finally, the elephant seals, *Mirounga*, more generally called sea elephants are the largest of the seals; adult males reach a length of about twenty feet and a weight of three and a half tons. They are named not only from their enormous size, but from the presence in the males of a proboscis or short trunk that hangs down over the mouth, and can be inflated together with a pouch on the snout when the animal is stimulated. At the breeding season the bulls are strongly territorial, and they inflate the trunk when threatening or fighting with each other. It appears to have no meaning to the females as a secondary sexual character, and is a handicap rather than an asset in fighting; it probably serves as a signal of aggressive intent, increasing the apparent size of the head, and drawing attention to and emphasizing its most dangerous part, the mouth with its large canine teeth. The trunk has been described as a resonater increasing the sound when bulls roar, but this function seems doubtful for the windy bellowings come from the throat with the mouth held widely open. Elephant seals congregate in large rookeries for breeding in the spring, and again later in the season for moulting. While ashore for these purposes they do not feed. The hair is coarse, and at the moult the hairs are shed together with large areas of the superficial layers of the skin to which they remain attached. Apart from the breeding and moulting haul-outs elephant seals are pelagic, though where they go is unknown for they are practically never seen at sea though stragglers turn up from time to time on coasts thousands of miles from their breeding grounds. They are believed to feed on fish and cephalopods – their stomachs often contain large quantities of squid beaks; their teeth, apart from the canines, are small and peg-like, but as with the ziphiid whales, a feeble dentition seems to be adequate for catching squids.

The southern elephant seal, *M. leonina*, was formerly very abundant on the shores of all the subantarctic islands and Tierra del Fuego, but was so catastrophically reduced by commercial sealing during the nineteenth century that hunting it no longer paid; in many places it was exterminated. After elephant-sealing was abandoned the herds began to recover in numbers, and the increase continued when controlled sealing, in which only the excess population of bulls is cropped, began in 1910. The present total population is nearing three-quarters of a million. The northern elephant seal, *M. angustirostris*, inhabits the islands off the coast of California and Lower California, and was brought even nearer to extermination through uncontrolled sealing in the mid-nineteenth century; it is now completely protected and has recovered in numbers to about 15,000 It is in general similar to the southern species, but has a rather longer trunk and some differences in details of the skull.

A striking feature in the biology of nearly all the seals is the short length of the period of lactation, and the great rate at which the pups gain weight, as though nursing the young renders the animals especially vulnerable, and must be completed without delay. The process lasts only a few weeks, and the mother feeds little if at all during this time. Although the pup shows a spectacular gain in weight it grows little in length, but becomes extremely plump owing to the thickness of the blubber that it accumulates. Seal milk has the consistency of thick cream, and contains about fifty per cent fat; lactation thus consists essentially in the mother making milk from her blubber, and transferring it to the pup, which at once converts it into its own blubber. A grey seal pup weighed daily was found to gain about three pounds a day for three weeks, while the mother lost about twice that weight a day – from a birth weight of about thirty pounds the pup increased to over ninety in twenty-one days. The pup has generally moulted by the end of lactation, and is well provided with reserves stored in the blubber to carry it over the period of weaning, and learning to find food for itself. The whole process seems to be a wild physiological scramble to get the animals back into the sea as quickly as possible.

ORDER SIRENIA

The order Sirenia contains the few species of sea cows, which are appropriately named for they are the only wholly aquatic mammals that are exclusively herbivorous. They are large dolphin-shaped animals – some reach a length of ten feet and a weight of 1,000 pounds – with fusiform bodies, flipper-like fore limbs, no hind limbs, and the tail expanded into a horizontal swimming organ analagous to the flukes of the whales. The skin is thick and rugose, naked or sparsely covered with scattered hairs, and a moderately thick layer of blubber covers the body beneath it. The eyes are small, and there is no ear pinna or dorsal fin. Sirenians are slow-moving placid creatures, very unlike the swift and active whales, dolphins or seals. They live in the warm waters of tropical Atlantic, Indian and western Pacific oceans, and are coastal, estuarine or fluviatile in habits. They never leave the water for they are unable to progress on land or even to haul out on to it. They are often said to have given rise to the legend of mermaids, though it is not clear why the fertile human imagination should not have thought up mermaids from its own inventiveness without a model, just as it has produced the yeti, the Loch Ness monster, and many others. The mammary glands are placed on the chest, but the story that sirenians suckle their young floating vertically in the water, with the baby clasped to the breast by the flipper, is an error; they both float horizontally and the young sucks as best it can without help. The most peculiar feature of the external anatomy is seen in the head, which is attached to the trunk by so short a neck that no division is apparent. The eye is small and the ear-hole minute, and the large truncated muzzle, with valvular nostrils at the top, overhangs the comparatively small mouth. The muzzle is a most peculiar structure, which has been described [230] as resembling 'a truncated proboscis rather than a modified upper lip'. It is rather flattened in front and is drawn out at the sides into fleshy lobes bearing short thick quill-like bristles; the lobes or flaps overhang the mouth on each side and can be moved towards and away from each other. When sirenians feed they use the very mobile side flaps for seizing the food and passing it into the mouth, where it is grasped between the rough horny plates that cover the front of the palate and lower jaw anterior to the cheek teeth.

The nature of the bones, too, is peculiar in the sirenians; they are extremely dense and heavy, especially those of the skull and ribs. Harrison and King [214] record that 'the vertex of the skull is the densest piece of bone (pachyostosis) we have ever tried to saw, making one wonder at the activity, or lack of it, of the parathyroids', and Gunter [190] sawed through a rib 'both crosswise and lengthwise. Except for a tiny lengthwise perforation, the bone is as solid as a piece of ivory'. This density has caused much speculation among zoologists, for owing to a lack of osteoclasts no marrow cavity is formed in the ribs or limb bones, and unabsorbed primary bone persists long after it would have been replaced in other mammals. Harrison and King [214] point out that similar conditions are found in human bones in certain endocrine disorders known as congenital athyroidism and osteopetrosis.

It certainly seems relevant that in manatees the thyroid appears histologically to be inactive and in a storage state. Sirenians are also the only marine mammals that eat sea vegetation, some of which contain quantities of iodine. The heavy bones of sirenians could have arisen secondary to an inherent hypothyroidism that developed in an early phase of imperfect adaptation to marine life.

They correlate this suggested hypothyroidism with the heavy bones, sluggish behaviour and low metabolic rate of the animals. On the other hand, as discussed below, the marine populations of sirenians, as well as the freshwater ones, eat vascular plants, not seaweeds.

Another point arises: in other mammals the red blood corpuscles are produced in the red marrow of the ribs, and thus Gunter [190] very pertinently asks, 'Where do the red blood cells form in this animal which, due to its mode of life, possibly has need of a greater supply than the ordinary mammal?' He suggests however that a sirenian has a low metabolic rate 'and possibly does not require a large supply of haemoglobin'. The question has not been answered. It may be that the haematopoietic function of the spleen is enhanced, for Hill [229] found that although the spleen in the dugong is not particularly large, there is 'a tendency to exhibit spleniculi' – small auxiliary masses of spleen tissue lying near but separated from the main body of the organ. Many writers have suggested that the density of the bones is correlated with the sirenians' aquatic life by increasing the specific gravity of the body. This argument seems unconvincing as sirenians live in shallow water and do not dive deeply like seals or whales, and furthermore they have to come to the surface at least every ten to fifteen minutes to breathe. Moreover, if dense bones are desirable in a diving mammal, why do not all the other diving mammals have them?

The order Sirenia contains two families, the Trichechidae and

Dugongidae, with four living and one recently extinct species between them. The anatomy of the sirenians has long ago been investigated and described by the zoologists of the last century such as Garrod [160], Murie [341] and Owen [363] and their findings have been confirmed and amplified by modern dissectors such as Dexler and Freund [109], Quiring and Harlan [385] and Hill [229], but studies on their general biology and behaviour are few and incomplete – they are shy and difficult animals to observe. Conservationists have recently been concerned for the preservation of the sirenians, and the attention thus drawn to them may lead to further investigations by zoologists.

FAMILY TRICHECHIDAE

The Trichechidae contain one genus *Trichechus* with three species, the manatees, which inhabit the coastal waters of both sides of the tropical Atlantic and the river systems flowing into it, so that they are present in the far interior of central Africa and South America. One of the American species is found well outside the tropics on the northern Caribbean coast to Florida, and beyond to Georgia. The tail of the manatees is peculiarly flattened and expanded at the sides to form an oval or truncated extremity without a caudal peduncle. The three species are distinguished mainly by details of the skull anatomy, but the Amazonian species differs in having no nails on the flippers whereas the other two have four small flattened nails. The name manatee is said by some authors to be derived from the Spanish *mano*, hand, and *tener* to hold, in allusion to the alleged way the female clasps her young with her flipper to suckle it. Apart from the fact that the female does not clasp the young, as shown by Moore [334] and others, Simpson [444] found a Carib word manati that means 'a woman's breast', and adds

the character of the sirenians that has most struck observers, and particularly primitive peoples, all over the world is that their mammae are extraordinarily like those of humans and unlike those of almost all other mammals. . . . It is clear that the Spanish colonists in the West Indies did accept 'manati' as the name for the animal and that we took 'manatee' from them.

In Brazil and Mozambique the manatee and dugong bear the Portuguese names *peixe boi* or *peixe mulher* meaning 'ox fish' and 'woman fish', but Cabrera and Yepes [65] suggest that the name manatee was introduced by the first negroes brought to America, because the African manatee bears the name mandí or mantí in the Mandingo language.

The peculiar dentition of the sirenians has been discussed in Volume 1;

in the digestive system the stomach, which is large, bears near its oesophageal end a glandular mass variously described as a glandular pouch or a gland with a large lumen. The secretion from the gland obviously helps in digesting the vegetable food, but nothing is known of the process, or whether bacterial fermentation occurs as in the ungulates – the sirenians are not known to regurgitate and ruminate their food. Barrett [22] says of the freshwater manatees of Nicaragua that when they are feeding 'little chewing is done. Probably the peculiar gland emptying into the stomach assists in digesting the coarse, unmasticated roughage'. His subsequent remarks, however, seem to contradict his statement that the food is unmasticated, for he says, 'The noise made by the flapping of the huge upper lip and the crunching of the large teeth can be heard distinctly on a still night 200 yards or more away. The sound made by a dozen or more manatees grazing on the "yerba Guinea" is much like that of horses grazing in a pasture'. His observation that the breath of both dugong and manatee smells 'almost precisely like that of a cow, though the former animal may never have tasted true grass' points to the possibility that there may be some bacterial digestion. On the other hand the captive manatee studied by Parker [365], who was timing its breathing rate at night, often blew its breath directly into his face. 'It was far from aromatic ... but had a most repulsive smell resembling somewhat that of phosphorus. This may have been due to the fact that she with the other two in the pool were fed regularly on cabbage.' In the coastal freshwater lagoons and the rivers of the Caribbean area and the Guianas manatees feed on vegetation growing on the banks, as well as that growing in the water; in particular they eat the stem tips of grasses of the genus *Panicum*, which grow on the banks but have stems floating on the surface up to ten feet from the side. Barrett says that in Central America these grasses form the sole diet of the animals, but Bertram and Bertram [37] report that in the Guianas they are unselective, and feed on almost any vegetation that is soft enough. Reference was made in Volume 1 to the use of manatees in the Guianas for clearing choked waterways of weeds, including the rampant water hyacinth – though they are selective in eating this plant, for they eat only the tops and not the roots, according to Moore [334]. In the sea the manatee appears to eat only the eel-grass *Zostera*, a flowering plant, and does not feed on seaweeds (Plate 21).

Manatees are gregarious, and are usually seen in small parties, but in Nicaragua they gather into loose herds of up to fifty animals during the day and scatter to feed at night. Although they are so shy that the sound of rain falling on a bucket in a canoe, or of a fisherman spitting over the side, is sufficient, according to Barrett, 'to keep the whole herd submerged from sight for hours; yet while they are grazing the hunter may go up and slap them on the back unnoticed'. They can, however, become adjusted to the presence of man for Moore [333] records that in south Florida one of the greatest concentrations of manatees lives in the Miami river 'in the very heart of the city. . . . Certainly no other city in the world can boast a population of free, wild manatees. . . .' In the northern part of their range on the Florida coasts, from which they sometimes stray to Georgia and even North Carolina, the manatees tend to move south in winter though they make no regular migrations. In years of exceptional cold some are killed by the low temperature, as in the exceedingly cold weather of January 1940, recorded by Cahn [66] and Krumholz [278], but others find refuge in the many warm springs discharging into the rivers. Moore saw concentrations of more than ten manatees 'on very cold mornings gathered in the outflow of a factory outlet where it vents warm water into the river beneath the Miami Avenue Bridge'. On the other hand in 1951 the manatees in the newly-established Everglades National Park 'reflect their recent unlawful persecution by meat-hungry commercial fishermen by being very shy' – fishermen who 'rather freely risked the $500 fine to slaughter these animals for food'. Moore thought that more animals escape observation from boats driven by noisy outboard motors 'by remaining hidden on the bottom of these turbid rivers and bays than attract attention to themselves by frightening up and swimming violently away'. The Bertrams [36] appear not to have considered this point in their enquiry into the manatees of the Guianas for they state that 'the increasing use of powered boats, including dug-out canoes with outboard motors, may be scaring the animals away from some reaches of the rivers and so restricting their distribution'.

Little is known of the breeding physiology of the manatee, or even if there is a breeding season, for recently born young three to four feet long have been seen in Florida waters from December to August. Neither is it known if the manatee is monogamous, or whether there is any social organization, territorial or courtship behaviour when breeding. Moore several times saw wild manatees nuzzling each other in 'a rather bovine fashion' – a mother with her young – and two that appeared to be adults 'rose and put their muzzles together above the surface . . . the "kissing" took place only once during an hour of observations'. It is impossible to say whether this was more than a casual encounter. Young manatees have been born in captivity on several occasions; the mother is said to support the young at the surface for up to an hour after its birth.

The intelligence of the manatee is, according to Moore [334], ordinarily underrated. Large animals captured as adults have been described as stupid and uninteresting, but they were kept in large aquaria 'where they could easily remain away from intimate contact with their keepers'. Younger animals kept in smaller pools behaved quite differently; one 'came at call, rolled over, and curled its tail up to thrust the tip into the air', and another 'came at call, rolled over, swam away toward the far end of the pool and back, offered its right flipper to shake hands, and reared up against the side of the pool . . . to "rub noses" with its keeper. . . . Each act was separately called for by voice and motion and rewarded with a lettuce leaf'. Schevill and Watkins [423] have tested manatees to find out whether they use echolocation as do some other aquatic mammals. They found that the sounds uttered are not particularly loud, but are squeaky and rather ragged. There is no evidence that they are used in echolocation; they are communicative rather than navigational.

Most of what is known about wild manatees has been learnt from *T. manatus*, which lives on the very doorstep of the American zoologists. Surprisingly little is known about *I. inunguis* of the Amazon, which is found from the mouth of the river as far up as Iquitos in Peru, or about *T. senegalensis* of West Africa, which is found up the Niger above Timbuctu and in many other rivers, estuaries and coastal waters. The distribution of these species is known approximately, and much is said of the value placed upon them as a source of food and oil by native peoples, but little indeed has been recorded of their lives and habits. The manatee living in the Orinoco is *T. manatus*, not *I. inunguis*, which is confined to the Amazon basin.

FAMILY DUGONGIDAE

The family Dugongidae contains two genera, *Dugong* and *Hydrodamalis* – the latter is the extinct Steller's sea-cow, *H. stelleri*, the discovery and extermination of which in the Bering sea is mentioned in Volume I, together with the little information about it that was recorded in the eighteenth century. A report in recent years that some animals of this species had been seen in the Arctic by a party of Russians has been shown to be based on a mistake. The dugong, *D. dugon*, is at once distinguished from the manatee by the shape of the tail, which more nearly resembles the tail flukes of a whale in shape, and has a central notch. In addition the second pair of upper incisor teeth is present; in the females these teeth do not erupt, but in the males they project as tusks for a few inches from the corners of the mouth, though most of their length of up to a foot is embedded in the socket. The dugong is widely distributed in coastal waters from

Madagascar and the tropic northwards on the east coast of Africa, the Red Sea and the coasts and islands of the Indian Ocean, through Malaysia, the East Indies and New Guinea to the Philippines and Solomon Islands, and along the north coast of Australia.

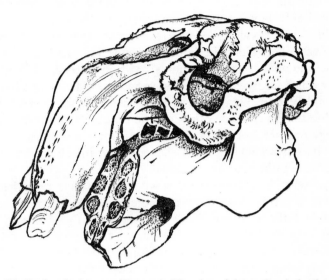

Figure 20 Skull of male dugong (*Dugong*). The tips of the tusks are broken off.

The dugong is commonly spoken of as feeding on seaweeds, which are understood to be marine algae, but there seems to be no definite record on the matter – the few samples of stomach contents that have been examined consisted of eel grass of species allied to *Zostera*, which are flowering plants, and Troughton [478] speaks of the animal using the flaps of the upper lip 'to grasp the Zostera-like marine grasses and draw the food into the mouth'. It is very probable, therefore, that the diet of the dugong differs little from that of the manatee, and does not include marine algae. Barrett, already quoted as saying that the dugong may never have 'tasted true grass' says further 'what sort of marine plants are eaten remains a mystery; the Persian and Arabian Gulfs and the Red Sea appear comparatively free of large seaweed masses' and he suggests that an unidentified weed seen floating near the surface of the Red Sea may be the food. Even if dugongs eat not algae but eel grass, the statement that they never taste true grass remains correct. Jarman [254] found that in the sheltered bays of the Kenya coast, where dugongs are still plentiful, the animals graze on underwater meadows of plants belonging to the orders Potamogetonaceae

310

and Hydrocharitaceae, orders closely allied to the Zosteraceae. 'There were no signs, and the fishermen emphatically denied, that the dugong ate any of the algal seaweeds'. The stomach of the dugong carries a glandular mass similar to that of the manatee, and it may well be that the diets and digestive processes are similar. If that should turn out to be the case, theories of hypothyroidism due to a diet of algae rich in iodine are untenable. On the other hand, Cave and Aumonier [73] found that the histological appearance of the thyroid gland did suggest that there might be a functional hypothyroidism; if there is, it cannot be due to the effects of a diet of algal seaweeds.

Although the structure of the dugong has been described in much detail by the anatomists already referred to, practically nothing is to be found in the literature about the natural history of the animal – nearly every author who mentions the species writes of the various ways in which it is hunted and killed, and tells how delicious is the meat, how useful the oil and other products. Throughout their range of distribution the sirenians are highly esteemed as food by whites as well as blacks, not only for the excellent quality of the meat but also for its abundant quantity, and the oil rendered out of the blubber is prized for use in cookery. In some parts of Asia the meat is believed to have aphrodisiac effects, and is in consequence particularly valued; everywhere the fat and oil are thought to have medicinal properties for the treatment of aches and pains and tuberculosis as, indeed, is that of many unusual or bizarre creatures, from adders and rattlesnakes to shearwaters and gorillas – myths which the credulous of all races are ever eager to swallow. The pressure of human predation, however, has led to an increasing scarcity of the animals in many places, and their imminent extermination is feared by some alarmists.

The latest scientific enquiries, however, show that the continued existence of the sirenians is not so precarious as some fear. The manatees of the southern United States are now in no danger – as we have heard, they even live in the heart of the city of Miami – and Bertram and Bertram [36] said of the manatees in the Guianas 'they are not yet very high on the danger list'. The Bertrams [37, 38] have also looked into the status of the dugong and say that although an increasing scarcity is assumed because of the reduced populations in the Red Sea and on the coasts of Ceylon and Malaysia, their investigations have led to certain encouraging conclusions. They write, 'in parts of the Australian northeast the dugong is not to be regarded as a rare animal. In places its numbers may even be increasing now. . . . The world's stock of dugongs is certainly much greater than that of manatees'. The capture of dugongs is not easy and 'because of its difficulty, and because of the progressive movement of the native peoples concerned on to a wage basis and the use of purchased food-stuffs, the hunting pressure on the dugongs would seem to be diminishing in many places'. Sirenians are protected in many parts of their range; the $500 fine for killing them in the United States has been mentioned, and they are legally protected in the Guianas. The African manatee is protected in some parts of its habitat, but it is being ruthlessly killed by the negroes. The dugong is protected on the coasts of East Africa, and in Australia it may be taken legally only by the native blackfellows. Unfortunately much of the legal protection is ineffective; often its existence is not known, even to government officials, and it is widely ignored to the extent of openly offering sirenian meat for sale in the markets. As Carleton Ray [395] points out, none of this protection is of value unless it is enforced, and unless the places where the animals live are effectively policed.

In view of the value of sirenians as food in a world afflicted with a population explosion, and the usefulness of the other things they produce, it is not enough just to forbid hunting and killing them, and to see that the ban is observed. They should be conserved, and the maximum crop that the population can give without depletion of the stock should be taken. In order to do so a great deal more must be learnt about the biology of the animals – we have seen how little is known about their natural history. The biologists, if given the chance, could within a few years gather the necessary knowledge to form the basis of a management policy to secure a valuable contribution to the food of those living in the tropics, and the perpetuation of an order of mammals of the greatest scientific interest. The cost of one moon-shot would cover the investigation many hundred times over, and the investment would pay continual dividends both material and intellectual; a futile hope indeed when we remember what has happened to the whales, in spite of the scientific advice offered to vested interests.

Chapter 11

Elephants. Hyraxes

ORDER PROBOSCIDEA

THE LARGEST land mammals form one of the smallest orders, for the Proboscidea are represented by only two living species, the elephants of Asia and Africa, placed in separate genera *Elephas* and *Loxodonta*. These two species are the last of a very long line in a family that included a large number of variations on the elephant theme, particularly in the production of tusks of many shapes and sizes in the upper or lower jaws, or both. The large size of the bones of the animals has favoured their preservation as fossils, and consequently palaeontologists have had the material with which to work out the course of evolution of the proboscideans, and their branching into several families, from their early ancestors in the late eocene. The history of the order is set forth by Osborn [357] in his classic monograph. At least one of the extinct elephants, the mammoth, was contemporary with palaeolithic man.

The Asiatic elephant *E. maximus* which inhabits, or inhabited until recently, much of India and Ceylon extends eastwards to Malaysia, and Sumatra, but is not native to Borneo, where it has been introduced. The African elephant *L. africana* formerly inhabited the whole of Africa south of the Sahara, but has been exterminated in settled countries so that its range is now much reduced; in early historic times it extended into North Africa. The species differ in the greater height, larger tusks and ears, and straighter back of the African elephant, in the narrower and more numerous plates of the molar teeth in the Indian elephant, and in the presence of a single finger at the end of the trunk of the latter and of two in the former. The tusks, which are incisor teeth, are present in both sexes of *Loxodonta* though smaller in the females; in *Elephas* they are often large in the males though they may not grow larger than tushes, and in the females they are nearly always rudimentary.

For many centuries tame elephants have been valued in the east as beasts of burden and status symbols; they are still used for work and show, but are increasingly replaced by the internal combustion engine. They are the only domestic animals caught as adults and individually tamed, not because they cannot be bred in captivity but because of their slow rate of growth, for they are not able to do useful work under the age of at least eight to ten years. Several methods of capturing wild elephants are traditional in the east, the best known being that of driving them into a stockade where they are subdued by hunger, thirst, tethering and severe disciplining. Other methods include the use of pitfalls, decoying, snaring, running down, and baits drugged with opium. Tame elephants play an important part in the capture and subduing of wild elephants whatever method is used; their role is similar to that of the Judas rams of the Chicago slaughter-houses. It is strange that man can not only dominate such huge and potentially dangerous animals, but can also make them help to bring their wild relatives into subjection. Many authors have given descriptions of the capture and taming of elephants; Deraniyagala [108] who was present at a 'kral' in Ceylon said 'the litheness of so large an animal as an elephant with its apparently stiff body is extraordinary. . . . The contortions of an elephant when first noosed are almost unbelievable. . . .', and he gives a number of lively sketches illustrating them, including one of a captive landing 'a right upper-cut upon a pushing decoy' so effectively that 'the thud of the blow and the clash of the recipient's lower jaw against his upper' could be heard thirty yards away. He also notes that a new captive 'can use its teeth very effectively on the two extremities of its perfidious tamed relatives', and gives sketches of wild elephants trying to bite off a trunk-tip and a tail during the struggle. Deraniyagala draws attention to a peculiar point: the wild elephants crowded in a stockade give off a cloud of steam, clearly visible in his photograph, but 'the tamed elephants which work as hard or harder in hauling the captives do not do so. . . . Although this vapour emanates so freely it is strange that the perspiration does not trickle down as in the horse or man'. It is, in fact, not strange, for the skin of the elephant is not covered with sweat glands, and consequently his suggested explanation that the vapour is 'due to heavy endocrine gland secretion induced by terror' cannot be valid. It is more likely to be caused, as he suggests alternatively, by the elephants withdrawing regurgitated water from the throat with the trunk and spraying it over themselves. This habit is well known in both the Asiatic and African elephants, as a way of cooling the skin.

In the elephant the only skin-structure resembling a sweat gland is a complex gland on the temple between the eye and ear. This temporal gland consists of a mass of tubules, similar histologically to sweat glands, surrounded by connective tissue and opening onto the skin through a single common duct. The function of the gland and its secretion is not known; it has been suggested that it has a sexual function and that it is especially active at the time of rut in the male. Evans [139] writing of the Asiatic elephant says, 'Male elephants and very rarely females, on attaining maturity, are subject to peculiar periodical paroxysms of excitement, which seem to have some connection with the sexual functions to which the name *musth* is applied. . . . The temples become puffy due to the swelling of the temporal glands . . . later an oily discharge exudes from the hole or duct over the gland. . . .'. During musth the animals are disobedient, destructive and dangerous, and have to be chained; musth may last a few days, weeks or months, and generally occurs every year. It is not certain that musth is a sexual phenomenon, for it is not always soothed by the company of female elephants – it is sometimes exacerbated – and its correlation with the activity of the temporal gland remains obscure. In the African elephant the gland appears to be active at all times of the year in both sexes, as is shown by the streak of secretion spreading on the skin below the duct. The secretion appears to be oily, as Evans said, and thus there is some doubt that the gland, as described by Eales [127] who, however, examined only a foetus, does indeed represent modified sweat glands, from which a watery secretion would be expected. Schneider [427] examined the histology of the temporal gland and found that it is similar in both species of elephant, and that the cells lining the tubules contain much lipoid – hence the oily nature of the secretion. The gland thus differs from sweat glands in not secreting sweat, but resembles them in its apocrine method of secretion. The sebaceous glands at the mouth of the duct might contribute to the oily secretion – like sweat glands, sebaceous glands are absent from the skin of the elephant except on the eyelid, at the opening of the temporal gland, and in the outer ear-passage. Perry [373] notes that because there are no marked symptoms of oestrus in female elephants it is often stated that 'In the elephant it is the male, and not the female, that comes on heat'. He adds, 'This is nonsense. Mating occurs only at oestrus . . . and the relation of "musth" to breeding is still obscure'.

Wild elephants, unlike many large mammals, are in no danger of extermination. Gee [161] points out that in Assam elephant numbers are controlled, in order to prevent damage to crops, by licensing hunters to shoot the rogues, and leasing the right to conduct elephant catching operations in areas 'where elephants are known to be on the increase or causing serious damage to crops'. This policy ensures that the stocks are maintained, and a useful yield obtained from them. Similar measures in other parts of India and in Ceylon and in other neighbouring countries are enforced. Asiatic elephants were for a century or more essential to the teak industry in Burma for extracting logs from the forest and for heavy work in the timber yards. Evans [139], whose knowledge of working elephants was never surpassed, points out that elephants are by nature nocturnal animals, and should not be worked in the heat of the day. He considered it imperative 'for humanity's sake that employers should detail, whenever possible, a European assistant to frequently . . . inspect and report upon the health of the animals, their physical capability for work, and above all the quality and amount *by weight* of fodder supplied. . . . The healthy elephant with good treatment, liberal allowance of fodder, can work regularly for six or seven hours a day without injury'.

The anatomy of the Asiatic elephant has been described by many zoologists whose findings are well summarised in Grassé [183], and the physiology has received some attention particularly from Benedict [30] who among other things constructed a huge respiration chamber for measuring the basal metabolic rate. Elephants when at rest are never still; they move trunk, tail and ears incessantly, and sway the body from side to side with a weaving movement that brings the weight successively on to each of the four legs. Benedict and Lee [31] however point out that 'If an elephant could be placed within a huge glass cage and, by means of a reducing glass, the ensemble be made to appear to have the size of a half-pint fruit jar containing an elephant the size of a mouse, the elephant would seem extraordinarily tranquil when compared with a normally quiet mouse'. They conclude that only when the animal is weaving can the muscular activity be sufficient to have a measurable effect in increasing the metabolism. The animals sleep for only a few hours each night, and 'one marvels at the endurance of the elephant in its ability to be for hours in incessant motion with no rest . . . the constant muscular activity is accomplished with but small drafts upon the energy to be derived from body material or food'. Apart from these studies the natural history of the animals in the wild has been the subject of little published work in recent years. The African elephant, on the other hand, has been investigated more thoroughly owing to the recent greatly increased interest in obtaining basic information for the management of wild populations of mammals, especially in East Africa.

In the early days of elephant hunting in Africa during the last century white hunters pursued their quarry not for amusement but to make a living, and the skilful ones were very successful in amassing a competence.

As the country became settled, and the elephants shot out or driven away, they were forced to hunt in ever more remote areas, and professional ivory-hunting gradually died out, though some hunters were still working into the end of the 1920s. When big game hunting became a fashionable pastime the expenses of a hunting trip could often be covered by the sale of the ivory obtained, and the hunting pressure on the elephant by professionals and amateurs led those who framed the game protection laws to give special consideration to the elephant. The cost of a licence to kill even one elephant was so high that many sportsmen refrained from elephant shooting, especially after the invention of artificial substitutes for ivory drastically reduced the value of the tusks. The value is, however, still high enough to tempt native poachers into illicit killing; game departments are perpetually alert to catch poachers and confiscate their illegal ivory, from which the authorities derive a useful revenue. Elephants increased in numbers under this protection and in many places they became a nuisance by damaging crops, sometimes devastating and destroying a whole native plantation in a night. As Laws [291] summarises the matter,

The elephant is one of the problem species in National Parks, Game Reserves and Forest Reserves throughout Africa, not because its numbers are declining, but because some populations are increasing and nearly all are more restricted in range than formerly owing to human settlement. This means that population densities are usually higher than in the past and this often leads to degradation of the habitats and ecosystems.

The superabundance of elephants produced the paradox of the game departments assuming the duty of destroying the excess of the protected animals, and for many years about 2,000 elephants were slaughtered annually in several east African countries. The opportunity of examining a large random sample of wild elephants was recognised by some zoologists, and in collaboration with the elephant control authorities, a study was started on the reproductive biology of the animals, and the results were published by Perry [374]. He found that in Uganda breeding occurs at all times of the year, and gestation lasts about twenty-two months. After the birth of a calf and a lactation anoestrus of similar length, the female experiences a number of short oestrous cycles until she becomes pregnant again, and lactation continues through the subsequent gestation period. During pregnancy the ovaries contain numerous corpora lutea, and it was deduced that about mid pregnancy they are replaced by a second set, some formed from unovulated and some from ovulated follicles; the occurrence of secondary corpora lutea would thus resemble the similar phenomenon

317

found in the mare. Both sexes reach maturity at about ten years of age, and the females then produce a calf about every four years until extreme old age. Perry concluded that at puberty the tusks of females stop growing but those of the males do not; very large tusks are generally the result of rapid growth rather than of great age, and their occurrence may be genetically controlled.

Since these pioneer studies, research on wild elephants has been greatly stimulated by the establishment of field laboratories in Africa, which have attracted numerous biologists to the study of the fauna and flora – and by the allocation of funds to enable their work to be done. Scientific investigation has thus replaced the casual and anecdotal observations that were almost the only source of information about the animal. Buss and Smith [63] made further studies on the reproduction of elephants in Uganda, and although they confirm most of Perry's results they concluded that the corpora lutea present in the ovaries during early pregnancy are not replaced but are maintained throughout gestation and for about two months after the birth of the calf. Like Perry, they found that all females were either pregnant or lactating, that many females begin to breed at about seven years of age, and that all females breed by the age of eleven. By watching the behaviour of wild elephants they found that cows in oestrus are receptive to several males successively, that there is no courtship behaviour, and no prolonged male-female relationship. Herds often contain several bulls, but although one of them may be dominant and drive off his subordinates, they mate with an oestrous female after him; there is little fighting between bulls for females, and what occurs is not serious or productive of more than trifling damage. Short [440] observed a female elephant throughout oestrus, and found that she was mated by many different bulls. At first they did not compete for the cow, but later they began sparring and 'one eventually appeared to establish mastery, driving the other bulls away. It was interesting that this master bull showed no discharge from his temporal glands'. The ovaries of this cow were examined and the recently formed corpus luteum was assayed for progesterone; very surprisingly none was found in this young and presumably active tissue. Short and Buss [441] had already examined older corpora lutea, and Short concludes, 'it is remarkable that we have never succeeded in detecting progesterone in the corpora lutea of elephants at any stage in the cycle or of pregnancy'. It is highly improbable that the corpora lutea of elephants do not contribute some kind of hormone with an action similar to that of progesterone, for their histological appearance suggests active secretion – the results of further researches are awaited with interest.

318

The problems produced by an over-dense population of elephants are shown by the conditions in the Tsavo National Park in Kenya in 1962. Glover [166] found that up to 10,000 elephants inhabited the Park, which covers an area rather larger than Wales, but that nearly all were within fifteen miles of permanent water. The whole region is arid, and was formerly covered with low bush and trees, which have been destroyed in many places, especially near the few rivers, and have been replaced by grasses of poor grazing value. The devastation of the habitat has been caused not only by the elephants in search of food, but by fires started by 'poachers and other intruders'. During the long dry seasons 'much of the most nutritious food is provided by legumes, herbs and woody vegetation. In fact, over-crowding near water and its ecological consequence, the destruction of bush, has produced elephant "slums" '. The park could probably safely carry about half the number of elephants that live in it, but the animals go on breeding and the mounting numbers present a very difficult problem to the authorities, who are, naturally, anxious to save the area from becoming a dust bowl (Plate 23).

Buss and Savidge [62] studied the elephants in the Murchison Falls and the Queen Elizabeth National Parks in Uganda, and found that the animals moved from grassland to woodlands during the dry season and back to the grass with the rains. In the different areas of their work they found that the population increase was inversely proportional to the density of population, and that where elephants were thickest on the ground the reproductive rate was lowest. The reduction was brought about by an extension of the interval between births to about three times that found by Perry fifteen years before. Changes in the size of populations are caused not only by the breeding rate but by immigration. There had been no elephants in the Serengeti area for more than forty years until 1955, when they invaded the Serengeti National Park. Lamprey and his colleagues [286] found that in 1967 they had increased 'mainly through immigration, to their present level of approximately 2,000 and have remained at this level for the last three years'. It might be thought that the return of elephants to a National Park would be welcome, for Fosbrooke [150] showed that they were there up to 1882. But unfortunately elephants can be very destructive to the environment, and they destroyed the larger trees at the rate of at least six per cent a year. In the wooded areas of the park the elephants eat the branches, foliage, and bark of the trees, and not only break off branches but push trees over and uproot them. Lamprey and his partners observed that

elephants seem to take pleasure in breaking and uprooting large trees. No

319

trees seem too big for them. . . . Many trees which have been freshly uprooted or broken show no signs of being eaten, but in others a large proportion of the foliage has been eaten, leaving broken and stripped branches around. In many instances bark is stripped off and thrown aside.

A film of a bull elephant pushing a tree over shows that the African elephant does not push with its forehead as does an Indian elephant when, for example, it is working in a Rangoon timber yard. It lifts its trunk up vertically and places it against the tree so that the push is given with the underside of the animal's trunk.

The elephant populations in national parks raise difficult problems of management for the authorities. If the function of a park is to preserve the flora and fauna in a natural state, should the elephants be interfered with? They and the results of their actions are part of the eco-system, and their influence on the vegetation is part of the natural course of events. On the other hand the parks form sanctuaries to elephants from the harrassment they receive outside, and they no longer have, as formerly, great tracts of country into which they can move when they have rendered an area an unsuitable habitat for themselves. In addition, the scenery of the parks is one of the amenities, and devastated woods are not generally admired by visitors. Control by reducing the numbers is inevitable in order to preserve the habitat, not only for the survivors but also for the numerous other species of animals of all sorts that live there. Although elephants are so destructive to trees, grass nevertheless forms the greater part of their diet even in the Tsavo Park, as observed by Napier-Bax and Sheldrick [344], who found that trees are attacked mainly at the height of the dry season. It was in the great drought of 1961 that the elephants really started wrecking their habitat: 'Near the Galana river there was no grass, most of the trees had been smashed and the herbs and shrubs had been eaten down to stumps. The large-scale destruction of baobabs and the ringbarking of *Acacia elatior* by elephant was particularly noticeable'. The destruction was such that there was no browse left for the black rhinoceroses, many of which died of starvation, and other bush dwelling species disappeared. Elephants eat a wide variety of food plants, but are selective and avoid some species, or eat some only at certain periods of their growth – normally the leaves and bark of trees are secondary, not bulk foods. Dougall and Sheldrick [118] watched a tame ten-year old elephant, weighing about 4,000 pounds, feeding during a whole day when released in the Tsavo Park; they found that he ate 64 different species of plants from 28 botanical families, but that 74 per cent of his browsing came from eight families. Chemical analysis of the latter gave values for protein, fibre and trace elements much like those for good European meadow hay, but a higher amount of calcium.

The elephant is not a good converter of food, for much that it eats is not digested. Benedict and Lee [31] say of the Indian elephant that the 'appetite is insatiable' and the animal is continually searching for food, but that only about 40 per cent of what the elephant eats is digested, an amount lower than that in any other animal they had studied. This may be correlated with the way an elephant eats; after it puts food in the mouth 'there is a remarkably small amount of chewing'. On the other hand, there may be some digestion of the food by intestinal bacteria; the rumbling of gas in the intestines of elephants can be heard at some distance 'which is perhaps surprising, as it is practically impossible to hear the heart-beat with a stethoscope'. Benedict and Lee collected intestinal gas in a basket-ball bladder connected to a rubber tube 'inserted well up in the capacious rectum', and on analysing it found nearly 50 per cent of methane, 20 per cent of carbon dioxide, 2·5 per cent of oxygen, and the remainder nitrogen. They also found that methane production over twenty-four hours averaged 655 litres from an elephant weighing 3,672 kg, and conclude therefore that 'in the elephant there is an almost perfect condition for anaerobic fermentation'.

Benedict and Lee also point out that although it is well known that not more than two molar teeth are in use at any one time, and that the teeth towards the front are lost as those at the back erupt and move forward, it is mistakenly and 'commonly believed and not uncommonly stated that the elephant's teeth are lost by being pushed forward and ejected *in toto*'. The teeth do move forward, but owing to their oblique position in the jaw they wear away from before backwards, so that the front laminae are worn down to the base although the hind ones have not long been in use. The tooth is consequently lost by the remains of the laminae breaking off in small fragments as they reach the fore end, and generally being swallowed with the food – there is nothing resembling the shedding of milk teeth in our species. Benedict and Lee exploded another popular myth about elephants; they showed that elephants are not afraid of mice or rats, and were not 'the slightest bit interested in or disturbed by the presence of mice'. They point out, however, that elephants are very easily alarmed by any un-accustomed noise 'such as the barking of a strange dog. If a rat scurried over a newspaper lying on the floor, the noise would challenge the attention of the elephants, and if they were lying down, would undoubtedly cause one or more of them to give an alarm, to which all the elephants would react by trumpeting'. These were Indian elephants belonging to a circus.

Elephants can react dramatically in unfamiliar circumstances if their behaviour patterns are disrupted. Winter [511] describes the actions of a herd of elephants after he had shot three out of a herd of about thirty during elephant control on Mount Elgon in western Kenya. The animals were up wind, and had not seen their attackers: 'They milled around trumpeting and shrieking fearfully, throwing up showers of grass and stones in their fury, and then with incredible determination and complete disregard for their own safety attempted to lift their dead companions'. The attribution of complete disregard for their own safety seems hardly justified, because they were unaware of the presence of the hunter and his game scouts; furthermore it is doubtful if any wild animal can have a mental concept of death such as we have – to the living the dead elephants probably appeared to be behaving unacceptably in suddenly lying motionless. Be that as it may, they entwined 'their trunks with those of the dead animals and tried to lift the carcases'; one cow that tried to lift a dead one by kneeling and putting her tusks under its belly snapped off her right tusk above the socket so that it flew through the air and landed thirty feet away. Shortly afterwards 'the herd moved off making a terrific commotion, smashing the bamboo and small trees as they went'. They returned and left three times, stampeding through the forest, trumpeting and screaming incessantly before finally leaving, when they could be heard screaming and calling far away in the forest. This behaviour seems to show that the apparently rather loose organisation of an elephant herd, in which the animals, except for mothers with up to three successive calves, do not seem to be particularly closely attached to each other, may in fact have much more cohesion and be less casual than is generally assumed.

In studying the composition of a population of mammals it is important to know the ages of the animals examined. A method for finding the age of an African elephant by examination of the teeth has been developed by Laws [291]. The teeth consist of a number of plates or lamellae of dentine, each covered with a layer of enamel, bound together with cement to form a single tooth. Six, rarely seven, molar teeth come into use successively, as mentioned above, and each can be identified by its measurements and by the number of lamellae, so that through knowing which teeth are in use, and the amount of wear they show, the age of the animal is approximately known. From the information given by the teeth, criteria depending upon the size of the animals are derived for estimating the age of an elephant seen in the field. Laws found that the upper age limit is about sixty years, but, by examining the teeth of animals that had died natural deaths, the mean expectation of life is less than fifteen years. His estimate of tusk growth differs from that of Perry, mentioned above, for he found that the growth rate of the tusks in females is linear throughout life, but that in males it increases, so that the tusks have a mean combined weight of 240 pound at sixty years. 'Big tusks are generally the result of prolonged growth: extremely big tusks probably result from prolonged and above average rate of growth'. He, like others, found that in areas that are over-populated the age of females at puberty is increasing and the interval between births is lengthening – 'a depression of the reproductive rate which is almost certainly density dependent'. If the information provided by studies on the ages of elephants that die naturally is integrated with the results of population counts from the air and on the ground 'a much more complete picture of elephant population dynamics than we have at present' will be obtained.

In a later paper Laws [291] checked the accuracy of his results by taking the dry-weight of the eye lenses, a technique that has been used for age estimation in several species of mammals; the dry lens-weight alone, as shown in his graphs, can be used to assess the approximate age. Laws, Parker and Archer [294] also used the method devised by Smith and Ledger [450] of estimating live weight from the weight of a single hind leg, a convenient method with large animals once the labour of obtaining the basic relative weights of leg and the entire carcase has been done. Krumrey and Buss [279] have also worked out the relationships between body height, total length and weight in the female African elephant, so that the weight of an animal can be determined from measurement without the labour of weighing it. They calculated the growth equation for female elephants up to twenty-five years of age, estimating the age by the state of the teeth, and found two among their series of fifty-six wild elephants that were at least sixty years old.

Over a dozen subspecies of African elephants have been manufactured by systematists who have overlooked the individual variation of the animals; as with so many species that have been subdivided through an examination of bits of dead animals rather than a knowledge of them in life, most of them are now discredited. The elephants living in the forests of West Africa and across the continent to the Nile valley are, however, constantly different from the bush elephants elsewhere, and are regarded as being a valid subspecies, *L. africana cyclotis*. The most noticeable difference of this subspecies from *L. a. africana* is in size, for the males are generally slightly less than eight feet in height and the females less than seven. The shape of the ear, too, differs; in the bush elephant the bottom corner forms a pointed lobe, but in the forest elephant the outline is rounded so much that there is no well defined lobe. The forest elephant has five nails on the fore foot and four on the hind, against the bush elephant's four and three; the number of nails, however, is not constant in either subspecies – Morrison-Scott [340] considers that the lesser number is probably correlated with age and weight, for some young bush elephants have five nails on all feet. Another subspecies, the 'pygmy' elephant, received a latin name; it was based on a captive specimen under four feet high which later confounded its describer by growing up. As Morrison-Scott says, 'the pygmy legend had taken firm hold' and it has not entirely let go yet; confusion is perpetuated by people sometimes using the name for the forest elephant. Even among zoologists there are a few prone to give some credit to native stories and hunters' yarns about an aquatic pygmy elephant, though no one has produced any concrete evidence that it exists. Blancou [43] who reviewed the matter thinks that although there is little documentation the evidence may yet be discovered, and concludes his essay with the comforting thought that if this form exists it is in no need of conservation because '*L'éléphant nain est généralement redouté des Africains pour son réputé mauvais charactère, peu intéressant pour sa viande, encore moins pour son ivoire, et il hante des terrains absolument répulsifs pour les chasseurs*'. As Hilaire Belloc said, 'Oh, let us never, never doubt what nobody is sure about'.

The forest elephant also differs from the bush elephant in being more amenable to captivity and training for work. Since the beginning of the century an elephant catching and training station was maintained in the Belgian Congo, where methods similar to those used in India were adopted, but it was not very successful, and is believed now to have been abandoned. Hediger [219], who visited the station in the Garamba National Park at Gangala-na-Bodio in 1948, said that its commercial returns were then minimal, and that as a working animal in agriculture and forestry the elephant of the Belgian Congo had no practical significance. It was the small forest elephant that was domesticated and used in war by the ancients. The subspecies was formerly more widely distributed across the continent and also extended into northwest Africa; Gowers [180] argues convincingly that it was the forest elephant that the Ptolemies and Carthaginians trained for war. Gowers and Scullard [181] also show that from numismatic evidence it is probable that Hannibal had at least one Indian elephant in his squadron of African elephants, and that it may well be that it, or they, 'survived the rigours of the Alps better than the main bulk of the Africans and that the sole surviving beast on which Hannibal himself rode through Etruria was an Indian'.

ORDER HYRACOIDEA

The order Hyracoidea is a small one containing a single family, the Procaviidae, with three genera and eleven species of small furry animals about the size of rabbits or hares. They are so unlike all other mammals that they are placed in a separate order: they are often said to be the nearest relatives of the elephants, and to a certain extent this is true, for they both have many subungulate characters. The impression thus conveyed to the non-specialist is unfortunate, for the close relationship applies only to very remote ancestors on both sides, and the popular paradox that 'hyrax is the nearest relative of the elphant' is misleading. Hyraxes superficially resemble large guinea-pigs or marmots in conformation; the colour of the soft fur is brown above, lighter or white on the belly, and with a spot, white, yellow or black according to species on the middle of the back; the spot marks the site of a skin scent-gland. They are stocky little animals with short legs, short rounded ears, and no visible tail. There are four short digits on the front feet, three on the hind; with one exception all are protected with flattened nails usually described as hoof-like, but the inner, anatomically the second, digit of the hind foot bears a stout claw said to be used in grooming the fur but more probably important in maintaining a grip when the animals are climbing. The palms and the soles are covered with soft naked pads which are said, though as Grassé says no proof has been provided, to be used as suckers when the animals are climbing trees or rocks. Sclater [430] stated that the centre of the sole is raised by muscular action to produce a partial vacuum, and maintains suction so effective that a hyrax shot dead when clinging to a vertical rock does not drop off but remains fixed by the suckers; he does not explain how the muscular tension can be maintained in a dead animal. The teeth are peculiar: there is a wide gap between the incisors and the cheek teeth resembling that of the rodents, and the single pair of upper incisors grow from persistent pulps, but, unlike those of rodents they are triangular in section, not chisel-shaped, and furthermore they are separated by a central gap. The two pairs of lower incisors are chisel-shaped, slightly serrated, and procumbent; the cheek teeth bear a remarkable resemblance to those of a rhinoceros in miniature. The animals are exclusively herbivorous.

Three genera of hyraxes are recognised, *Procavia*, *Dendrohyrax*, and *Heterohyrax*; the third is treated by some systematists as a subgenus of the second. Hyraxes inhabit nearly the whole of Africa south of the northern tropic, and *Procavia* extends north into Syria, Palestine and southern Arabia. Although there are only eleven species of hyrax, a very large number of subspecies has been described – they will probably be much reduced when the systematics of the order has been critically revised. Hyraxes live among rocks, cliffs, scrub and more open country, and some species are arboreal; it is only in the last few years, however, that scientific observations have been made on the habitats, behaviour, and general biology of several species in East Africa, and have provided exact information in contrast to the casual and anecdotal observations previously available. *Procavia* and *Heterohyrax* are rock hyraxes, generally known as dassies in South Africa; Sale [411] found that in East Africa the two genera show no basic differences in behaviour and selection of habitat, but that a subspecies of *Procavia* living high on Mount Kenya differed in habits from those of the lowlands. Rock hyraxes do not burrow, but live in colonies in natural crevices of rocks or among boulders. The availability of suitable shelter therefore determines the distribution of the animals. In South Africa predators have been exterminated in some places and the numbers of hyrax have increased so much that the excess population has been driven out of the rocks on to the plains, where the animals have taken to living in the abandoned burrows of other animals, and in the crevices of stone walls and other man-made structures, thus showing an ability to adapt their behaviour to the necessities of the environment.

The geology of East Africa, with the frequent occurrence of rock outcrops, provides innumerable sites suitable for the establishment of colonies – Sale considers that a family group of some five adult animals is about the smallest that can form a colony. The gregarious nature of hyrax is emphasised by Sale who says, 'It is extremely rare to find a group living in a single isolated hole or crevice. . . . Unless a cavity has a similar shelter within about ten metres it will not be used as living quarters by hyrax, although they may take temporary refuge in it during flight from an enemy'. Hyraxes dislike strong winds, and choose the sheltered side of an outcrop for their dwellings, but they show a wide tolerance to altitude, temperature and rainfall. Rock hyraxes are diurnal; they come out of their holes in the early part of the day to bask in the sunshine on exposed rocks, which reach so high a temperature that it is surprising that they can tolerate it – they seem to like being almost roasted alive. The temperature within their holes, however, remains much cooler, and the humidity higher, with a comparatively limited diurnal range. The eye of hyrax has a unique character, perhaps correlated with the basking habit; the iris bears a peculiar lobe which is believed to shade the retina from the damaging effect of the sun's rays – native legend has it that the hyrax is the only animal that can look directly into the sun (Plate 22).

In many places hyraxes do not drink for comparatively long periods but obtain their water requirements from the vegetation on which they feed. The physiology of their water metabolism does not appear to have been investigated, but it is probably unusual, for not only can they withstand high sunshine temperatures, but their urine appears to be unusually concentrated. Hyraxes are latrine users, the animals of a colony going to particular places for defaecating and urinating – the urinating places become thickly encrusted with the solids of the urine and form conspicuous white deposits streaming down the surface of the rocks.

Turner and Watson [479] found that the colonies of *Procavia johnstoni* living in the granite kopjes, surrounded with dense woody thickets, of the Serengeti show a constant daily pattern of activity. They write that

at dawn they emerge from the crevices and climb to high rocks with an eastern aspect where they spread themselves to take the maximum radiation from the early morning sun. In the first stage of sunning all hyrax are grouped very closely together, touching other animals on all sides, the young often climbing onto the backs of the adults to burrow into the fur.

Later, the animals spread out and start grooming themselves and stretching, and then come down from the rocks and start feeding; they retire to the shade of the thicket during the heat of the day and then feed again between four and six o'clock in the late afternoon before retiring to their holes for the night. These animals fed almost exclusively on grasses and did not browse on herbs or shrubs; on the other hand in Kenya Sale [409] found that *Procavia habessinica* and *Heterohyrax syriacus* fed mainly on woody plants, especially shrubs of all sizes, preferring the leaves and young shoots, though during a drought they ate the bark. The animals commonly climbed into shrubs to reach leaves inaccessible from the ground, going as much as three metres up the stems. These species, like that on the Serengeti feed for less than an hour in the morning and again in the afternoon. They eat a very wide range of food plants according to the season, but usually concentrate on a few species at any one time. Sale [409] records that *P. habessinica* feeds with impunity on *Phytolacca dodecandra*, a plant highly poisonous to most animals, and found one colony that fed 'almost exclusively on the fleshy leaves and shoots of this rambling shrub . . . the plant contains an alkaloid and a high proportion of haemolytic saponin'. He suggests that the ability to eat this plant may be correlated with the presence of two unique caeca in the region of the large intestine; the poison may, however, be neutralised by some stomach secretion, or there may even be an immunity in the blood and body fluids.

Hyraxes are gregarious in feeding, as they are in their other activities. Sale describes group-feeding, in which all the animals start browsing together; if the feeding place is at a distance from the holes the colony goes to it in single file along a well-worn track. When feeding the group fans out in the manner of a herd of grazing antelopes thus having 'wide visual cover in case of approaching predators'. Group feeding goes on for about twenty minutes, seldom more than half an hour, and ends gradually 'as individuals or small groups of animals wander back to the holes'. Rock hyrax thus eat a lot in a short time, and Sale remarks on the speed and directness with which group-feeding starts, and the intensity with which it is maintained. The animals are most vulnerable to predators when in the open away from their holes, 'hence it is of survival advantage to get the business of feeding over as quickly as possible'. If one of a feeding group sees or hears potential danger it gives a short squeak which alerts all the members of the group; if the danger approaches a 'high intensity alarm note' is given with the lips wide open and pulled back tightly against the jaws. Immediately the group dashes for shelter, all the animals repeating the alarm signal, those in bushes leaping down to join the stampede. They take shelter in the nearest hole, whether or not it is home; the dominant animal enters last and turns at the entrance to face the danger, 'showing its tusks (upper incisors) and growling fiercely'. *P. johnstoni* in the Serengeti has a similar reaction, but has the advantage of a sentry system. Another species, the tree-hyrax *Dendrohyrax brucei* lives in the kopjes with *Procavia*, but after the morning sunning, when *Procavia* groups spread out grazing it climbs quickly into its selected tree or shrub to browse, and the two species are then completely separated. Turner and Watson found that the

alarm call of the tree hyrax has the immediate effect of sending all hyrax (of both species) running for cover . . . the separation of the two species in feeding is obviously of great benefit to the rock hyrax, which avoids competition for food and is moreover enabled to make efficient use of grazing up to 60m from the kopje by virtue of the system of tree hyrax sentries.

They add that the tree hyrax gains little from the separation except in avoiding competition with the heavier and more aggressive rock hyrax.

The habit of feeding by day in this species of tree hyrax appears to be unusual, for tree hyraxes are generally nocturnal and come under notice more through their whistling calls at night than through being seen. Other species of tree hyrax live in smaller groups and shelter in hollow trees, rather than sharing crevices in kopjes with rock hyraxes. Owing to their more solitary and nocturnal habits they are less common in museum collections, though I have seen them taken in quantities great enough to make fur coats and capes in the souvenir shops of Moshi and Tanga. Tree hyraxes are surprisingly agile in climbing about the branches of trees, and Turner and Watson note how they run down vertical tree trunks at great speed.

Sale [412] has investigated the interesting feeding method of hyraxes; the upper incisors are defensive tusks, and the lower ones are used as a comb for grooming the fur, and so are quite unsuited to grasping or cropping food. Food is consequently bitten off by the cheek teeth and taken in at the side of the mouth 'like a carnivore gnawing a large bone which remains projecting from the side of the mouth'. The length of the row of cheek teeth thus allows a large amount of food to be cut off at each bite so that 'for a herbivore of its size the hyrax can take in food at a very great rate and this must largely account for the relatively brief feeding time'. When feeding hyraxes do not pick up things with their fore feet; at most the feet are used for pushing down the stems of tall herbs to bring the shoots within reach, or for pinning down a food object to prevent it rolling from the teeth. Hyraxes chew their food with a rapid side-to-side motion of the jaws, but they do not ruminate, in spite of the biblical reference to the cony chewing the cud. The structure of the stomach is simple and entirely unlike that of the ruminants; Sale has found no evidence for rumination, although others have stated that they thought that it occurs. The mistake has probably arisen from the habit of hyraxes 'chewing air' – producing a chewing motion of the jaws while there is no food in the mouth, particularly as a displacement activity when confronted with something that is strange to them. Sale suggests that perhaps refection may take place as in the rabbit, when the animals are in their holes, but no proof or evidence has been produced to show that it happens. Hyraxes do not take food into their holes but eat it immediately as they gather it. Sale concludes that the dry matter content of a food determines the amount of it that a hyrax will eat. Foods with a low protein content probably demand a lower water intake than those rich in protein. The dry matter intake of an adult *Procavia* was found to be 33·6 gm per kg body weight per day, 'which is low for an animal of this size and may be connected with the relatively inactive life and poor temperature regulation of the rock hyrax'.

The habits of the subspecies of rock hyrax that lives in the alpine zone of Mount Kenya differ considerably from those of the lowland subspecies. Coe [76] studied *P. johnstoni mackinderi* in the Teleki valley on the mountain, where the animals live in the hollows and crevices among the boulders of the lateral moraines of extinct glaciers between the altitudes of

329

10,500 and 15,500 feet. In the forest zone below this level the tree hyrax, a subspecies of *D. arboreus* is the only species, but in the Ruwenzori range where *Procavia* is absent another subspecies of *Dendrohyrax* has occupied the habitat in the alpine zone exposed by the retreating glaciers. Here it lives much as does *Procavia* on Mount Kenya, except that it has maintained its largely nocturnal habits. On Mount Kenya the activities of hyraxes depend upon the illumination – they come out by day only when the sun is shining, and go back to their holes if cloud covers the sun for only a few minutes. Coe watched a colony during a period of bad weather and found that the hyraxes 'remained below ground for three days, and only emerged when the weather improved'. The hyraxes were also active on moonlight nights and were heard calling as soon as the rocks among which their holes lay were illuminated. Coe concludes that the activity is thus dependent upon illumination and not upon the heat derived from the sun shining on the animals. The mountain hyraxes showed a preference for grasses and mosses as food, but also ate some higher plants with succulent leaves. In spite of the high water content of the mountain fodder the hyraxes drink regularly, going as far as 700 yards to do so. 'The animals use regular tracks which go straight from the colony to the water below: they are about a foot wide and are worn deep into the ground by continual use'. Coe found that colonies are made up of small family groups, each with a dominant male, who protects a party of females and their young. When the sun shines into their hole the male comes out first and is shortly followed by the others; if alarmed he gives the warning signal and is the last to enter when they rush home. Coe remarks that like all species the mountain hyraxes are good jumpers and that he has seen them leaping down from rocks, covering a distance of about twenty feet without ill effect.

Many species of hyrax in the tropics have no well defined breeding season, but Glover and Sale [167] find that mature males of *Procavia* and *Heterohyrax* have only limited periods of sexual activity. They suggest that since the family group consists of a male and several females and their young, the surplus males must form bachelor groups, and they found from observation of captive animals that the presence of an old sexually active male inhabited the activity in the gonads of the subordinate males. In the *Dendrohyrax* of Ruwenzori O'Donoghue [352] found that breeding is not restricted to any particular season, but that births are most frequent 'in the middle of the year'. The structure of the placenta of hyrax has attracted the attention of embryologists such as Wislocki [513, 515, 520] and Sturgess [463] who studied the placentation of the South African species *P. capensis* in the hope that their results might throw some light on the affinities of hyrax and the origins of the order. Wislocki found that the placenta most closely resembles that found in the insectivores, rodents, and the primate *Tarsius*, and has no resemblance to that of the ungulates. He described the avascular allantois, and showed that it is peculiarly drawn out into four sacs at the points where the allantoic vessels penetrate the chorionic wall. The allantois is annular and surrounds the amnion like a broad girdle; it resembles a flattened quadrilateral pyramid or tent, with the points in contact with the chorion. When he wrote in 1928, 1930 and 1940 the details of the placentation and of the membranes in the elephant were not known, but in 1953 Perry [374] showed that in the elephant too the allantois is similarly drawn out into four sacs, a resemblance that may point to a phylogenetic affinity. The serological experiments made by Weitz [503] support the suggestion of such an affinity, for they showed a 'very marked relationship between the sera of hyrax and elephant'.

Details of the sexual rhythm, frequency of breeding, and the lengths of gestation and lactation have yet to be determined precisely. Gestation lasts about seven months, and the young are precocious – at birth they are comparatively large, have the eyes open, are fur clad, and are active at an early age. The litter size is generally one to three, but six is recorded – the Mount Kenya rock hyraxes have only one young at a birth, and it has been suggested that this low reproductive rate is correlated with the high population density, and also perhaps with the severe environment in which the temperature falls to freezing or below nearly every night – an environment which must throw a strain on an animal with poor temperature regulation.

Hyraxes utter characteristic calls in addition to the warning and alarm notes and the aggressive growling already noted in the rock hyraxes. The nocturnal calling of the tree hyraxes has led to the animals being known as 'bush babies' in some parts of South Africa, according to Roberts [406]. Rock hyraxes when sitting on a boulder or rocky ledge utter a high-pitched mewing note of considerable carrying power; Coe heard animals answering from more than a mile away. The use of this calling is not apparent; it can scarcely be of territorial meaning in a creature so tied to a restricted range, nor is it likely to be sexual in view of the structure of the social groups.

All the authors quoted, who have studied hyraxes in the wild at first hand, are agreed that the leopard is the main predator on the hyrax in East Africa, and that large snakes, hunting dogs, and large birds of prey also eat them. The habit of group-feeding, and the alarm signals, help to protect the hyraxes from leopards, as does the selection of holes as dwellings. Sale points out that holes with entrances large enough to admit a leopard are never used except as temporary refuges. On the other hand, holes of little extent within are often used if the entrance is too small to admit a leopard, and 'one can sometimes observe a group of hyrax huddled inside such a hole at very close quarters'. They appear to 'feel quite secure' even when so near to such a large enemy as man. On Mount Kenya, however, where leopard are less common and hyrax are very plentiful, they are not always so careful in selecting a hole with a small mouth. In captivity Sale found they showed a similar complacency when they were threatened by a dog outside their cage; they quickly realised they were quite safe and took very little notice, 'sometimes even deliberately sitting with their noses against the mesh while the less intelligent dog charged them ferociously.' The group-feeding habits, general alertness, the quick dash into shelter inaccessible to large predators, and the adapatability to accepting the different food stuffs available, combine to make hyraxes common and abundant animals in habitats that suit them throughout Africa.

Although hyrax fur has some commercial value the animals are not of great economic importance. Turner and Watson mention that a substance known as hyraceum was formerly used in medicine and pharmacy as a substitute for castoreum, and that the 'rather shiny deposits of hyraceum' are derived from the urine of hyraxes at their micturating places. It is difficult to imagine any similarity between the two substances, except that they were probably equally without any therapeutic value. The dried dorsal scent gland, or its contents, seem to be a more probable identification for hyraceum, and would certainly bear a closer resemblance to castoreum than do urine deposits.

These excellent researches on the biology of hyrax – there is room for no more than a summary here – are a splendid example of the first-class work now being done by zoologists of both colours in East Africa. To zoologists of an older generation, who themselves have done something towards starting such investigations, and who for long felt that they were very lonely minuscules on a vast page, it is a source of much pleasure to see their early efforts carried forward by their successors on an ever increasing scale of scope and excellence. At one time we felt that the work would not be started until it was too late – now we at last have the satisfaction of seeing research and scientific conservation pushing ahead hand in hand to a greater knowledge and appreciation of a fascinating fauna.

Chapter 12

Rhinoceroses, Tapirs and Horses

ORDER PERISSODACTYLA

THE ORDER Perissodactyla contains the odd-toed hoofed animals, in which the weight of the body is carried chiefly or entirely by the third digit of each limb; the limbs are termed mesaxonic because the axis passes through this digit. The digits are shod with hooves, the front walls of which are closely bound to the underlying terminal phalanges, whereas the sole is separated from the bones by a plantar or palmar cushion of fibrous and elastic tissue. The perissodactyls are exclusively herbivorous, and the cheek teeth form a continuous broad row in correlation with a diet of grasses or leaves. The skull is elongated by the comparatively large development of the facial part necessary to accommodate the long tooth-rows.

The perissodactyls were once a large order containing twelve families, most of which are now extinct leaving only three with living representatives. All three families contain species that are now rare and on the verge of extinction. Of the two suborders the Ceratomorpha contains the Rhinocerotidae and Tapiridae, and the Hippomorpha contains the Equidae.

FAMILY RHINOCEROTIDAE

The family Rhinocerotidae contains five living species partitioned among four genera, inhabiting southeast Asia, Burma, Malaysia, some of the East Indies, and most of Africa south of the Sahara. Although the familiar figure of rhinoceroses is that of large clumsy-looking animals of great weight, they can move with surprising speed and agility when necessary. The front feet have three or four digits, the hind three, each with a small nail-like hoof in front and a common sole behind. The skin is very thick and sparsely haired; the horns are entirely dermal growths set above a thickening of the nasal bones, and of the frontal bones in those species

333

with two horns. The horn consists of a mass of horny fibres agglutinated together with keratinous material; the fibres, however, are not true hairs because they grow from skin papillae and not from follicles.

Rhinoceroses are mainly nocturnal and solitary; for such ponderous and seemingly invulnerable animals they are surprisingly timid, in spite of the fact that the black rhinoceros of Africa sometimes charges blindly if alarmed. The Asiatic species are now rare, their numbers having been reduced by excessive sport-hunting, destruction of habitat by increasing settlement, and by hunting to obtain the horns which are and always have been greatly prized in the East for their supposed aphrodisiac qualities when powdered and made into medicine.

Three species inhabit Asia, *Rhinoceros unicornis*, the Indian rhinoceros, *R. sondaicus*, the Javan rhinoceros and *Dicerorhinus sumatrensis* the Sumatra rhinoceros. The reduction in their numbers is dramatically shown in the maps published by Talbot [466] in which their distribution a hundred years ago is indicated by cross hatching covering thousands of square miles and their present localities by a few scattered dots. The species of the genus *Rhinoceros* are the only ones with a single horn; in both of them the thick skin is thrown into a heavy fold over the fore and hind quarters so that it resembles armour plating, a resemblance heightened by the presence of rounded stud-like thickenings that look like rivet heads.

The Indian rhinoceros is the largest species; it reaches a height of six feet at the shoulder and may weigh up to two tons. Little is known about the ecology of this, or of the other Asiatic species. The Indian rhinoceros formerly inhabited a wide tract of country in northern India and Nepal from the foothills of the Hindu Kush to the border of Burma; it now lives in eight reserves or sanctuaries in India, five of them in Assam, and in the National Park in the Rapti valley in Nepal, about 400 in the former and rather fewer in the latter, a total of less than 800. In Assam rhinoceroses live in swampy country with dense stands of tall elephant grass, but Gee [162] found that in Nepal during the cold months they live mostly in thick tree- and forest-scrub, and in the rainy season in grassland or forest. In Assam they feed both by day and by night on grass, reeds and twigs; they are solitary, and spend much time in wallows especially in hot weather. They return to the same place for defaecating and accumulate large piles of dung. They are not strongly territorial, although individuals do not appear to wander from their habitual home range. Talbot found that 'wallows were shared by as many as five rhinos at the same time. Other rhinos wandered through the areas at will and the use of the dung-hills seemed to be a matter of chance, determined by which hill they were nearest at the time. These dung hills are quite considerable structures, some of them measuring over fifteen feet long and up to four feet high'. Hutchinson and Ripley [249] on the other hand, thought that *R. unicornis* was solitary and inhabited definite territories during most of the year, but that it wandered more widely during the breeding season. They supposed that the territory centred round a pond or wallow, and had a central dunghill and minor ones on the edges – a much more restricted pattern than Talbot found.

In the sanctuaries where they have not been disturbed for many years the animals are fearless and indifferent to the near presence of man, but in Nepal where they have always been shot at and driven away from cultivated areas Gee found 'they have become nervous, frightened of the sight of human beings, and almost entirely nocturnal'. At the Kaziranga sanctuary in Assam Talbot found that a few old rhinoceros lived outside the park near the paddy fields to the south. 'Instead of becoming dangerous rogues, these individuals have become extremely docile, paying little attention to livestock or to the native life which goes on nearby. . . . There are very few cases where rhinos have run amok, causing injury or destruction'. On the other hand a female with a young calf generally reacts aggressively if disturbed, and will charge at once. It is peculiar that when attacking the rhinoceros does not use its rather blunt horn but the very sharp large procumbent canine teeth of the lower jaw – there is a single large incisor but no canine in the upper jaw, and a small one in the lower. In using the lower canines the animal throws its head upwards and slashes its victim – Talbot saw an elephant thus gashed to a height of seven feet from the ground.

Durer's famous illustration of a rhinoceros cut on wood in 1515 shows this species; it was used or copied in the works of Gesner and Topsell. The animal was the first of its kind to be seen in modern Europe, and was imported from India by King Emmanuel of Portugal for presentation to the Pope, but was lost in a shipwreck on the way to Rome. It is a remarkably accurate picture of the creature, in view of the fact that Durer had never seen it, but made his woodcut from a sketch by another hand, and a description in a letter from Lisbon. It shows, however, a second small horn on the neck above the shoulders, evidently derived from a misinterpretation of the sketch, a mistake that could readily be made, for in Walker's [495] photograph of an Indian rhinoceros the ear of the animal happens to be in such a position that it could easily be mistaken for a second horn on the neck by one who had never seen a living rhinoceros.

R. sondaicus, the Javan rhinoceros, is rather smaller than *R. unicornis*,

335

though Talbot, with field experience of both, could detect little difference between them in size. It differs in the details of the skin-folds and in the size of the horn which is less than a foot long and is very small or absent in the females. It formerly inhabited southeast Asia from Sikkim to Vietnam, Burma, Thailand, Malaya, Sumatra and Java; between two and four dozen animals living in a reserve on the Java side of the Straits of Sunda are all that remain today. This species appears always to have inhabited the lowlands of its former range, unlike the two-horned rhinoceros which inhabits densely forested mountainous areas unfrequented by man. This habitat may, as Talbot says, be an enforced one, as the animals would have been driven away from the lowlands by human disturbance if they ever lived in them. The distribution of *D. sumatrensis* was very similar to that of *R. sondaicus* but did not extend so far west and north, or include Java, but it extended farther east by including Borneo. Today the numbers are woefully reduced, but the animals still exist in small areas scattered throughout the former range, from the north of Burma to the south of Sumatra, and tracks of them have recently been seen on Mount Kinabalu in Borneo.

This, the smallest species of rhinoceros, is hairier than the others, especially when young; the skin bears coarse bristles and the ears are fringed. A very hairy specimen was wrongly regarded as a separate species, *R. lasiotis*, to the confusion of the nomenclature. The front horn reaches a length of up to one foot or a little more but the hind one only about five inches, though exceptional longer ones have been seen; the horns are smaller and the hind one rudimentary or absent in the female. Hubback [243] who had hunted the Sumatran rhinoceros extensively in Malaya emphasises the extraordinary agility of the animal, which is able to climb up thickly forested mountain sides so steep that a man has great difficulty in following – the rhino pushes through the thickest tangles of thorns, climbs over fallen trees – 'nothing is too difficult for him'. He states that the rhinos are solitary and feed by browsing, never by grazing, and that the main food supply is the twigs from young trees that they break down; they invariably feed only in virgin forest or very old regenerated jungle. 'When feeding and quite undisturbed a rhino will continually squeak and talk to himself, making some of the noises through his mouth and some through his nose'. The animals dig wallows with horns and feet, and use them two or three times a day; when the mud is thin after rain 'he thickens it up a bit by breaking down the bank with his horn or forefeet'. Hubback adds that they are fond of walking and swimming in the rivers. They frequent salt licks, and polish the nearby rocks by licking them too, not for anything they obtain from them but, according to Hubback, to get rid of leeches from the mouth and lips.

This species appears to be not territorial, for it does not make dunghills; but perhaps disturbance prevents territorial behaviour – Hubback followed the tracks of one for forty days and was close several times, heard it three times, but never saw it. As, however, the total of forty days was spread over five hunting trips the animal was, presumably, constantly inhabiting a definite area of country. The eyesight of this species is poor, but the senses of smell and hearing are acute; Hubback mentions how when disturbed and trying the wind with its enormous nasal cavities it curls up its pointed upper lip, and on getting the scent of man 'wastes no time in getting away, and . . . voices his fear in no uncertain manner'. He snorts 'like an engine blowing off steam', and then goes off at full speed 'in any direction – if facing you he is liable to run straight in your direction – making a noise something between the bark of a dog and the quack of a duck'. On the other hand rhinoceroses can be docile in captivity; one caught young and brought up for seven years foraged and wallowed in the jungle during the day, but came home for supper every evening. 'When the rice was ready one of the household would call with a loud and shrill, hoh! hoh! hoh! and the rhino would answer from the jungle and come back at full speed for its evening meal. It slept under the house'. Hubback found a rhino trail between limestone boulders 2,000 feet up on a mountain; the boulders were polished and the rock of the path worn down. 'How many thousands, nay, tens of thousands, of rhino's feet must have passed along that trail! None pass now, because there are none to pass'.

In contrast to the Asiatic species the African rhinoceroses are much better known biologically, partly because one, the black rhinoceros is still numerous, but mainly because a number of fully trained and experienced field biologists have been working at the research stations and university colleges of East Africa during the last ten to fifteen years. Two species of rhinoceros live in Africa, the black rhinoceros *Diceros bicornis*, and the white rhinoceros *Ceratotherium simum*; both are dark slaty grey-brown, and neither of them black or white. Both species have two horns, which vary greatly in size and proportions; in both there are no incisors or canine teeth, and the skin folds do not divide the surface into armour-like plates. Both species were formerly widely distributed in suitable habitats through-out Africa south of the Sudan but are now much diminished in numbers and distribution.

The black rhinoceros is still found in many places and is common in parts of East Africa, in spite of the useless slaughter of about 900 of the animals during land clearance for the ground nuts fiasco. Goddard [168, 169, 170] whose researches are the main source of our knowledge of the biology of the animal, found that black rhinoceroses are extremely limited in their home range, in which they spend their entire life. The area varies in different places according to the availability of food and surface water, and ranges from less than three to over twenty square miles. Although they are solitary they do not hold well defined territories, for the home range of individuals may overlap to a considerable extent. The area of the home range used is greater in the wet season when food is plentiful than in the dry season when the animals stay in the part of the range nearest to water. Many rhinoceroses are very regular in their movements about their home ranges, and can generally be seen in the same places at the same times daily. When they meet their neighbours in overlapping parts of their ranges they are not generally aggressive, but having accepted each other's presence they are indifferent or tolerant. On the other hand, if a stranger tres- passes, as when visiting a salt lick, 'the resident rhinoceros invariably attacks and is extremely vocal. The head is lowered, eyes rolled, ears flattened, tail raised, and the animal curls its upper lip, emitting a scream- ing groan'. The stranger remains silent and on the defensive, and tries to repel the charges of the resident – the animals use the front horn for goring and for clubbing the opponent on the sides of the head. The African rhinoceroses thus use the horns for fighting as, unlike the Asiatic species, they have no canine tushes for slashing. Aggressive behaviour of this kind occurs between males, and less intensely between males and females; there is little aggression but great caution when two females meet. This explains the common aggressive reaction of black rhinoceroses when they are disturbed by human intrusion on their range – they treat the trespasser exactly as they would a stranger of their own species.

The black rhinoceros is most active at night, and between early morning and late afternoon spends the day sleeping in sand or dust-filled depressions often in full sun even though shade is available nearby. Rhinoceroses sleep lying on their fronts, and rise for about ten minutes once in every hour and a half. The daily routine consists of a walk round the home range for feeding, a sleep in the accustomed bed during the day, and a visit to a mud wallow in the late afternoon. Like the Asiatic species the black rhinoceros makes dung piles, but, unlike those species, it does not let them accumulate into high mounds; it scatters the dung with sharp kicking motions of the hind legs as soon as it is deposited, scuffles among it, and sweeps it from side to side with the front horn using a threshing movement of the head. Goddard found that shuffling through the dung smears the hind feet so that a trail is left when the animal walks about its range – he experimented with a drag and found the animals could follow exactly a complicated zig-zag trail. Thus, although the dung piles are not territorial markers, their indirect use is to define and advertise the home range both to the owner and to strangers.

The black rhinoceros is a browser, and gathers food into its mouth with the pointed prehensile upper lip; Goddard found that in all types of habitat rhinoceroses eat a wide variety of the available plants but that they are highly selective for certain herbs and shrubs – his list of food plants runs to 191 species of forty-nine botanical families. Green succulent herbs are always preferred, though some dominant herbs soon become unpalatable and are neglected if legumes such as clover are available – the latter is particularly sought after in the wet season. It is peculiar that rhinoceroses eat plants that are known to be highly poisonous to some other mammals; these species include *Datura stramonium* and *Phytolacca dodecandra*, the latter also eaten with impunity by hyrax. The finger-euphorbia is highly palatable and is eaten in great quantity; it forms about a quarter of the diet in the wet season and nearly three quarters in the dry, in places where it is available. Goddard relates that the animal uses its horns with great dexterity to obtain the higher branches of this tree.

Straining upwards and placing its fore limbs on convenient lower branches, it wedges the higher branches between the anterior and posterior horns and simply walks backwards. I have seen higher branches up to seven inches in diameter snapped by this method. When the branch falls almost all the branchlets are consumed by the animal. The bark is removed with the tip of the anterior horn and the front premolars.

Some animals in the Ngorongoro caldera were not adversely affected when their home ranges were completely burnt out; they did not leave, and continued browsing on the charred shrubs, even after the rains stimulated new green growth. Similarly in overgrazed areas which afford poor pasture for other animals rhinoceroses find a good habitat, and browse on the shrubs that quickly appear in such areas.

Breeding occurs at any time of the year, and there is no formation of a close pair-bond; the animals are polygamous and polyandrous. Within a few weeks of the birth of her calf a female comes into regular oestrous cycles about every thirty days. Gestation lasts fifteen to sixteen months, and a calf is born about every twenty-seven months, hence 'it is apparent that successful conception does not occur until about a year after parturition, despite the cow's regular oestrous cycles and apparent receptiveness to the bull'. Lactation is not interrupted by oestrus or by pregnancy for the cow continues to suckle a calf until shortly before the next is born; cows must therefore be in almost continuous lactation. Shortly before the new calf is born the previous one is driven away by its mother; it then usually takes up with another animal, immature like itself or adult, and remains in companionship for some time before establishing a home range of its own. When a female is in oestrus several males may be attracted to her, and a good deal of charging, snorting, squealing and jousting takes place with little damage done; the female will accept any or all of them, depending on which can drive the others away. The male follows close behind the female as she wanders about; she urinates frequently and each time the male sniffs the deposit and shows the flehmen reaction, curling the top lip upwards in a snarl-like expression, and testing the receptiveness of the female. He advances with a stiff-legged gait and prods her with his horn and the pair face each other, jousting with their horns. Sometimes the female charges the male aggressively, whereupon he gives way but runs in a narrow circle to return; such rough horse-play may continue for some time. The bull often mounts the cow, and the mounting may be repeated as many as twenty times during several hours before copulation, which lasts about half an hour, occurs. Goddard thinks that this behaviour has contributed to the legends of the aphrodisiac value of rhinoceros products. Some authors, such as Ritchie [405], stated that after mating the cow generally again became aggressive and charged the bull, but Goddard's observations did not confirm this. The pair may remain together for several months, but the bond is not strong or permanent; it 'is governed to a large extent by the size of the home range of the male, and the extent to which the female in oestrus uses that home range'.

Lastly, Goddard points out that the behaviour of the black rhinoceros has an important bearing on the conservation of the species. Its sedentary habit and attachment to a small home range make it an easy subject to preserve in parks and other limited areas with suitable habitats, but these very characteristics make it extremely slow in extending its range into suitable places, when protection is given, in areas where it was formerly abundant before it was exterminated.

In spite of its thick hide the rhinoceros is attacked by blood-sucking arthropods, and is bitten by ticks, tsetse flies (*Glossina*), and horse flies (*Tabanus*). Two species of bird, the red- and the yellow-billed oxpeckers or tick-birds (*Buphagus*) run about on the animals' skin eating the attached ticks, as they do on most large animals, except elephants but including domestic stock. They do not only remove the ticks but feed also on the scar-tissue, blood, and exudate from scabs covering abrasions; in doing so

340

they keep the wound open and prevent it from healing. Black rhinoceroses very commonly carry a running sore behind the shoulder, kept open by tick birds. Rhinoceroses and other animals show complete indifference to tick-birds running over their bodies and pecking at their sores. Tick-birds also eat biting flies attacking their hosts, and drongos (*Dicrurus*), too, have been seen catching flies pestering rhinoceroses. The alarm notes of tick-birds when disturbed by man are said to give rhinoceroses warning of approaching danger. Parsons and Sheldrick [367] have found that two stomoxyd flies, related to the common stable-fly, feed on the blood of rhinoceroses. Although these flies have a stabbing proboscis they seem to be unable to pierce even human skin, but insert the proboscis into the underlying tissues through abrasions, scratches, or the bites of other blood-suckers in order to get a blood meal. After digesting the blood, copious black excreta are produced after even a single meal. Parsons and Sheldrick saw black rhinoceroses covered with masses of these flies, which rose in a cloud when disturbed; the excreta covered the hide of one animal so thickly that it was 'black from forward of the shoulder to behind the hips, and appeared as though tarred ... they must cause it much discomfort at the height of the dry season, when wallowing is impossible'. The authors suggest that in drought conditions, when food and cover are reduced partly by elephant damage, the swarms of flies may contribute to the stress and death of the rhinoceros. The rhinoceros itself is a contributor to the nuisance, for the flies lay their eggs and pass their larval stage in the rhinoceroses' dung deposits.

The white rhinoceros, the largest land mammal after the elephant, is today much more limited in distribution than the black species. It is distinguished by the broad level upper lip, adapted for grazing, and by its neck-hump accentuated by the position in which the head is held with the nose close to the ground. It is gregarious and lives in small herds – solitary individuals are rare – and its tolerance and non-aggressiveness towards man, in contrast with the behaviour of the black rhinoceros, is no doubt correlated with this social structure. The exception is when a female is accompanied by a small calf; she may then be aggressive towards an intruder even of her own species, apart from human disturbance. Foster [152] saw a female lower her head and utter a deep growl when a young male approached her and her calf. 'A larger male ran up and with a side swipe of its horn sent the smaller male running off into the bushes making excited bird-like chirps'. White rhinoceroses graze on grasses, and consequently are plains and not bush animals. They wallow in mud, and make dunghills though they do not scatter the dung as does the black rhinoceros; but the habits of the animals are not known in detail, and studies on this species similar to those of Goddard on the other are yet to be made. The comparatively placid white rhinoceros was easily shot by the early white settlers and hunters in Africa, and long ago became a very rare species exterminated from most of its range. A number exist in reserves in South Africa, including the Kruger National Park, a former habitat into which it has been introduced, and in northern Uganda and adjacent parts of the Congo and Sudan. Wherever it is strictly protected from poachers it seems to be increasing in numbers, so much so that transfers of the animals can be made from such areas for liberation in places from which they have long been absent. The use of modern immobilising and tranquillising drugs, injected from a distance by means of a gun or crossbow, has made these desirable operations possible. As with other species of rhinoceros attempts have been made to recognise subspecies, and to separate the northern and southern populations taxonomically; the pendulum however has now swung in the other direction, and some systematists even think that two genera for the African rhinoceroses are not needed, and that both species should be placed in the genus *Diceros*.

FAMILY TAPIRIDAE

The family Tapiridae contains one genus, *Tapirus*, with four species, one Asiatic and the others South American. The peculiar discontinuous distribution is due to the family formerly being widespread in the northern hemisphere; the extinction of many forms has left the survivors which moved into the tropics separated on the opposite sides of the world. Tapirs are stout-bodied animals of moderately large size with short legs and tail, rounded ears, small eyes, and the nose and upper lip prolonged into a short but very mobile proboscis. They are entirely herbivorous, shy and unaggressive animals. They have a full dentition of forty-four teeth, with the third incisor caniniform and larger than the true canine. The four toes of the front feet and three of the hind are shod with small hoofs and supported by a common sole over a thick plantar or palmar cushion. The eastern and western species differ conspicuously in colour, but the young of all species are unlike the adults and bear a pattern of light longitudinal stripes and spots on a darker background. The structure of tapirs has been investigated in considerable detail by the anatomists, but no serious studies appear to have been made on the biology of the animals – a number of scattered notes on various aspects of their natural history have appeared, but most of the information about them comes from hunters and collectors bent upon their destruction rather than on learning about their lives.

The Malay tapir, *T. indicus*, the largest species, has an unusual colour pattern; the head, shoulders, limbs and belly are black, the rest of the body, the back and flanks, is white. This colour pattern makes the animal very conspicuous in unnatural surroundings, but it may be disruptive and thus help to conceal it in the wild; if it is, one may wonder why natural selection has not favoured a similar pattern in the American species, which are all uniform dull brown or greyish. The striped pattern of the young of all species certainly seems to be disruptive and obviously conceals them among thickets and tall herbage. Sanborn and Watkins [414], who collected tapirs in Siam for the Chicago Natural History Museum in 1949, say nothing on this point, but from their account of hunting the animals it is apparent that tapirs live in jungle so dense that they are invisible even at close quarters; perhaps the colour pattern is not important in concealing them – we can only wonder what other meaning it may have, if any. The hunters spent many days trailing tapirs, and found that 'it follows a zig-zag course during its nocturnal feeding, taking a few leaves from one bush and then moving on to another. It appears to keep moving, never staying long in one spot and never eating all the leaves on a bush'. This implies that the animal is not territorial, but it is probable that there is some sort of home range, as with the black rhinoceros. In Siam the tapirs feed upon several different kinds of bush, at least one of them thorny; while feeding 'they often leave the jungle to cross open areas'. The hunters found that success in killing tapirs depended upon taking advantage of the animals' habit of 'making a hairpin turn down wind from its trail' before bedding down for its daily rest in some thick jungle, and that when disturbed it 'invariably back tracks on its trail when escaping'. The skin of the tapir is so thick, especially on the head and back of the neck, that it is very difficult to preserve it successfully in the hot and humid climate. Sanborn and Watkins estimated that the young are born in November or December, but they could not determine whether the animal breeds every year or every other year; they thought it probable that breeding occurs every other year because some mothers were accompanied by large young ones, a fact which suggests that lactation lasts for a considerable time. Since the gestation period [54] is thirteen months this is inevitably a reasonable suggestion. The authors conclude that tapirs were most plentiful in the south, 'perhaps because the Malays associate them with the pig and will not touch them. It is not hunted in northern Siam and those killed are either shot for sport or in ignorance of what the animal is'. A mother is sometimes shot in order to capture the young for the animal dealers.

The Malay tapir was formerly widely distributed from Burma to

Indo-China and throughout the Malay peninsula and Sumatra; it is becoming increasingly scarce, mostly through human encroachment into its habitat, for it is generally not hunted by the native populations partly because of superstition, and partly because it has no commercial value, is shy, and its meat is not eaten. Harper [205] quotes a letter written to him by F.N. Chasen stating that although in 1937 it was still not uncommon in the Malay peninsula and Sumatra it was 'much persecuted by menagerie keepers. No zoo anywhere in the world is considered complete without a pair of Malayan tapirs and for every animal that survives in an exhibition, several die in Singapore, or in transit'. The tapir was alleged to live in Borneo, but has never been found there; its portrait has, however, appeared on the postage stamps of Saba, then North Borneo.

The South American tapirs are found from southern Mexico to the south of Brazil; *T. bairdi* in Central America, *T. terrestris* to the south, and *T. pinchacus* in the Andes of Colombia, Ecuador and Peru. They are slightly smaller than the Asiatic species, and the first two have a pronounced crest on the neck topped by a bristly mane of short hairs. *T. terrestris* and *T. bairdi* live in the forest and feed on a great variety of plants, roots, twigs, leaves, and fallen fruit. They are nocturnal, but not, as sometimes supposed, amphibious although according to Cabrera and Yepes [65] they generally live near rivers or streams and bathe daily; they also often take to the water if hunted by jaguar or man. On the other hand they are not confined to wet places, for Miller [321] found tapirs in very dry parts of Matto Grosso in Brazil, where water was so scarce that he had difficulty in finding enough for camp use. Ingles [250] watched *T. bairdi* in Panama enter a lake, submerge and come out in a shallow place forty feet away with 'only the snout, eyes and ears . . . above water when the animal had its feet on the bottom'. He was able partly to confirm the native belief that tapirs walk on the bottom under water, for he observed one that could be 'clearly seen as it stirred up the muddy bottom under about five feet of water', though he could not determine whether it was walking or swimming along the bottom. Tapirs appear to be territorial, according to the observations of Hunsaker II and Hahn [247] on a group of eight captives in a zoo. Nevertheless these authors found a special hierarchy among this group, with the oldest male and female dominant, but this may have been the effect of captivity, because in the wild they are solitary or are at most found in pairs. They found that in this group both males and females were aggressive when their territories were occupied by other tapirs, a characteristic 'to be expected in solitary animals'. Tapirs wear regular paths between the water and accustomed feeding places – presumably home ranges – and

344

use similar tracks through the bush for feeding, though when disturbed they can flee through dense cover, pushing through thick tangles with lowered heads. Indeed Miller says they crash blindly off and sometimes collide with trees with sufficient force to knock themselves down.

Whether or not the eyesight of tapirs is acute, recognition of individuals by sight is precluded by the dense herbage of the habitat, and the voice is used in communication. The sounds made are generally of low intensity; a squeal of sliding frequency and short duration is made while feeding. Hunsaker II and Hahn suggest that the squeal serves to keep members of the population in contact with one another, though they do not say why contact should be necessary in a solitary animal – perhaps tapirs are not so solitary as is sometimes supposed. A similar sound of greater length and intensity is made in pain or fear. The species recognition signal is a clicking noise produced by the tongue and palate, and is uttered when animals come close to each other. When startled, tapirs snort much as do rhinoceroses. It is said that natives imitate the sliding squeal call to 'attract tapirs into traps'; Tate [468] was hunting with an Indian who, on locating a tapir, called to it 'first with a short shrill whistle and then with two short clicking noises. The animal ran again, then stopped, whistling and clicking in answer. A second tapir, which we had not noticed before, then commenced to squeak. After a while, however, they tired of this sport, and we heard no more of them'. In some parts of South America tapirs are hunted for food, but in others, as Miller records of the Bororó Indians of Matto Grosso, they are, with the capybara, *santo*, and are never killed or sought for food. The mountain tapir, *T. pinchacus*, lives at altitudes of over 10,000 feet in the Andes; it is rather smaller and darker than the other species, and has longer hair; little seems to be known of its natural history. Several subspecies of doubtful validity have been designated under *T. terrestris*.

FAMILY EQUIDAE

A single genus, *Equus*, forms the family Equidae containing seven species, the horse, asses and zebras. The animals of this family are the remnants of a once large and widely distributed one with many genera and species; with the exception of two species of zebra, all the living species are now rare and seem to be close to extinction unless artificially preserved by man. A remarkably complete fossil record has enabled the evolution of the equids to be traced from small Eocene browsing ancestors with four toes on the front foot and three on the hind, to the modern grazing forms with a single functional digit on all limbs. The reduction in number of digits accompanied an increase in size, in length of leg, length of face, and specialisation of the cheek teeth. It is peculiar that throughout this immense period of time the evolution of the horses took place in north America, with occasional side branches of the family tree extending into Eurasia and latterly South America – but that horses became extinct in North America towards the end of the pleistocene, leaving the modern horse to survive in Eurasia, whither its ancestors had moved from America. Horses were not seen again in America until they were introduced after the Spanish conquest. The living equids are essentially grazers on open plains and steppes; they are gregarious and polygamous; family groups consist of up to a dozen mares with their stallion, and surplus stallions form bachelor groups. The equids have comparatively simple stomachs, and digest the cellulose of their food with the aid of ciliate protozoa and bacteria in the enormous caecum in which carbohydrate fermentation takes place.

Wild horses, *E. caballus*, existed in Europe in prehistoric times, and in historic times they have inhabited a vast tract of Asia from the northern and eastern shores of the Caspian Sea to the steppes of Mongolia east of the Altai mountains. They have been exterminated throughout most of this range, but a few groups may still exist in Mongolia, though it is probable that they are not pure blooded and have interbred with feral domestic ponies. The mongolian animal, known as Przewalski's horse, is a thick-legged coarse-headed pony-size horse, dun in colour with mealy mouth and dark hog mane. Similar horses formerly inhabited Asia to the west, and on the Kirghiz steppes just north of the Caspian Sea they were known by the Kirghiz Tartar name of 'tarpan', a name that has given rise to much confusion and the belief that the tarpan was a different species. Przewalski's horse is thus no more than the tarpan of Mongolia, and is the species that once extended from its present habitat right through Asia to Europe where it formed the prey of neolithic man. Confusion has been added by the claims of a German zoo to have 're-created' the former wild horse of Europe in the form of a 'tarpan' by crossing various strains of domestic horse and introducing Przewalski horse blood. The Union of German zoos unanimously rejected these claims about 1952, and accepted that there is, and has been, only one form of wild horse in Europe and Asia from neolithic times to the present, and that the last living members of it are the Mongolian wild Przewalski's horses. This species is the one from which domestic horses are derived in all their variety of size, build and colour.

There is, however, a peculiar point: the chromosome number of the domestic horse is sixty-four, whereas that of the Przewalski horse is sixty-six. Frechkop [159] has claimed that this difference in chromosome number shows that the two are different species, but the argument is refuted by the fact that crosses between the two are perfectly fertile, as are back crosses to any degree of complexity, although the chromosome number of the first cross is said to be sixty-five. In comparison it is pertinent to note that the chromosome number of the ass is sixty-two, and that crosses with the horse have a chromosome number of sixty-three, and are invariably infertile. According to Zeuner [529] the last wild horses were captured in Poland in 1812, and in the Ukraine in 1851; it is a pity that we shall therefore never know the chromosome number of the tarpan of eastern and south-eastern Europe. The overwhelming balance of evidence, however, shows that the domestic horse is derived from the once widely distributed wild horse of Eurasia, and forms with it a single species, *E. caballus*. Benirschke [33] points out that the chromosome number of the different species of living equids decreases with increasing distance from the original introduction of the stock across the former Bering land-bridge, so that the Cape or Mountain zebra, the most remote, has only thirty-two, but he does not suggest that the decline in numbers shows a line of descent. Indeed, Hopwood [236] has shown that the zebrine horses are older and more primitive than the caballine ones, and were the first wave of invasion from the east – the caballine horses arrived subsequently. Both groups evolved independently from the *Pliohippus* and *Plesippus* group. The long separation however, does not prevent interbreeding, for horses can be crossed with zebras to produce hybrids nicknamed zebroids – a crossing of sixty-four chromosomes with forty-four. There are less than 150 Przewalski's horses now living in captivity, descendants from some of the twenty-eight animals imported into Europe in 1901. They have, however, been crossed with domestic horses, and few are of 'pure wild blood'; but if only animals with sixty-six chromosomes are used for breeding, the original form may perhaps be preserved.

The domestic donkey it derived from the wild ass *E. asinus* of north-east Africa, where it inhabited arid regions and low rocky hills, living in small family parties. It formerly lived in Somaliland, Eritrea and the Sudan, but its numbers are today much reduced and its area of distribution restricted. Several subspecies have been named – in domestication the ass has not produced the wide range of variations shown by the horse, though modification has gone far in producing the extraordinary and bizarre Poitou jack, once so highly prized in France for breeding mules. The Asiatic wild ass, *E. hemionus*, is believed to have been domesticated in antiquity; it was widely distributed over the arid areas from Syria to Mongolia, but exists now as a few scattered populations that show local differences and are designated as subspecies. Such are the kiang of Tibet, the onagers of Persia and India, and the kulan of Mongolia; little is known of their biology (Plate 24).

The remaining equids are the zebras, with characteristic pattern of black and white stripes; they have facetiously been called 'horses in fancy dress', but anyone who has been in Africa and heard their nocturnal braying knows that they are really donkeys. The original zebra of South Africa, *E. zebra*, is now scarce – it is often called the mountain zebra, and the population living in south-west Africa is regarded as a subspecies, Hartmann's mountain zebra. It is distinguished by the pattern of short cross stripes on the rump, usually called the 'gridiron', and a small angular dewlap under the neck. Grévy's zebra, *E. grevyi*, is the largest of the zebras, and is distinguished by its stripes being more numerous and narrower than in other species; they are arranged vertically instead of horizontally on the sides of the rump which is almost unmarked above the tail. This species inhabits the southern parts of Somaliland and Abyssinia, and northern Kenya.

Burchell's zebra, *E. burchelli*, is by far the commonest of the zebras, and in its various subspecies is still numerous in many parts of Africa south of the Sahara. Its pattern of stripes shows an interesting geographical cline in passing from north to south, and subspecific names have been given to some of the races, though there is no sharp division between them and they merge into each other. In the extreme north of the range the stripes are bold and clear-cut and extend transversely down the legs to the hooves. Further south stripes of a paler brown colour appear between the main stripes. These shadow stripes occur at first on the hind quarters, and at the same time the stripes on the legs reach no lower than the knees and hocks. The farther south the more the shadow stripes creep forward and the more the leg stripes recede up the legs until in the extinct race inhabiting the former northern Cape Province and Orange Free State, the original Burchell's zebra, *E. b. burchelli*, the shadow stripes extend on to the withers, the legs are practically unstriped, and the body stripes do not extend on to the belly, which is white. Finally there is the extinct quagga in which only the head and neck were fully striped and there was slight striping on the withers. The rest of the body was dull brown and the legs were white, as though the leg striping had completely retreated, and the shadow stripes had filled in the white stripes between the dark ones and merged into a uniform brown, an appearance heightened by the indefinite body stripes over the withers.

It might thus be thought that the quagga was the most southerly form of Burchell's zebra, and showed the extreme development of the confluent pattern of stripes; this view would undoubtedly be accepted by systematists were it not for a scrap of information recorded before the quagga and the true Burchell's zebra were exterminated. In 1836 and 1837 Harris [207, 208] went on a hunting expedition from the Cape as far north as the Tropic of Capricorn; it can be inferred from his writings that the quagga and Burchell's zebras both lived between the Vaal and the Orange rivers, but that they did not mix and were thus truly distinct species. He says the quagga existed from the Vaal river south into the Cape Colony, but that Burchell's zebra was found in the interior north of the Orange river. The Vaal is a tributary of the Orange river, and if he were speaking in general terms this might not have implied an overlap in the habitats of the two animals, but he does say that the quagga never mixed with the herds of Burchell's zebra and this could mean that they lived in the same places. On the other hand he says that both animals habitually associated with the then plentiful herds of gnu, Burchell's zebra with the northern species, the brindled gnu, and the quagga always with the southern one, the white-tailed gnu. As both animals are extinct the question whether they were distinct species or merely subspecies can probably never be answered, but Harris' statement does not necessarily refute the suggestion that the quagga represented the extreme southern end of the Burchell's zebra cline of patterning extending from the north. In passing, it is interesting to note that Harris described the quagga as 'reddish brown, irregularly banded and marked with dark brown stripes, stronger on the head and neck, and gradually becoming fainter until lost behind the shoulder . . . belly, legs, and tail, white'. Of Burchell's zebra he says that the general ground colour of the head, neck and body was sienna, 'capriciously banded with black and deep brown transverse stripes . . . belly and legs pure white'.

Many attempts have been made to show that the striped pattern of zebras is disruptive, and serves to conceal the animals from predators. In certain conditions of lighting, and when the animals are among shadowing vegetation or seen against irregular backgrounds, the pattern certainly can be disruptive. Zebras, however, are typically animals of the open plains, where at a distance they appear greyish and the stripes cannot be distinguished; seen against the light they appear dark, but in the opposite aspect they appear light. Furthermore if the stripes are important as a concealing pattern, why do they progressively become less distinct as one traces the cline southwards? In addition zebras do not appear to rely upon any concealing colouration for their safety – they graze in the open and

349

appear to be totally indifferent whether they are seen or not, and to rely on the senses for warning to escape quickly if they are alarmed. As Harris told, Burchell's zebras habitually associate with herds of brindled gnu, and this they do throughout their range where the two species exist side by side. Herds of the two species together are subject to exactly the same ecological conditions, and if stripes are necessary for the preservation of the one, it may well be asked how it is that the more or less uniformly dark gnu manage to survive. On the whole, the stripes seem to have little survival value as protection from predators, and some other function must be attributed to them. The pattern on the face, however, may have some social meaning within the species; the muzzle is black, and the curved and angled stripes on the sides of the face may well draw attention to the mouth – and zebras bite. The facial stripes, indeed, give a rather snarling expression to the face, which to a human observer has a distinctly un-friendly and repelling look. An unusual colour anomaly has been seen in a zebra in Rhodesia in recent years – the animal is black with white spots, the black stripes having encroached upon and broken up the white ones. This extraordinary individual – which looks like a child's toy – has been photographed from the air, and does not appear to be an outcast from the herds of its normally coloured relatives or in any way handicapped in earning its living.

Although much has been written of the distribution and habitats of zebras, only the northern race of Burchell's zebra – *E. b. boemhi*, Grant's zebra – living on the plains of Tanzania, has been the subject of biological research. Klingel [270] was able to tranquillise, mark and release over 100 zebra, and follow their movements for up to two years. The pattern of stripes is as individual as human finger-prints, and consequently photographic records allowed all the members of a group containing a marked animal to be followed, so that over five hundred animals were kept under individual observation. Klingel found by using this method that zebras live in family groups which are extremely stable in their composition. Each group consists of a stallion with up to half a dozen mares and their foals, and may total fifteen animals in all. Changes in the group are rare, for the mares stay with their stallion and very seldom transfer to another group. Both mares and stallions sometimes disappear, presumably the victims of lions or other predators, and the disappearance of stallions showed a most interesting detail of social structure. Five groups from which the stallions disappeared were later taken over as a whole by another stallion. 'This demonstrates that the family members are not held together by force by the stallion, but form a stable group even without him'. When the foals grow up there are some changes in the family group, for the young mares are abducted by other stallions when they come on to heat at the age of one to two years. The young stallions leave when from one to three years old or more, and generally form bachelor groups of up to fifteen; small groups of from two to four were found to be very stable, and to remain together for several years. They do not start families of their own until they are five or six years old; they become sexually mature before this age, but the competition of older stallions forces them to remain single for several years.

Although the mares first come into oestrus at the age of twelve to fifteen months they are not fertile until two years old; the age at first foaling observed by Klingel was three. The gestation period lasts about a week over a year, and although foals are born at any time of the year the majority are born between October and March. The sex ratio at birth is equal between the sexes, and so it is inevitable that many stallions must remain bachelors all their lives. Some do not stay with bachelor groups but become solitary, though some of the solitary stallions appear to be old animals that have given up, or been driven away from, their families. King [269], supported by a financial grant from the Horserace Betting Levy Board, studied the reproductive physiology of both Burchell's and Grévy's zebras in East Africa, and found that on the whole it resembles that of the horse. There was some indication, but not proof, that the ovary produces accessory corpora lutea early in pregnancy. In some Burchell's zebra mares an infertile foal-heat is followed by up to four months anoestrus, but a similar lactation anoestrus was not found in Grévy's zebra. The family groups of Grévy's zebras appeared to be less well defined than those of Burchell's zebra, though stallions do consort with groups of mares, and other stallions form bachelor groups. Some of the solitary stallions had a home range of about a square mile but although they were large powerfully built animals their sexual activity, as shown by the level of testosterone in the testis, was only a third of that in younger animals. It is possible that these stallions mate with any mare that comes into oestrus within their 'territory', and may have to fight with the stallion of the group to which the mare belongs. This would account for the heavy scarring on the necks of such animals, though King saw no serious fighting among wild stallions and has seen a stallion allow another to drive off his mare after he had served her several times in quick succession. He remarks that 'stallions did not appear to fight over a mare in oestrus . . . although they often had scars on the neck'. He concludes that there is still a great deal to be learnt about the reproduction of both species, as indeed there is about many other

351

aspects of their lives. These researches in East Africa are a splendid start on the study of the natural history of these handsome animals, and will no doubt be followed by many more now that the biology of African animals is at last being investigated scientifically.

Chapter 13

Pigs, Hippopotamuses, Camels and Llamas

ORDER ARTIODACTYLA

THE ARTIODACTYLA or cloven-hoofed ungulates form a much larger order than the perissodactyls, and contains nine families with about seventy-five genera and 170 living species. In the members of this order the axis of the foot passes between the third and fourth digits, which in the more specialised species are the only ones. In other species the second and fifth digits are present either as functional toes or as reduced dew claws that do not normally touch the ground. With the exception of the pigs and hippopotamuses the artiodactyls ruminate; in many of them the head carries horns or antlers, sometimes in one sex only. Representatives of the order inhabit all the continents and many of their islands, except Australia. The order is divided into three suborders, the Suiformes, the Tylopoda and the Ruminantia.

SUBORDER SUIFORMES

FAMILY SUIDAE

In the first suborder the family Suidae contains the eight species of pigs native to the Old World. The coarse hair is bristly, and in some species is sparse; apart from the familiar truncated snout with the nostrils opening on its disc-like end, the development of the canine teeth as tusks is characteristic. The canines of both jaws curve outwards and upwards, and in the full grown males project from the mouth as formidable weapons. The disc of the snout is partly supported by a pair of pre-nasal bones which lie between the nostrils and strengthen the cartilages that form the disc; the appropriate facial muscles control its considerable mobility. The disc is a sensitive tactile organ but is also robust, so that it is used in digging when pigs search for animal or vegetable food beneath the surface of the ground. Food is located by scent, and the raised rim of the upper part of the disc is used in exploration, but once it has probed into the ground the facial part of the snout is used with a ploughing action for rooting. The

Y

combination of sensitivity and strength in the nasal disc is surprising; a captive bush-pig put into a sty with an apparently smooth concrete floor investigated its quarters carefully and discovered a minute hair-crack in the surface – it at once set to work on the crack and in a few hours had enlarged it enough to start breaking up the floor. The snout in domestic pigs has similar strength so that they can be restrained from rooting only by a ring through the cartilage to make the activity too painful.

The genus *Sus* contains three species, of which the wild boar *S. scrofa* is by far the most widespread; it inhabits most of Europe south of the Baltic, North Africa, and thence across southern Asia to China and Japan. A large number of subspecies has been described from different parts of this vast range; some of them were formerly thought to be distinct species. There is a considerable degree of variation both geographically and individually in size, and in the amount of bristles in the skin; in some there is a dorsal mane or crest, in others conspicuous whiskers on the lower part of the cheeks. Wild pigs usually live in family parties which often join to form larger bands. They are crepuscular and nocturnal and lie up by day in thick bush, and frequently wallow in wet mud – the use of a coating of mud in helping to regulate the body temperature has been mentioned in Volume I. They are omnivorous; although they eat plants and their roots they take much animal matter ranging from earthworms to carrion. The wild boar is the ancestor of the domestic pig, and its ready acceptance of almost anything as food has made it well adapted for domestication in all parts of the world. The young of the wild boar, like those of most suids, are marked with irregular longitudinal stripes of lighter colour on a dark background; it is peculiar that this character has been lost in the numerous domestic breeds.

The bush pig or river hog, *Potamochoerus*, of Africa south of the Sahara is the dandy of the suids; it is rather smaller than *S. scrofa* and has softer bristles. The general body colour is rich red with a white crest along the back, white eyebrows and snout, voluminous bushy white whiskers at the sides of the face, and pendulous tufts at the tips of the ears. In addition there is a wart below each eye, larger in boars than sows. The depth and shade of colouring vary considerably between individuals, as does the amount of hairy ornament. Bush pigs are nocturnal, and appear to have habits similar to those of the wild boar; they, too, are sometimes destructive raiders of cultivated fields, and the rooting of a small party may leave as much as half an acre of land looking as though it had been ploughed as a result of their night's work. The giant forest hog, *Hylochoerus*, of east, central and west Africa was discovered only in the early years of this

354

century, or rather distinguished from the bush pig with which it had previously been confused. It is a large species with long coarse black hair which becomes sparse in older individuals. The nasal disc is comparatively large, and below the eyes on the almost bare face there is a large lump or enormous wart of thickened skin, above which lies a pre-orbital gland. In spite of the huge snout the giant forest hog does not dig but browses on the soft tips of broad leaved plants, and sometimes young fallgrass. It lives in family parties in dense forest and appears to be more active by day than the common bush pig.

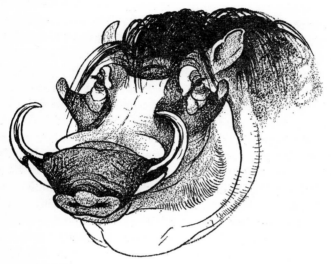

Figure 21 Wart hog (*Phacochoerus*).

The wart hog, *Phacochoerus*, is widely distributed throughout Africa south of the Sahara and abundant in places from which it has not been banished by white settlement. It is rather sparsely haired, but has a crest of lengthy drooping hairs along the ridge of the back and on top of the head which, together with the fantastic ornamentation of the face, gives the boar a truly ludicrous, some might say hideous, expression. The bulbous eyes are set high up on the long face with bags and wrinkles below them, and a large wart on each side; a second pair of warts adorns the lower part of the face, and two pairs of huge tusks curve upwards from the mouth. Unlike the other pigs the wart hog is diurnal. Ewer [140] has studied its feeding habits; it does not dig but feeds almost exclusively on short grass which it crops with a back and forth motion of the head and jaws. It is extremely selective and takes only the tender tips of short grass

355

shoots, avoiding broad-leaved weeds and long grass. The function of the large tusks, which are not used for digging, appears to be to protect the face when the animal is pushing into thorny scrub in search of short grasses growing at the base of the taller herbage. The upper tusks of the boars are less curved than those of the sows, and flare outwards much more widely. They are used in fighting during sexual rivalry; the fight consists of a pushing match with lowered heads and apposed tusks – when the weaker animal submits it breaks away and the dominant tries to gash him on the flank as he turns, usually without inflicting a serious wound, though the flanks may carry many scars of minor lacerations. Ewer was able to show that the characters of the skull of the wart hog, bush pig, and forest hog are correlated with the different foods and ways of feeding in the three species. If disturbed it runs away at a fast trot and only gallops if pressed, carrying the tail erect with the terminal tuft streaming out like a flag. When feeding it commonly kneels on its wrists to bring the head close to the ground; it bears thickened callosities at the point of contact. The callosities are not the result of wear on the skin of the wrist for they are present at birth and are genetic in origin. The wart hog generally finds temporary shelter in an aardvark burrow when pursued, skilfully turning and backing in at the last moment so that it presents its tusks towards a predator. Hubbard [244] observed that the wart hog, unlike most game animals, chooses the hottest time of the day for wallowing and drinking; 'early in the morning and from four o'clock on in the afternoon they root and feed'. The wart hog is often said to have poor sight, but Guiraud [189] found its sight very sharp, 'Ces animaux étant d'un naturel curieux il suffit de détruire la silhouette humaine en marchant à quatre pattes ou par tout autre artifice, pour pouvoir à bon vent, les approcher d'assez près'. Ansell [14] points out that, as far as present knowledge goes, most of the larger African mammals, both herbivores and carnivores, have no definite breeding season, and that young are born at any time of the year although in some species there may be a seasonal maximum. Among the ungulates he found that the wart hog in Zambia seems markedly seasonal in its breeding, with a peak in July and August, but that in the more northern parts of its range it is not seasonal. The young of the wart hog are not striped, thus differing from those of other pigs except the babirusa.

The babirusa, *Babyrousa babyrussa* is the only other Old World species of suid; it inhabits Celebes and some of the neighbouring islands, and is perhaps even more bizarre than the wart hog. The body is so sparsely haired that it is nearly naked, the skin is wrinkled, especially over the forehead, and the tusks, which are largest in the males, extend upwards. Those of the lower jaw project from the mouth, but those of the upper grow from upwardly turned sockets so that as the teeth erupt they penetrate the tissues and skin of the snout and come out on the upper surface. The tusks grow high over the face, curving backwards, and sometimes reach so far as to touch and injure the region above the frontal bones. These extraordinary tusks are believed to be sexual characters, but their function if any is unknown for little scientific study has been made of this or other species of pigs.

FAMILY TAYASSUIDAE

The peccaries are the wild pigs of America, and are found from the southern United States to central Argentina. They form the separate family Tayassuidae, distinguished from the suids, among other characters, in the canine tusks of the upper jaw projecting downwards, not outwards and upwards. The family name has no connection with that of the Suidae in spite of the spelling; it is derived from the Amerindian vernacular name 'tayassú'. The single genus *Tayassu* was formerly named *Dicotyles* in allusion to the navel-like opening of the dorsal skin gland, but the law of nomenclatural priority decrees that the barbarism be perpetuated. The trivial names of the two species are as bad: *Dicotyles torquatus* and *D. labiatus* are now *T. tajacu* and *T. pecari*. Tayassú is the Tupí-Guaraní name for peccary, and pakirá the Carib. Fifty years ago Lydekker [307] objected to the use of ugly, ungrammatical, and absurd terms, and accepted the laws of priority only with protest, and 'in one case ... I could not bring myself to replace a classically-formed name by one of these ill-sounding barbarisms'. Most people are now so ignorant of the classics that perhaps it does not matter that they thus mangle them.

Peccaries are small pigs with dark brown to blackish bristly coats, the collared peccary with a whitish collar round the neck, the white-lipped peccary with white patches on the sides of the jaw; the latter is considerably the larger species. The two species are sympatric over most of their South American range, but the white-lipped peccary does not extend north of southern Mexico, whereas the collared peccary's range continues north into the southern United States. Both occupy varied habitats, but the white-lipped is more an inhabitant of tropical forest; they live in bands ranging from a few individuals up to several hundred. Both kinds of peccary are omnivorous. Miller [321] studied peccaries in an area of southern Matto Grosso, Brazil, where both species were present. He found collared peccaries in small bands but never in large herds, and was surprised to discover that here they lived in burrows, either those of the giant armadillo or natural cavities under tree roots enlarged by the animals themselves. When pursued they took refuge in these dens, about eighteen inches in diameter.

On Barro Colorado Island, Panama, Enders [133] found peccaries very numerous, especially in dense cover and near water; their trails led either to a stream, a wallow, or a palm tree. 'The ground beneath many of the palm trees had the appearance of a pig-pen because of the activities of the peccary, which in common with the squirrels and the coati are very fond of the pulp covering the seeds of the palm nut. . . . Some of these trees had been rubbed smooth by the animals using them as rubbing posts'. He says that one smells peccaries more often than one sees them, for their scent remains in their haunts;

When in doubt as to the users of a trail, it is only necessary to smell a few twigs at the level of a peccary's back to catch the odor, if the animals have been along recently. A young captive did not give off much odor, but when petted would erect the hair about the dorsal gland, trying, at the same time, to rub the gland against the petter. The odor of the gland was characteristic and difficult to remove from the hands.

Epling [136] found that the scent gland, embedded in the dermis and subcutaneous fascia of the lumbar region, is a compound 'storage gland complex' consisting of numerous sebaceous and sudoriferous glands that empty their secretions into common storage sacs, of which the primary sac opens at the summit of a nipple-like elevation. He describes the scent as resembling that of the skunk, though he refers to the peccary's alternative name of 'musk hog', and says that the animals are alleged to be able to eject the scent at will as a protection; he thinks that this would be possible because the connective-tissue capsule of the gland is attached to the skin-muscle. Enders' peccaries fed both by day and by night, and were active especially on moonlight nights; they rooted here and there as they moved along either singly or in groups numbering three to ten.

The larger white-lipped peccary was not so numerous in Matto Grosso, but nevertheless Miller came across it several times in both small groups and large herds. He followed the trail of a large band for several hours, and came up with the animals on a fairly open savannah.

As we galloped up, the bulk of the herd which numbered at least two hundred individuals, gathered in a compact mass among some scattered bushes and low growing trees. They kept up a constant rattling of their tusks and a low eerie moaning. Occasional little parties charged out at us but these our horses easily evaded. The decided aggressiveness of these animals was very apparent and the repeated warnings from my hunter that I should not dismount, were hardly needed.

The dogs, which were accustomed to hunting collared peccaries, refused to approach within several hundred yards – yet young peccaries were easily tamed and made affectionate pets. Enders, who had read of the ferocity of peccaries, says that at first he was afraid of them, but after he had seen collared peccaries several times bluff for a moment and then retire he was no longer worried about attack. He was convinced that 'if one did not get between sections of a divided herd there was little danger from peccaries; this impression was rudely shaken later'. He came across a large peccary, which may have been of the white-lipped species, but instead of bluffing and running it advanced upon the trail 'and kept on coming, and passed within thirty feet of me'. Fortunately the animal then bolted, but its approach was so deliberate that 'in my haste to change the light charge in my gun to buckshot, the desirability of an accurate identification was overlooked'. The stories of the ferocity of herds of peccaries are evidently true, at least for the white-lipped peccary. Enders records that the flesh of the collared peccary is fairly good eating, but stringy and dry; nevertheless five natives ate the entire carcase of a forty-six pound female with 'a liberal amount of fat on her shoulders and in the omentum' in two days.

FAMILY HIPPOPOTAMIDAE

The family Hippopotamidae contains two species, *Hippopotamus amphibius*, and *Choeropsis liberiensis* the pygmy hippopotamus, and is solely African. The hippopotamus, weighing two or three tons, with huge barrel shaped body, short legs and tail, heavy head with small ears, bulbous eyes, and huge mouth, is familiar to all. The body is so sparsely haired as to appear naked, and all four toes of the limbs touch the ground when walking; they are loosely bound together forming a foot shod with nail-like hooves. The lower canine teeth are a pair of large upwardly curved tusks, the upper are smaller; the outer upper incisors curve downwards like the canines, but the inner pair project forwards, and the inner pair of the lower jaw are much larger forwardly directed tusks. The arrangement somewhat resembles the dentition of the pigs, with tusk-like canines and procumbent lower incisors; the resemblance is heightened by the broadening of the maxillary part of the skull that bears the sockets for the upper tusks – the animals would more appropriately be named 'river pigs' than 'river horses'. The hippopotamus can remain underwater without breathing for up to about five minutes, and when submerged keeps the ears and nostrils closed. The ears, eyes, and nostrils are set periscopically so that the animal can float with the rest of the body below the surface. The peculiar coloured sticky skin-secretion, popularly called 'bloody sweat' has been described in Volume 1.

The hippopotamus is widely distributed in rivers, lakes and swamps throughout Africa south of the Sahara, but has disappeared from the more settled areas, and from the Nile below Khartoum. Its biology has been studied in some detail during the course of the control measures mentioned in Volume 1, but elsewhere little scientific work has been done on its ecology. In the Victoria Nile, and in Lakes George and Edward, and the Kazinga channel between them, the animals live in schools, usually of about ten but sometimes containing as many as 100. The schools consist mainly of cows and calves, juveniles of both sexes, and a few adult bulls; a few solitary bulls are generally found near them. During the day the animals live in the water or sleep on sandbanks and the shore, but at night they come out to graze on the banks; the food consists mainly of grass grazed on land, and not as sometimes stated of water plants. Hippopotamuses often go a mile inland when grazing at night, and use well worn tracks which cut deep into the banks at their landing places. They are, nevertheless, very sedentary, and each school frequents its own water-territory and grazing ground without trespassing on that of others, though there are seasonal movements of populations. Near the lakes hippopotamuses also live in numerous pools or wallows at some distance inland from the shore; the wallows are particularly favoured by the bulls. Laws and Clough [293] who investigated the breeding biology of the hippopotamus, said that although no detailed observations had been made on social organisation and behaviour 'the greater size and higher incidence of broken canine teeth in the males suggests that some form of dominance hierarchy may be maintained. The present indications are that if territories are maintained, they are aquatic and variable'.

In the Upper Semliki between Lake Edward and the River Karurume, Verheyen [488] found that over a distance of thirty-two kilometres there was a minimum of over 2,000 hippopotamuses, an average of one to every fifteen metres, with a concentration of one to every five metres in the upper five and a half kilometres of the river. In these conditions each group of females and young is attended by up to six adult males of approximately equal rank in the hierarchy, each dominant only in the neighbourhood of its own bit of river-territory. The males maintain their territories and their social rank by threat and fighting, most commonly by threat alone followed by withdrawal of the weaker. Fighting of different degrees of severity occurs, and wounds that appear to be very severe are inflicted but, as Verheyen points out, the animals are well adapted to fighting and its consequences by their very thick skin on the back, flanks, rump and tail, and by the astonishing speed with which the wounds heal. Nevertheless, when the animals are overcrowded, fighting sometimes ends with the death of one combatant; while making his study Verheyen examined the bodies of five animals that had been thus killed out of the population of 2,000. In the Kruger National Park the hides of practically all old bulls are, according to Stevenson-Hamilton [458] 'seamed with the scars of war'. When the animals defaecate on land they rapidly oscillate the tail so that the semi-liquid mess is splashed about; this is believed to be a method of marking territory. Laws and Clough observed that mating takes place mostly in the water, in wallows, lakes, and rivers, and that the cow is almost submerged when mounted. 'Births usually occur on land but they have also been observed both in and under water'. The young is suckled under water, and sometimes is carried on the mother's back when it is small; the period of lactation probably lasts at least ten months, although the calves start eating grass as well as milk from the early age of six to eight weeks.

The Uganda hippopotamuses had, under complete protection from molestation, increased to numbers greater than the habitat could support without serious destruction. Clough and Laws estimated that the Queen Elizabeth and Albert National Parks alone contained about 26,000 hippopotamuses, and that their nocturnal grazing coupled with the considerable amount of faecal waste discharged into the water during the day results in a continual one-way drain of nutrients from the terrestrial ecosystems. On the other hand, Bere [35] points out that the presence of great numbers of hippopotamuses is not entirely harmful to everyone, because the fisheries of Lakes Edward and George are amongst the most productive in the world and support a thriving and developing industry; the animals provide a continuous application of organic manure which encourages the growth of the plants on which the fish feed, particularly the palatable *Tilapia*. The presence of schools of hippopotamus near the shores also helps to protect the breeding grounds of the fish from disturbance by over-eager fishermen. 'Indirectly, therefore, the grass torn from the banks by the hippos demanding jaws reappears as a highly productive crop in the nets of the prosperous local fisherfolk'. He adds, however, that 'probably half the hippo would do the job equally well'.

In other parts of Africa hippopotamus are regarded as destructive crop raiders; in Gambia the extension of rice-growing – rice has been an important crop there for at least two and a half centuries – is thought by the natives to be limited by the depradations of the hippopotamus. Clarke [75] who investigated the alleged damage found that although they do indeed damage rice crops the amount of destruction is greatly exaggerated. When crops were damaged the natives were able to call upon government help, and hunters were sent to shoot the intruders; the carcasses, too heavy to be removed, were left to the natives who, suffering from a diet deficient in meat, were delighted to have a feast. The damage to rice caused by baboons and other monkeys, bush pig, and weaver birds was overlooked, although even separately these animals might do as much damage as hippopotamuses – but they are less palatable compared with the greatly prized hippopotamus meat. Even when the rice was grazed, and shoots three feet high, above the one to two feet of water covering the paddy, were grazed down to one to four inches above the water, it regenerated almost completely by ten days later, 'the tops of the plants being four feet above the water and only with difficulty distinguishable from the originally ungrazed plants'. Elsewhere in Africa damage is caused to many other crops including sweet potatoes, not only by grazing but also by trampling. The hippopotamuses in the Gambia river form an isolated population, but though isolated it may well be recruited from elsewhere; the animals sometimes travel long distances overland, and are also able to swim in the sea from one estuary to another and thus enter the different river systems. Clarke found that damage can be greatly reduced by the erection of fences, not massive structures as might be expected, but quite flimsy posts and rails two or three feet from the ground; the animals apparently do not step over such barriers, which they could easily push down – but do not. The complexity of the problem of control is emphasised by Clarke who says that if extensive fencing of rice plots is started the trees of the savannah belt may have to be felled to provide the materials. 'The growth of grass which would follow such deforestation might be sufficient to check erosion, but there will be the temptation to put more cattle into such cleared areas, thus reducing even the grass cover'. Once again, if one feature of the ecology is artificially altered, many unintended results are produced.

Laws [292], by examining the growth and wear of the teeth from about 3,000 hippopotamuses in Uganda, concluded that the expectation of life for a three year old hippopotamus is perhaps as much as twenty-four years, and that the greatest life span in the wild is about forty-five years; the mechanical senescence of the teeth is the main cause of death at ages above thirty years. He was able to check his estimates of the ages of animals in several ways, one of the most interesting being by the amount of rinderpest-neutralising antibodies in the blood serum; a severe epizootic of rinderpest affected the animals in 1932–3, and the highest immunity occurred in animals of estimated age thirty years or more when examined in 1963 – the survivors that had recovered from the disease thirty years before. The age at puberty is nine years in the female but about eighteen months younger in the males; reproduction follows a basic two-year cycle. As in some other mammals, accessory corpora lutea and luteinised follicles are produced in the ovaries during the gestation period of about 240 days. Like some other diving mammals the hippopotamus has a greatly reduced heart rate when submerged; Elsner [132] found that the rate dropped from about ninety to less than twenty a minute during dives in a young captive specimen. Lastly Hediger [220] confirmed the reports from several parts of Africa that a fish associates with hippopotamuses. He identified the fish as a cyprinid, *Labeo velifer*, and found that it cleans the skin of the mammal by eating the deposits of vegetable detritus and mud sticking to it.

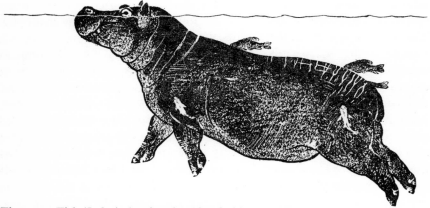

Figure 22 Fish (*Labeo*) cleaning the skin of a hippopotamus.

The much smaller pygmy hippopotamus *Choeropsis liberiensis* is restricted to some forested parts of West Africa. It is solitary and much less aquatic in its habits than the larger species, but little is known of its life.

The pigs, peccaries, and hippopotamuses form the suborder Suiformes; they differ from all other artiodactyles in that they do not ruminate, but their stomachs show varying degrees of complexity resembling the stomachs of those that do ruminate. In the suids the stomach is drawn out to form a comparatively small pouch near the junction with the oesophagus, so that the stomach can be termed 'two chambered'. In the peccaries the stomach is divided into three distinct chambers, and in the hippopotamuses it is enormous and divided into three chambers with complex foldings in the lining – two huge diverticula lie on each side of the anterior chamber, the rear end of which extends so far back that it is lodged in the pelvic cavity. Although the stomach is three-chambered, the last chamber is divisible into two parts by the nature of its lining, only the posterior part of which secretes gastric juice. Thurston, Noirot-Timothée and Arman [474] found that the anterior chamber and the diverticula, particularly the lower part of the left diverticulum, contain great quantities of ciliate protozoa, some of which ingest small fragments of vegetable matter and break down the cellulose. Large amounts of volatile fatty acids that are produced through the action of the ciliates show that carbohydrate fermentation takes place. The ciliates are more diverse in species, and less highly evolved, than those in the rumen of ruminants – the contents of the diverticula of the hippopotamus are drier and less finely divided than those in the stomach of a ruminant. In the rumen of the ruminants a similar fermentation process takes place through the action of ciliates – the rumen is the first of the four chambers into which the stomach is divided in all ruminants except the tragulids, which have three.

Not all the artiodactyls that ruminate, however, are classified as ruminants; the camels and llamas ruminate, but they are not included in the suborder Ruminantia and are placed in a separate suborder the Tylopoda. The tragulids, mentioned below, are regarded as the most primitive of the ruminants, and early traguline stock is considered to have given rise to the other members of the suborder. The tylopods are separated from the ruminants because they arose from a common stock with the tragulids during the Eocene, but evolved in parallel thereafter, and thus form a group very distinct from the Ruminantia living today.

SUBORDER TYLOPODA

FAMILY CAMELIDAE

The suborder Tylopoda contains only one family with living species, the Camelidae, with two genera. The fossil remains of numerous genera and species of extinct camelids are known, but the survivors – four species – are, as Simpson says, uninterestingly few, and of these few the two species of camel and two subspecies of llama exist solely as domestic animals. They differ from the ruminants in being digitigrade, and in retaining the third incisors of the upper jaw. The last two phalanges of the digits are placed almost flat on the ground when the animals stand, and are supported by an elastic cushion covered with a common horny sole; the terminal phalanges bear nails on their upper sides but are not encased in hooves.

The side digits II and V are absent and consequently there are no dew claws. As in the ruminants the front of the upper jaw is toothless owing to the absence of incisors, the place of which is taken by a hard pad of the gums. In the camelids, however, the third upper incisor is retained as a caniniform tooth at the side of the mouth, so that the animals appear to have four upper canines. The lower canines retain their canine form and are separated by a gap from the procumbent lower incisors. The first chamber of the stomach in the camels bears a series of cellular diverticula that were formerly thought to be cells for the storage of water, but are now known not to have this function. The anatomical and physiological adaptations of camels to life in arid regions have been discussed in Volume I. The work of Schmidt-Nielsen and colleagues [425, 426] has shown that camels can withstand water loss through desiccation of up to twenty-five per cent of the body weight, and that they conserve water by reduced evaporation and by the production of concentrated urine. They can tolerate a range in body temperature of 12° F, and store heat during the day by allowing their temperature to rise instead of keeping it down by sweating and thereby losing water; the heat is lost during the night. They can get sufficient water from their food during the winter to meet their physiological needs.

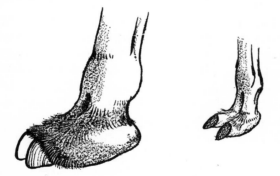

Figure 23 Foot of (a) guanaco; (b) camel.

Although the two species of camel, the Arabian or one-humped camel, *Camelus dromedarius*, and the Bactrian or two-humped camel, *C. bactrianus*, look so different they are closely related; indeed anatomically there is little difference between them except for the humps, and the voluminous winter coat in the Bactrian camel. The one-humped camel, however, has two humps, or at least a rudimentary first hump over the shoulders in front of the 'single' one, from which it is separated by an internal fold of fibrous tissue – the two are fused into one. As a consequence some zoologists have proposed that there is but a single good species with two subspecies. This suggestion is supported by the fact that the two kinds readily interbreed; the hybrids have only one hump. On the other hand, the two-humped camel flourishes in the highlands of Asia where the seasonal climatic variation is extreme; it has a very thick winter coat, and moults in spring to almost complete nudity. There are no wild one-humped camels, and probably no wild two-humped; small numbers of two-humped camels live at large on the edge of the Gobi desert in the foothills of the Altai mountains, but it is doubted whether they are truly wild animals as it is equally possible that they are feral, and descended from escaped domestic camels. Their alleged differences, especially in having smaller humps, appear to be trivial. However that may be, the original home of the wild species, according to Zeuner [529], was parts of North Africa and the Arabian peninsula for the one-humped camel, and Iran and Central Asia to northern China for the bactrian camel; both have been domesticated for probably 2,000 years before the Christian era. Camels are used as beasts of burden, for drawing the plough, and as a source of wool, milk and meat. The one-humped species is very widely spread in domestication from India through North Africa to southern Europe and the Canary Islands – it has also been introduced into the southern United States, South Africa, Australia, and elsewhere, and has become feral in some places. The two-humped camel has stayed at home and is little used outside its original range mentioned above. As with other domestic animals, numerous breeds of the one-humped camel are found in different places; they differ considerably in conformation, as for example riding camels and baggage camels.

Camels are usually docile, but are liable to sudden fits of bad temper, especially the males; they are quiet and obedient but when handled or loaded they make peculiar gurgling, groaning and bellowing noises as though protesting at having to work. While making these sounds the back end of the soft palate is inflated from behind and blown into the mouth; during sexual excitement this peculiar structure known as the 'palu' extends outside the mouth like a large toy pink rubber balloon (Plate 25). The skin scent gland on the occiput has been referred to in Volume I. The callosities on the legs and the chests of camels, which look as if produced by long wear, are congenital and are present in the foetus long before birth. The red blood corpuscles of camels are unique among mammals in being oval and not circular; this fact has somehow become distorted so that it has been stated that the erythrocytes are unlike those of other mammals because they are nucleated – if true this would indeed be remarkable. The paces of the camel have been described in Volume I, but the peculiar rack gait is not its only one; the photograph illustrating the note by Montagu [330] on the feral or perhaps wild Bactrian camels of Mongolia shows the animals fleeing in haste from a pursuing motor-car at a gallop. One animal, at least, plainly has the left hind and right fore legs out-stretched forwards, so that it appears to be using a diagonal gallop – the left fore is flexed and the right hind touches the ground vertically.

The remaining genus of camelids, *Lama*, is confined to South America; although throughout the Tertiary the centre of evolution of the tylopods was in North America no camelids now live in that continent whence they moved south to South America, and west to Asia across the Bering land-bridge. Members of the genus *Lama* resemble small humpless camels, but have comparatively longer ears, and semi-digitigrade feet with the digits less closely bound together, and with comparatively larger claws. The guanaco – pronounced wanáhko – *L. glama guanicoe* and the vicuña *L. vicugna* are the two wild species; the guanaco is found from southern Peru to the extremity of Tierra del Fuego, the vicuña lives in the high plateaux of the Andes from the south of Ecuador to northwest Argentina. The guanaco inhabits semi-desert districts both in the mountains and the plains, and occurs almost from sea-level up to 13,000 feet of altitude in the mountains, whereas the smaller vicuña is common only in places above 14,000 feet, whence it seldom goes above 16,000 feet or below 12,000 feet. Both species are able to flourish in regions where the vegetation is scanty, and the temperature has a daily range of up to 50° F, and often falls to zero.

In high mountains the atmospheric pressure is low and respiration is consequently affected because of the decreased partial pressure of oxygen; this matter is discussed in Volume I. In man and many other mammals acclimatisation to this condition takes some time and is effected by increasing the number of red blood corpuscles during the period. Hall [194] found considerable differences between species in the affinity of the haemoglobin in the blood for oxygen, but that it does not increase in individuals with changes in altitude; the increased oxygen capacity of the blood in such animals as man, the rabbit, and sheep, when taken to high altitudes is thus attained by increasing the number of erythrocytes. In the vicuña and llama the oxygen capacity of the blood alters negligibly with difference in altitude, because their blood possesses haemoglobin 'with a much higher affinity for oxygen than those of rabbits or sheep'. In addition, the red blood cells are more numerous; in the vicuña at 9,000 feet the blood contained 14·1 million red cells per cubic millimetre, whereas that of man had only 5·4. The size, too, of the red cells is smaller; 2·12 cubic millimetres x 10^{-8} as against 9·8 in man. Hall points out that the 'large number of very small cells in the vicuña naturally affords a larger surface area for the diffusion of oxygen. The oxygen can be more readily "picked up" in the lungs and more readily "unloaded" to the tissues'. The efficiency this gives to the respiration was shown when the expedition of which he was a member chased a herd of vicuñas across the desert; 'We drove two automobiles for some time at a speed of at least thirty miles per hour at an elevation of 15,000 feet without gaining on the herd. . . .'

The guanaco inhabits open country whether highland or lowland; Dennler de la Tour [106], who studied the South American camelids for many years, says that in the mountains it chooses the high pampas and plateaux, avoiding rocks and steep slopes. It never enters forests or woods but likes clearings in the wooded regions of the southern Andes and Tierra del Fuego. Guanacos live in small herds of four to ten females led by a male. Young males and older ones without harems unite into herds of a dozen to fifty. Darwin saw herds of as many as 500 in southern Patagonia, but today female herds usually number less than ten, and male herds not more than twenty-five. In a small family herd, while the females are grazing, the male stands at a vantage point on watch. 'Unlike the camel the guanaco is a good swimmer, which explains its presence in Tierra del Fuego and Navarino Island'. When in rut – from November to February – 'the male chases the female but dares not go far from the herd lest another male should try to steal one of his harem. When this happens the two males fight bitterly, often inflicting deep wounds with their upper incisors and lancet-shaped canines'. The same author [107] says that the herd structure of the vicuña is similar; this species 'likes the steep slopes of the mountain tops, minding neither snow nor ice, and does not feel uneasy in glacial winds or snow-storms. It enjoys humid meadows which lie adjacent to the eternal snow-summits of the Andes, many of which reach and pass the 6,000 metre mark'. Koford [272] closely studied the vicuña in the central part of its range, west of lake Titicaca, on the 'puna' which consists of 'high, rolling, semi-arid grasslands, barren pampas, and volcanic peaks, above the limit of cultivated crops'. The animals feed not only on grasses but also on broad-leaved herbs as their need for succulent food is apparently greater than that of the guanaco.

Communal dung piles, often a foot deep and fifteen feet across, are characteristic of vicuña pastures; 'on well grassed flats these piles are spaced, with striking regularity, at intervals of about fifty yards'. Koford found that the population density ranged from one animal to ten acres downwards, with an average of one to about fifty acres; on the basis of these figures he calculated that Peru harboured about 240,000, probably more than half the living vicuñas. Vicuñas live in groups of two kinds, family bands and troops of males. A family band consists of an adult male and several females and their young of the year; each occupies a territory which the male defends against other vicuñas, principally against males. Males that do not have family bands gather into leaderless troops which range widely. Vicuñas graze by day, and rest for about half an hour every one or two hours, and usually drink once a day. They graze continuously late in the afternoon on lower slopes where forage is best, and soon after sunset they walk upslope and lie down for the night. The male leads and defends the band; he grazes close to his females and runs out to drive off an intruder on his territory. When he approaches a female she 'walks quickly away in a subordinate attitude, head and ears low'. Although there is no dominance hierarchy among the females they are intolerant of close approach; 'if one comes within two yards of another, the latter kicks or spits at the offender'. The habit of spitting in defence or aggression is common to all the lamoids – a sharp expiration blows saliva and nasal mucus over the object of annoyance. 'In the event of a serious alarm, the male gives a high whistling trill, the females gather together, and the band flees, the male running behind the females, guarding them from the rear'.

The animals are strongly attached to their territories, and it is the territory rather than the band of females that the male defends. Unattached females are generally allowed to join a band, but one from a neighbouring band is chased away. When a territorial male was injured or sick all his females left him and joined one or two adjacent bands. The male troops consist mainly of one and two-year old animals with about ten per cent of adult males – they often number twenty to thirty, and sometimes seventy-five or more. 'The troop is an open society whose members may join and leave freely without more notice than a few stares from the others'. These troops are nomadic and do not hold territory, but as most of the good grazing is in the territories of family males, they are continually being chivvied and kept on the move when they trespass, first by one male then by another. The troop generally moves away uphill from the territories, but comes down late in the afternoon when they become bolder and resident males more tolerant. Encounters between family bands occur when they graze close to the territorial boundary, which is not marked but is known to the animals and is revealed by their behaviour. The males graze on the boundary side of their females and chase off any of the other side that may trespass, and if in doing so one male crosses the boundary the two males chase each other back and forth. 'Commonly encounters end with the males standing several yards apart, displaying, grunting, and spitting for a minute, before they return to grazing. On rare occasions adult males fight violently, wrestling and tripping each other with necks and heads, clashing with chest and forelegs, biting and shrieking'. The rut starts about a month after the young are born, mostly in March; the young stand up half an hour after birth and after about a week associate with other young which 'play vigorously among themselves, chasing each other, and wrestling with their long supple necks'. Koford concludes that the main advantages of territoriality are that it reduces strife among males, protects the feeding area for females and young, and prevents range damage from over-grazing.

Both the guanaco and the vicuña are valued for their soft coats and their meat, and consequently have been much hunted by man, from long before the Spanish conquest. In Inca times great drives to catch the animals were held at intervals which allowed the numbers to recover so that the stock was maintained. Since then the animals have been increasingly exploited, particularly after the Indians got possession of modern firearms. The skin of young guanacos, 'guanaquitos', is particularly prized; as a result the animals have been exterminated in many places, as must any species of which the young are constantly killed. They are also persecuted by sheep farmers who think, mostly incorrectly, that the presence of guanacos is inimical to sheep farming. As both guanaco and vicuña are polygamous so that there are herds of surplus males, it should be possible to control their exploitation without damage to the stock. The light, soft and silky fur of the vicuña is most highly prized and, owing to the great diminution in the number of vicuñas in many places the animals are protected in Peru, Chile, Argentina and Bolivia – but law-enforcement is often difficult in the sparsely inhabited puna, or neglected, though exports can be controlled. There are no vicuñas in Paraguay, however, and consequently no law protecting the animals; since World War II a poaching trade has been developed whereby vicuñas are illegally killed, and their skins flown into Paraguay whence they are, or were until recently, exported in great numbers to Europe, particularly to Germany.

The llama, *Lama glama glama*, and the alpaca, *L. g. pacos*, are domesticated breeds of the guanaco, the latter perhaps with some hybridisation with the vicuña. The llama is a beast of burden, but is also used for its meat, wool, hide and other products. The alpaca, which is rather smaller, has been selectively bred for its wool during many centuries – the best alpaca wool is long, silky and very valuable. The llama and guanaco weigh up to 200 pounds or more, the alpaca rather less, and the vicuña only half as much. Koford found that alpacas dominate much of the grassland immediately below the vicuña zone. When they approach vicuñas, the latter retire to poorer sites. 'One morning I watched a single alpaca charge again and again at two large bands of vicuñas and finally drive them from their territories to adjacent sandhills. Llamas, grazing freely or in pack trains, pass within site of vicuña bands nearly every day, and vicuñas retreat from them . . . all of the other lamoids compete with vicuñas for food; alpacas on the wetter sites, llamas on the drier sites, and guanacos in some of the most arid places'. No wonder the species needs protection.

Chapter 14

Deer, Antelopes, Cattle and Sheep

SUBORDER RUMINANTIA

THIS CHAPTER deals with the suborder Ruminantia of the artiodactyles, containing five families, two large and three small. Simpson [445] observes that the Ruminantia are the most widespread, most highly differentiated, and individually most numerous of ungulates, having replaced the Suiformes as the faunal dominants of the order since the Oligocene. In ruminants, with the exception of one small family, the stomach is four-chambered; in ruminating the food, swallowed without being masticated, is returned as cud to the mouth where it is chewed and re-swallowed. During the process of fermentation in the stomach the cellulose of the food is broken down by the cellulytic enzymes of an enormous population of ciliate protozoa and bacteria. The fermentation produces fatty acids which are absorbed through the walls of the rumen, and methane and carbon dioxide gases which are expelled by belching. A copious secretion of alkaline saliva containing sodium bicarbonate buffers the acid products of microbial fermentation in the rumen; Bott and his colleagues [50] point out that the sodium content of the daily secretion of saliva represents more than five times the sodium content of the blood plasma so that the cycle of sodium secretion and re-absorption is rapid. In sheep and goats the daily secretion of saliva amounts to ten to fifteen litres, and in cattle it may be five to ten times this quantity. Barnard [21] records that the pancreatic secretion in ruminants contains a high proportion of ribonuclease, the enzyme that breaks down ribonucleic acid. A large part of the protein in the food passes into the bacteria in the rumen and is incorporated in their RNA; they pass with the food to the fourth chamber of the stomach where they are killed and digested. The RNA thus released is degraded in the intestine by the pancretic ribonuclease, so that the protein in the food is the source of an increased nucleic acid-nitrogen metabolism peculiar to ruminants – the amount of the enzyme in the pancreatic secretion of ruminants is over a thousand times that of the secretion in man.

The upper incisor teeth are absent in ruminants, their place being taken by a hard pad of the gum opposed to the incisors of the lower jaw. The lower incisors are procumbent and often spatulate; the width of the row is increased by the lower canines, which are incisiform and lie at the outer ends of the row, so that there appear to be eight lower incisors instead of the six typical of most mammals.

Scent glands are present in the skin of most ruminants, and they may occur at one or more of several sites. The most easily seen are the orbital glands on the face below the eyes; frontal, occipital, inguinal, tarsal, interdigital, and other glands may be present. The secretions from the glands are used in marking territory, signalling alarm, laying trails in gregarious species, and in sexual display – no doubt other functions remain to be found.

FAMILY TRAGULIDAE

The family Tragulidae, containing the chevrotains, also inappropriately called mouse-deer, is considered to comprise the most primitive living ruminants, and to represent the basic stock from which the other ruminants evolved during the Oligocene. In this family the stomach is three-chambered, the third chamber, the omasum, psalterium or manyplies, being represented only by the passage between the second and fourth chambers. Chevrotains are small antelope-like animals, none standing more than a foot high, with slender legs and short tails; they are hornless but the upper canine teeth of males are large and form sharp-pointed tusks protruding from the mouth. The coat is brown with white spots or longitudinal stripes above, and white below. The animals are solitary, and nocturnal or crepuscular; they live in dense bush through which they run rapidly; they bear a striking superficial resemblance to the rodent agoutis of South America, a resemblance heightened by their habit of sometimes sitting on their haunches, a posture infrequent in ungulates. The family contains two genera, *Hyemoschus* with one species *H. aquaticus*, the water chevrotain of tropical West Africa, and *Tragulus*, with three species in India and Ceylon, Malaysia, and the East Indian Islands. Little is recorded about the habits of these animals in the wild; Sterndale [457], who kept some young Indian chevrotains (*T. meminna*) as pets running loose about his house, said that they were timid and delicate but became very tame. 'They trip about most daintily on the tips of their toes, and look as if a puff of wind would blow them away'.

A smaller species, *T. javanicus*, from Malaya has recently been kept and bred in the New York zoo where Davis [103] made observations on its breeding behaviour. The animals showed an arhythmic activity pattern, 'alternately resting and moving about around the clock'. Gestation lasted from 152 to 155 days, and there was a postpartum oestrus within forty-eight hours of parturition; if conception did not then occur there was a further oestrus fourteen days later. Breeding was not limited to any season. The mother cleans the newborn young, which stands up at the age of about half an hour; during the first few weeks she rests in a far part of the cage – it is therefore probable that the chevrotains, like many deer, bed their young down in a lair to which they come only for suckling. When suckling the mother raises the hind foot nearest the baby, and 'the fawn nuzzles under the horizontally held foot at the udder'. The young grow rapidly and are of adult size at five months of age; a female came into oestrus at four and a half months and gave birth 152 days later, and a young male 'was seen to copulate with the old female when he was 135 days old, taking turns with his father. No hostile or competitive activity was noted between the two males. . . . The young male's tusks were not apparent externally until he was ten months old'. Still less is known of the African water chevrotain; Sanderson [415] who collected mammals in the North Cameroons says that its habitat is lush herbage near rivers in high deciduous forest, and adds, disappointingly, only that 'Two of these animals were kept alive. Both devoured insects with rapacity'.

FAMILY CERVIDAE

The Cervidae are a large family containing seventeen genera and about thirty-seven species of deer; its members inhabit the whole of Europe and Asia, both the Americas, and arctic lands such as Spitzbergen, Greenland and Novaya Zembla, and have been introduced into New Zealand, parts of Australia and some islands. Antlers are the characteristic feature of deer; they are carried only by the males of all species, except one in which they are also present in the females, and two in which they are lacking in both sexes. The use of the antlers in display, fighting, and defence has been discussed in Volume I; their use is similar to that of the horns in other ruminants. Horns are permanent appendages of their owners, but antlers, which in some species exceed in size and weight all but the largest permanent horns, are grown and then discarded every year, a process that must throw a great strain – one might think a crippling strain – on the animal's metabolism. The deer has to obtain several pounds of calcium salts and phosphorus from the herbage or grasses of its food in order to build them in the course of four or five months, and repeat the process every year as soon as the antlers are shed. The extinct giant deer (*Megaceros*) of the

Pleistocene annually grew antlers spanning over nine feet and weighing nearly a hundred pounds. Striking as this is, it is really no more surprising than that a young animal can obtain the necessary salts to make the bones of its skeleton in the course of less than a year – although having made a skeleton it does not throw it away periodically.

Fully mature antlers consist of dead bone attached to short pedicels on the frontal bones; the base of each antler is enlarged, at its junction with the pedicel, as an irregular ring, the coronet. During the period of growth the antlers are covered with skin containing hair follicles and sebaceous glands and bearing short hairs that give it its name of velvet. Wislocki [516] and Waldo, Wislocki and Fawcett [494] found that a vascular layer under the velvet contains branches of the superficial temporal arteries which branch profusely into capillary retia over the growing tips. At the growing point new capillaries are continuously formed and the old ones thus left behind are embedded in the bone and drain into numerous thin-walled venous sinuses within it. As growth goes on the vessels nearer the base are constricted by the accumulating bone, and atrophy when the bone is reconstructed from the delicate spongy bone first laid down to dense compact bone. The circulation is then restricted to the vascular layer of the velvet, and venous return of the blood occurs there instead of inside the bone. This change takes place from below upwards until, when growth is complete, the antler consists of dead bone covered with living velvet from pedicel to the tips of the tines. At this stage the increasing level of testosterone in the blood supply brings about an ischaemia in the velvet which dies, whereupon the animal rubs and scrapes it off against favourite trees and bushes known as fraying-stocks. The dead or dying velvet evidently causes the animal some irritation that drives it to this self mutilation – the half shed strips of velvet quickly shrivel and die but the velvet is still soft and contains some blood when fraying starts.

It is peculiar that even the living velvet, through which a large volume of blood circulates, does not bleed profusely if it is injured; even if an antler in velvet is amputated the loss of blood is negligible, and consists only of a slow oozing of blood and never of spurting from the cut arteries. If the pedicel or the frontal bone below the coronet is badly injured, however, the flow of blood is much faster and more copious. The minimal bleeding is brought about by the instant contraction of the extremely thick muscular walls of the arteries of the velvet, which have comparatively small lumina. The vessels in the velvet are particularly vulnerable to accidental trauma and thus 'a mechanism for their prompt and effective closure is of considerable importance'. It appears possible that the death

375

of the velvet, after growth of the antlers is complete, could be caused as much by the fraying carried out by their owner as by a hormone-induced ischaemia.

The growth of the antlers is controlled by hormones; Wislocki [517] and Waldo and Wislocki [493] found that it is stimulated by a factor derived from the pituitary, but that growth stops when the level of testosterone in the blood supply increases as the breeding season draws near. When this level falls at the end of the season the pituitary hormone again stimulates growth, but as the antler has long been dead new growth can take place only at the top of the pedicel below the coronet. The first effect of this growth is to weaken the junction so that the antler drops off. Wislocki, Weatherford and Singer [521] describe how the soft spongy bone of the early stage of antler growth is transformed into dense compact bone from fibrovesicular chondroidal tissue, a method of bone formation not known elsewhere, and intermediate between the normal intramembranous and endochondrial types of ossification.

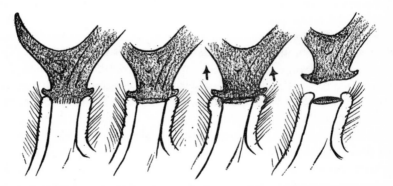

Figure 24 Diagram of vertical sections of antler during shedding. Dissolution of the bony connection – replacement by connective tissue – the upward growing skin pushing up the loosening antler – the antler detached. Blood-clot shown as black.

The process of antler shedding is equally interesting; Waldo and Wislocki [493] showed that the first stage consists in the absorption of bone at the top of the pedicel below the coronet, at the level where the break is to occur. As the bone is absorbed it is replaced by connective tissue which retains the antler in place on the pedicel, but with a much weaker attachment than before. At the same time the bone immediately below the coronet is resorbed so that the diameter of the butt of the antler is less than that of the top of the pedicel. While this is going on the skin and subcutaneous tissues surrounding the top of the pedicel start vigorous growth, and press against the under edge of the coronet so that an upward pressure is put on the antler, tending to lift it off the pedicel. The antler, now insecurely attached, oscillates slightly on its attachment when the animal moves its head and a slight haemorrhage occurs so that a small blood clot is formed between the connective tissue covering the pedicel and the base of the antler. The skin surrounding the pedicel continues to grow upwards and lifts the antler from its seating, until a slight contact with some object or merely a movement of the animal's head is enough to make the antlers fall off by their own weight. The antler comes away quite clean without any blood staining of its butt end, leaving the pedicel capped with a thin central blood clot and the growing skin of the pedicel rolled over its edge. The skin grows inwards over the pedicel, but, before it has closed the wound, growth is already slowly starting to form a new antler. After some weeks of slow growth active proliferation begins and the new antler is produced. Waldo and Wislocki point out that the function of the coronet is not, as had been suggested, to constrict the blood vessels supplying the velvet and thus to cause its death and shedding, but to act as a shelf against which the rapidly growing skin and underlying tissues of the pedicel can push to lift the antler from its seating long after the velvet has been removed and the breeding season is at an end. This interpretation is strengthened by the discovery that if the coronet is removed experimentally the skin of the pedicel grows upwards round the antler like a tube and fails to push it off.

The hormonal control of antler growth is complex, for female deer, apart from reindeer, sometimes grow antlers. It has generally been assumed that such growth occurs in abnormal animals in which the gonads are functionally deranged. But Wislocki [518] examined three white-tailed does which bore antlers, albeit small, asymmetrical, and in permanent velvet, which were quite normal sexually and had born and suckled calves. Gonadal abnormality is thus not necessarily or always involved. He was unable to give a complete explanation of the production of such anomalies, but points out that another assumption that has obtained some acceptance is without foundation, namely that lesions or atrophy of either ovaries or testes of one side results in abnormality or suppression of antler growth of the other side. This conclusion is only to be expected if the conditions are hormonally produced, for once hormones are in the blood they will circulate throughout the whole of the body.

The apparently wasteful process of growing and discarding antlers annually is an unsolved problem – presumably it is of biological value,

377

for if it were not the negative action of natural selection would long ago have eliminated it. A suggestion that antler growth absorbs surplus energy is not acceptable, and there is general agreement that it imposes a considerable physiological strain on the animal. Stonehouse [460] has proposed that the velvet of the growing antlers with its abundant blood supply acts as a heat dissipater in regulating the body temperature during spring and summer; the growing antlers do of course increase the surface area of the body, and their growing tips are higher in temperature than the shafts, but the suggestion has not been generally received with approval. Krog, Reite and Fjellheim [275], who investigated the response of the blood vessels of the antlers to adrenal hormones, found no evidence 'to support the hypothesis that antlers have a thermoregulatory function'. Geist [163] points out that the hypothesis seems plausible 'but rests on doubtful assumptions and fails to predict correctly'; he emphasises the uses of antlers as 'display organs, determinants of dominance and so on', and concludes that they evolved 'as social adaptations and are of great importance to the reproductive success of the individual'. All this can be agreed, but the reason, to speak teleologically, why antlers are grown and discarded remains an unanswered question.

Deer range in size from small to very large; a few species are spotted but most of them are some shade of grey or brown – the young of many species, however, unlike their parents, are spotted. In many deer there is a facial gland below the eye, and some species have scent glands on the legs or between the toes. All are herbivorous, and members of the family occupy a great variety of habitats from the arctic to tropical forests. As the family is a large one it is conveniently classified into four subfamilies.

The subfamily Moschinae contains a single genus and species, the musk deer, *Moschus moschiferus*, which differs widely from the other cervids. It is a small deer without antlers and with large outer digits on the feet. The upper canine teeth of the males are large tusks which project downwards outside the mouth, a character found also in the only other deer without antlers, the Chinese water deer of the subfamily Odocoileinae. It lives high in the mountains of central and north-east Asia up to the limit of tree growth. Musk deer are shy, solitary and nocturnal. The foreskin gland of the male is a large sac which contains about an ounce of sweet smelling secretion much valued in perfumery; thousands of musk deer are snared annually in order to obtain the gland, which when dried is known as a musk pod. The subfamily Muntiacinae contains only two genera of small Asiatic deer which live in forest and among thick vegetation. The several species of muntjac, *Muntiacus*, are solitary and nocturnal. They are peculiar in having very long pedicels for the short antlers, and in the canines of the males being protruding tusks. In the Tibetan muntjac or tufted deer, *Elaphodus*, the pedicels and antlers are shorter and are almost hidden by a tuft of hair.

The deer of the subfamily Cervinae are larger animals and include the only deer with spotted adult coats, the fallow, axis, and one race of the sika deer. The antlers of the fallow deer, *Dama*, a native of southern Europe and the Near East, are generally but not always palmate; those of the other genera vary much in size and complexity of branching into tines. In the axis and hog deer, genus *Axis*, of India they are three-tined, but in deer of the genus *Cervus*, which includes the red deer *C. elaphus*, the wapiti *C. canadensis*, and the sambar *C. unicolor*, they range from three-tined to the large many tined antlers of the wapiti. All these deer are gregarious, and both browsers and grazers. The name of the European elk, which belongs to a different subfamily, has been transferred to America as the vernacular name of the wapiti, which is closely related to the red deer – the true elk of America is there known as the moose. The name of the elk in some European languages is elan or eland, a name which was applied to the eland of Africa, an animal that is not a deer but a large antelope – pioneer settlers are seldom zoologists. With the exception of the wapiti all the members of this genus are European and Asian. The large Pere David's deer *Elaphurus davidianus*, with unusual antlers which look as though they are back-to-front compared with other deer, exists only in captivity as descendants from the herd once kept in the Imperial hunting park near Pekin; it is not known when it became extinct as a wild animal, probably some centuries ago.

The subfamily Odocoileinae contains ten genera, most of them found only in the Americas. With the exception of the reindeer or caribou *Rangifer tarandus*, the deer of this subfamily are less gregarious than the cervines. The genus *Odocoileus* contains the white-tailed or Virginian deer and the mule deer, both favourite quarry for North American sportsmen. *Alce alces*, the elk of Europe, the moose of America, the largest of the deer, with magnificent palmate antlers in the male, inhabits the forests of northern parts of the whole palaearctic and nearctic regions. The moose and the reindeer, *Rangifer*, are the only deer in which the wet muffle round the nostrils and upper lip is so reduced that the nose is dry and covered with fine hair – the large pendulous nose of the elk is as soft as the muzzle of a horse. This character is no doubt helpful to animals living in sub-zero winter temperatures. The reindeer migrates far to the north of the forests beyond the range of the elk; the migrations of the caribou, the American form of the reindeer, have been discussed in Volume I. It is the only deer in which the females carry antlers as well as the males – the response to hormone stimuli must be unlike that of other female deer. The reindeer, too, is the only species of deer that has been domesticated. The South American deer are small species, the largest being the swamp deer *Blastocerus dichotomus* inhabiting marshy savannahs from the Guianas to Uruguay; each branch of the two-branched antlers forks into two points. The pampas deer with the confusingly similar generic name *Blastoceros* formerly inhabited dry open plains from Brazil to Patagonia, but is now scarce; in Argentina it is reported to have been reduced to about fifty living in a swamp near Buenos Aires, from which a few were removed alive for preservation elsewhere when the animals were being massacred by sportsmen in 1968. Two species of Andean deer *Hippocamelus* live in mountain forests from Ecuador to Patagonia; the antlers are two-pronged. Four species of brocket, *Mazama*, live in the forests from Mexico to Paraguay; the antlers are simple spikes, and appear not always to be shed regularly every year. The two species of pudus, *Pudu*, of the west side of the continent range south to Patagonia, and have short antlers that are simple spikes.

Hydropotes inermis, the Chinese water deer, a small species without antlers, but with large canine tusks in the males, differs from other deer, which usually have one or at the most two young at birth, in producing a litter of four or more. The single species of roe, *Capreolus capreolus*, inhabiting most of temperate Europe and Asia is not gregarious; the short antlers of the male are usually three-tined. The roe is the only deer in which delayed implantation of the developing embryo is known to occur – it was in this species that the phenomenon was first mentioned by Harvey in the seventeenth century.

FAMILY GIRAFFIDAE

The family Giraffidae contains two genera, *Giraffa* and *Okapia*, the giraffe *G. camelopardalis* and okapi *O. johnstoni*, the first found over most of Africa south of the Sahara, and the second only in the equatorial rain forests of the Congo. Many species and subspecies of giraffe have been named, based mainly on the colour pattern, but also on the character of the horns, both of which are now known to show much individual variation. The subspecific classification appears to have little value beyond appeasing tidy minds, for widely different colour patterns can be found in a single group of giraffes. Krumbiegel [277] in his heroic attempt at a classification, divides the giraffe into two subspecies and thirteen sub-subspecies, and gives each a geographical location.

The arrangement of the blotched pattern is as individual as human fingerprints, and has been used by Foster [151] in tracing the movements and behaviour of individual animals – he was even able to recognise animals from photographs taken casually more than ten years before. Students of animal behaviour often have to use elaborate and troublesome methods of marking to recognise individuals, but with giraffes that is unnecessary. Among the many African ungulates the giraffe is one of the few that have been studied by competent field zoologists; in consequence we have a considerable amount of information about its life and behaviour. Innis [252] investigated the life of giraffes on the lowveld of the eastern Transvaal in an area of large-leaved deciduous bush, some of it dense and some more open and park-like. The animals browsed on a large number of different kinds of trees, but especially on those of the order Leguminosae; they were more selective when the trees were in full leaf. In a later study Dagg (née Innis) [93] found no correlation between the giraffe's preferences and the ash, crude ether extract, and crude protein content of the foliage analysed; the preferences are presumably based solely on palatability – a little of what you fancy does you good.

Giraffes can reach up to a height of five metres while browsing, and the trees show a marked browse-line at this height. In Kenya, Foster found that trees exceeding this height were browsed into an hour-glass shape, but those shorter into beehive shape; he suggests that the presence of hour-glass trees shows that giraffe were absent at some time in the past, for where they are present the trees are not allowed to grow taller than the beehive shape (Plate 27). Innis remarks that 'it was interesting to see all the taller giraffes feeding at one favourite tree whose lower branches had been eaten away, while the smaller giraffes who could not reach the foliage fed at shrubs nearby'. Giraffes spend the greater part of every day in browsing, and they chew their cud not only when lying down, but also when standing or walking about. When feeding a giraffe pulls a branch into its mouth with its eighteen-inch tongue and then pulls its head away so that the leaves are stripped off; 'it even browses from thorn bushes in this way, its thick bristly prehensile lips protecting it against wounds from the thorns'. Giraffe chew each bolus of cud on average forty-four times at a rate of one chew a second. The gaits of the giraffe, the rack-like walk and the gallop, have been mentioned in Volume I. The animals are much bothered by ticks which tend to congregate under the belly and round the genitalia where the hair is thinnest; tick-birds run about over their bodies removing the parasites, but giraffe also try to rub them off. 'To scratch the stomach they would stand over a six foot bush and rock backwards and forwards. If the ticks cause the back to irritate they back slowly in and out of the bush'.

Giraffe live in herds which are very loose associations, for individual animals often join and leave them, and there is no definite pattern of leadership although there is always one big male in every mixed herd, and he may be the dominant one. In addition to mixed herds there are herds of bulls, both adult and young either separately or together, and there is always a number of lone bulls which may be wandering in search of females in oestrus. There is no ritualised courtship; but a bull approaches a female and licks her tail or takes it between his lips. 'Without seeming to pay any attention, the female then urinates and the male collects some of the urine in his mouth or on his tongue'. He then raises his head and with closed mouth gives the flehmen expression; this testing probably informs the male whether the female is coming into oestrus. From Innis' observation it appears probable that the bull uses the rigid fore-leg gesture, described in Volume I, before mounting a female in oestrus. In the bull herds there is a peculiar manifestation of sexual play known as necking behaviour. Coe [77] who studied this behaviour closely in Kenya, describes necking as varying from two bulls standing head to head and swinging their heads so that the necks rub together, to the bulls standing head to tail and swinging the head to deliver heavy blows with the horns on the flank and loins; the second is often followed by flehmen and penis erection. Innis found similar behaviour in the Transvaal giraffes, and was the first to describe it. This homosexual behaviour – the word is used without the human connotation of stigma and abnormality – is considered to form 'an important sexuo-social bonding mechanism whereby a hierarchy is created amongst the males, and movement between strictly bachelor and mixed herds helps to maintain the contact between the sexes in this polygamous mammal' (Plate 26).

In the Transvaal, Innis found that males were more numerous than females, whereas in Kenya Foster found the reverse, and thought that the males tend to be overlooked because they live more in the forested areas than the females and young, which live more in the open. The males may in fact be fewer because they are more vulnerable to predation in forest where their view of the surroundings is limited. The home range of the animals has yet to be determined, but there are indications that the females and young cover an area of about fifty square kilometres – neither bulls nor cows hold territory. Not only is the herd structure a loose one, but the parental bond between mother and young is weak, as found both by Innis and Foster. The young start browsing in their first week, and are seldom seen being suckled after that; they wander away from their parents and join groups of other juveniles, or return to the mother's or another herd. Foster says, 'Apparently some newborn perish soon after birth as no trace can be found of them while their mothers continue to be recorded'. The adult herd structure is even more casual than the relation between young and parent, and Foster found that any large herd always contained some different individuals on consecutive observations. Juvenile males join the bachelor herds during their third year, and are not seen alone until they are adult. None of these workers record any vocal communication between giraffes, at most a warning snort was heard. Communication seems to be visual in that a giraffe learns of danger by the behaviour of its companions. Innis says that panic seems to be important in spreading an alarm; 'if one giraffe begins to run, the others stampede after it, sometimes in bush so thick that I doubt if they know what they are running from'.

Figure 25 Giraffes with three and five horns, and exostoses.

Both Dagg [95] and Spinage [454] have investigated the structure of the skull of the giraffe. The horns, which always remain hair-covered, are developed from ossification centres beneath the skin, and become fused with the underlying bone. The paired larger horns are attached to the parietals, the median unpaired horn to the frontal and nasal bones. There are often other smaller horns, paired on the occipital region, and above the eyes, and extensions of the median horn. As the animals grow older more bone is laid down enclosing the blood vessels in tubular cavities, unlike the process in deer where the blood vessels lie outside the bone under the velvet of the antlers. In old animals the exostoses are extensive and give the skull a very gnarled appearance. Spinage suggests that these arrangements are correlated with the necking habit, and that the secondary ossification protects the blood vessels from damage during heavy necking. He points out that 'a male wielding a skull of eleven kilogrammes of bone will have a considerable advantage over one armed with only seven kilogrammes'. The exostoses are not the result of trauma – they are a secondary sexual character of genetic origin and 'not a response to injury or stimulation from fighting'.

In many, but by no means all ruminants the liver is without a gall-bladder. Although a fair number of giraffes had been dissected there was doubt whether a gall-bladder is present in the giraffe until Cave [72] settled the point. By a queer chance the giraffe examined by Owen in 1838 – the first to be dissected in Europe by a zoologist – had a large gall-bladder, but later anatomists were unable to find one in their specimens. Cave found that a rudimentary gall-bladder is normally present in the foetus but that it usually undergoes physiological involution so that it is absent after birth; rarely it persists into adult life, as happened in Owen's abnormal specimen, which caused confusion for over a hundred years. The work of Lawrence and Rewell [289] and others [6] on the control of blood-pressure in the giraffe has been mentioned in Volume I; the arrangement of the blood vessels in the head and neck to cope with a rapid raising of the head from some seven feet below the level of the heart to nine feet above it is, however, merely an extreme case of the circulatory problem that faces all mammals, especially those with long legs.

There are no similar detailed researches to be discussed about the okapi in its wild habitat, although much has been written on its anatomy in addition to Lankester's classic monograph [287]. Unlike the giraffe it is extremely shy, and it lives in almost impenetrable forest. Nevertheless okapis can be maintained in captivity with little difficulty, and breed freely, so that a captive population is building up in the zoos of the world. They are docile and gentle when used to being handled. Great difficulty was found in keeping them at first because the wild-caught animals were infested with parasitic worms, which increased enormously in numbers when the resistance of the animals was lessened by the stresses of capture and strange climate and surroundings. Once a healthy stock was produced free from infestation its perpetuation has been successful. It is peculiar that because the okapi is a rare animal it has received so much publicity that its appearance is better known to most people than that of many commoner animals. A good many notes have been published on the breeding, management and behaviour of the okapi in captivity, such as the review by Hediger and others [222], but the particulars of its life in the wild remain largely matters of conjecture. Dagg [94] points out that the okapi has a rack pace similar to that of the giraffe, and that consequently the suggestion that the rack of the giraffe is correlated with its long legs and long neck is not upheld.

FAMILY ANTILOCAPRIDAE

The family Antilocapridae contains a single living member, *Antilocapra americana*, the pronghorn or 'antelope' of the western half of North America from southwest Canada to northwest Mexico. The family separated from the bovid stock about the beginning of the Miocene, and evolved many now extinct species in North America, to which continent it has always been confined. The pronghorn is thus not a true antelope; it differs from all hollow-horned ruminants in having deciduous horn-sheaths. Up to 100 years ago the pronghorn was very numerous on the semi-deserts and dry grasslands which are its habitat, but it was destroyed in thousands for food by early settlers, and by sportsmen, so that its extinction seemed imminent; at the eleventh hour, however, management in the National Parks dragged it back from its threatened fate, and it has increased greatly in numbers and is no longer in danger. The Americans are now justly proud of this unique member of their fauna; its portrait in one form or another has appropriately adorned the cover of the Journal of Mammalogy, the organ of the American Society of Mammalogists, since the first number was published in 1919 (Plate 28).

The pronghorn is a stockily built antelope-like animal weighing up to about 100 pounds, fawn and white in colour with black or dark brown patches on the face and throat, and a large white patch on the rump; several geographical subspecies are recognised. Pronghorns feed on the wiry grasses of the dry plains and on sagebrush, a species of wormwood, and do not thrive if fed lush green grasses; they drink, but appear to be able to go without water for long periods. They survive their harsh climate – hot in summer but cold in winter – by growing a dense soft undercoat which is lost by moult in the spring when the snow goes from the wind-swept plains. In summer they live in small parties, but in winter they form – or used to form – large herds once numbering thousands, to move south with the first fall of snow. The rut comes in the autumn, when there is much sparring but little serious fighting among the bucks; the successful bucks have small groups of three or four does but do not accumulate harems – in spring the herds disband and the small groups separate. The young, usually twins, are born in May and June, and after a few days are able to follow their mothers.

The pronghorn has the reputation of being the swiftest ungulate in America; it can run at forty miles an hour for several miles, making leaps of from ten to nearly twenty feet. When alarmed pronghorns erect the hairs of the white rump-patch, that is very conspicuous and visible at a long distance as large paired rosettes. The hairs are often raised and lowered quickly so that the patch gives a flash that attracts the attention of nearby animals and warns them of danger. Skinner [447] describes how, when 'the animals halt and face about, the signals disappear, and the otherwise neutral colour causes the animals to fade out as if by magic. But if there is still cause for alarm, the white signals flash out again and again long after the rest of the animal has become invisible'. When the rump hair is raised in signalling a dorsal gland is exposed 'releasing a musky odor noticeable to man's poor nose for quite a distance'. Somewhat similar methods of signalling alarm or other emotion are known in some of the true antelopes (Plate 29).

The most unusual features of the pronghorn are its horns and the method of replacing them, which were referred to in Volume I. Like the horns of the bovids, the horns of the pronghorn consist of cornified sheaths supported by bony cores growing from the frontal bones of the skull, but there the resemblance ends, for the nature of the horny sheath and its growth is quite different. The horns reach a length of a foot or sometimes more in the males, but are absent or only one to three inches long in the females; the upper part of the horn is curved and turned inwards, and at about the middle of its length it branches into a short forwardly directed prong – the horns of the females have no prong. The bony horn core is not forked, but is wider below the prong and may bend slightly backwards above its level. When the horn sheath is shed in the autumn the bony core is not exposed, for it is covered by the skin forming the new one, which pushes off the old one. In the young animal when the first horns start to grow the bony core grows upwards covered with skin which gradually becomes cornified from the tip downwards to the base. The first horns may attain a length of four inches but have no prong; they start growing in the spring when the animal is about ten months old, continue growing through the summer, and are shed in the autumn. The first rudiment of the prong shows in the next year, but the horns do not reach their full size until about the fifth year, the successive pairs growing larger than their predecessors.

The horn sheaths have a markedly fibrous structure, apparently derived from hairs embedded in a matrix, which soon decays in shed horns and reveals the fibres. When the horn-sheath is shed the bony core is seen to be covered with skin which at the tip is soft and fleshy and, according to Skinner, is fast hardening and changing to true horn. The rest of the skin bears long bristle-like hairs, and similar hairs extend up the horn from the base. The skin and bristly hairs 'agglutinate' into horn, 'the change slowly proceeding towards the base in which the remaining unchanged bristly hairs are embedded, even to the very tip of the prong and well up towards the main tip'. The process takes about three months for completion. Bailey [17] describes a half grown horn with hard horny tips reaching nearly to the middle, but with the skin below covered with long coarse hairs. 'As the horn thickens and hardens from the tip downward, this hairy covering is buried, rather than absorbed, and becomes a part of the horn'. On the other hand, Mearns [319] suggests that the new horn is covered with long hairs 'which soon disappear' and this may well be correct, for the process of alleged 'agglutination' is difficult to understand. Waldo and Wislocki [493] show how the old sheath is detached when it is to be shed. New growth starts in the underlying skin, particularly at the apex, and this loosens and pushes off the old sheath that encloses it. Frechkop [158] observes that the shedding of the horn-sheaths is perhaps more comparable with the loss of the velvet from the antlers of deer rather than the shedding of the antlers themselves. This suggestion cannot be accepted, for when the velvet of deer is shed all the living tissues die and leave only the bare bone, not an underlying layer of new velvet with its blood vessels and hair-covering; the process in the pronghorn is unique. What biological value there may be in the process of horn-sheath shedding remains as obscure as that in the shedding of antlers by deer.

FAMILY BOVIDAE

In the final family of the artiodactyles, the Bovidae, the horns, always present in males and sometimes also in females, consist of a bony core covered with a permanent horny sheath; this construction might be said to be the only sensible one if animals must have horns, and presents no puzzle to the enquiring human mind that wants to find a reason for everything. The bovidae are a large family of about forty-five genera – some systematists recognise over fifty – and about 115 species. It includes most of the game animals, apart from the deer, that provide those beautiful heads hung by sportsmen as trophies on their walls for moth and dust to corrupt. It is impossible here to do more than make some general observations on the bovids, though even if space allowed, information on their life as wild animals, based on scientific study rather than casual observation, is available for only a few species. The family is divided into five subfamilies separated by technical details of their anatomy that are described in the standard works of reference, and need not be repeated here.

The subfamily Bovinae contains the largest as well as some of the smaller members of the family. Bovines occur in Europe, Asia, America and Africa, all in warm regions except the bisons, the yak, and the extinct European wild ox. The seasonal migrations of the bison, *Bison bison*, or buffalo of North America have been discussed in Volume I; the European species *B. bonasus* is a more sedentary forest animal. So few of the latter are left, living under protection and management in Poland, that they can scarcely be regarded as truly wild animals, or be expected, as Jaczewski [253] shows, to exhibit the behaviour of their ancestors. Mohr [329] published some surprising photographs of bisons performing with tigers and other incompatible animals in a Russian circus. The massive African buffalo *Syncerus* of the plains is represented in the tropical forest by a smaller lighter-coloured subspecies, the bush-cow, with short curved horns.

The buffaloes of Asia, *Bubalus*, exist both as domestic races of long standing and as wild animals – they are named water-buffalo from their habit of spending much of the day submerged to the head in rivers or pools. The anoa and tamarau are dwarf species of *Bubalus* that live in the forests of Celebes and Mindoro of the Philippines respectively – they may be conspecific. The domestic ox, *Bos taurus*, is probably derived from the extinct wild ox of the forests of Europe, *B. taurus primigenius*, and the eastern humped domestic cattle probably from, or from crosses with, the wild humped species the gaur, *B. gaurus*, and banteng, *B. banteng* of India and southeast Asia. These humped species carry a large dorsal ridge-like hump so large in the first as to appear almost a deformity, but they have little dewlap. Together with the semi-domesticated gayal, which is probably derived from the gaur, they are often placed in a subgenus *Bibos*. The kouproh or kouprey of Cambodia, placed in a separate genus by Coolidge [82] but regarded as doubtfully valid by Bohlken [46], has been shown by Edmond-Blanc [129] to be a hybrid between local domestic cattle and, probably, the banteng. The yak, *B. grunniens*, of the highlands of central Asia, with thick woolly coat which drapes to the ground from the flanks, is now scarce as a wild animal but plentiful as a domesticated race. The nilgai or blue bull, *Boselaphus*, of India is a more lightly built animal with marked sexual dimorphism; the cows are hornless and dun coloured, the iron-grey bulls have short straight horns and a tuft of black hair hanging from the throat. The four-horned antelope of India, *Tetracerus,* barely two feet high at the shoulder, is horned only in the male; he bears a pair of short straight spike-like horns behind the eyes, and a second pair of smaller spikes above them.

The remaining two genera of the subfamily are entirely African in distribution. *Taurotragus* contains the eland, one of the largest and meatiest of the antelopes and capable of domestication, and the smaller forest-dwelling bongo, both with slightly spirally twisted horns in both sexes. *Tragelaphus* is a large genus, split into several by the older systematists, containing the kudus with long horns in open spirals, the bushbucks including the harnessed antelope, the nyalas, and situnga, animals of widely differing appearances, habits and habitats; many of the species, like those of *Taurotragus* are marked with narrow vertical white stripes. Wilson [508] found that in eastern Zambia greater kudu usually live in small groups and inhabit dense cover in rough and broken areas. They are almost entirely browsers, and although they have food preferences they take the foliage of a wide variety of plants, and are thus able to exist wherever there is sufficient quantity of browse. The calves are born when the grass is highest and are hidden by the parent among it for about two weeks. Owing to their habits and habitat they maintain their numbers in an area of tsetse control in spite of fairly intensive hunting for over twenty years.

The Cephalophinae are a small subfamily of small animals, but there are numerous species, the duikers, confined to Africa. They are small to medium-sized antelopes which live among dense vegetation. Both sexes generally bear short backwardly directed horns, often nearly hidden by a tuft of long hair. There are two genera; *Cephalophus* contains all the species except the grey duiker *Sylvicapra grimmia,* in which the females are usually hornless. Most of the species are some shade of reddish brown usually with a darker or lighter stripe along the back; one *C. zebra,* is marked with vertical dark stripes. In some parts of Africa attempts have been made to eliminate tsetse fly by exterminating the wild animals on which they feed. Wilson and Roth [510] studied the effect of tsetse control hunting on the population of duiker (*S. grimmia*) in Zambia. Two years intense hunting failed to give any appreciable reduction in numbers. 'Although general availability of duiker did not diminish, they became increasingly difficult to shoot because of behavioural adaptation and changing periods of feeding activity'. Furthermore the disturbance of the control area did not drive the duiker away, and the observers concluded that the shooting merely removed the annual increment due to breeding; though the operation was a failure as a measure of fly control, it pointed the way to possible exploitation of the animals by game-cropping. In an earlier paper Wilson and Clarke [509], basing their studies on similar material from the game destruction operations, found that duikers are mainly solitary though often seen in pairs, and that they probably have a territorial system. The females are slightly larger than the males, and breed at any time of the year, and hide their lambs under patches of thick grass – the lambs can run swiftly at the age of twenty-four hours. Duikers are almost entirely browsers, but take a wide variety of plants; they are fond of the fallen fruit of the duiker-tree (*Pseudolachnostylis*) and when chewing their cud spit out the seeds of the fruit so that small piles of them accumulate at the chewing lairs.

The subfamily Hippotraginae contains nine genera all of which are found only in Africa. Although none of them is as heavy as the eland they are large antelopes; horns are present in both sexes of all but three genera, and in all but one they bear transverse rings. *Kobus* contains five species including the waterbucks, puku, lechwes and kobs, which frequent marshy places and flood plains, sometimes in large herds. The horns, carried only by males, sweep backwards and upwards and turn forwards near the tips. The defence of small breeding territories by male Uganda kob and the associated behaviour have been mentioned in Volume I. There are two forms of waterbuck, *K. defassa* with white rump and *K. ellipsiprimnus* with a white ring on the rump; in some parts of their range the two overlap and interbreed producing offspring with many intermediate patterns – the specific names are therefore not justified. Waterbuck are territorial, and the males hold territory extending inland from thickets on the banks of streams, in which each herds a small group of females. In Kenya and Uganda they spend the night in the thickets and come out to graze on the grasslands during the day, but farther north where temperatures are higher the routine is reversed. Kiley-Worthington [265] found that the territorial sense becomes weaker the farther the males go from their home thickets and that the females are not tied to the male territories but wander from one to another if the male is not able to head them back. Nevertheless,

there is a more intimate and lengthy bond formed between the male and female than is usually found in more patriarchally organised antelope. This is illustrated by the long and often interrupted and repeated courting sequences, also the male obviously prefers the company of even non-oestral females or young, and will follow a group around within his own territory if only to graze and rest with them.

The horns of the three species of reedbuck, *Redunca*, similarly turn forwards, but are much smaller. These animals live in the neighbourhood of water but enter it much less frequently than those of the genus *Kobus*. The South African rhebok *Pelea capreolus*, the only species of the genus, differs in having a short woolly coat very unlike the smooth coat of most antelopes. The males carry short, almost straight, upwardly pointing horns. The animals live in small groups on rocky mountains on which they climb and jump with agility. The genus *Hippotragus* contains the roan and sable antelopes, large stoutly built animals with backwardly curved horns that reach a great size, and beautiful symmetrical sweepback in some sable antelope bulls. They are generally found in small groups; they live in varied habitats but seem to prefer open grasslands with scattered bush. A third species, the blue buck of the Cape, was exterminated about 1800. The genus *Oryx* has three species; the gemsbuck or beisa, and the Arabian oryx carry long nearly straight horns in both sexes, the scimitar-horned oryx, curved ones. The first inhabits dry grasslands and bush, the others the deserts of Arabia and North Africa. The Arabian oryx has become scarce owing to the use of motor transport which has made their remote and arid habitat accessible to irresponsible people armed with modern rifles. Some points in the mating behaviour of the oryx have been described and illustrated in Volume I. The closely allied addax, *Addax nasomaculatus*, also inhabits the Sahara desert, and is almost white, as are the desert oryxes, but bears horns twisted in an open spiral in both sexes. Addaxes are becoming rare; the development of oil and gas fields in North Africa has brought them within reach of people with rifles in light aircraft and motor vehicles.

The genus *Damaliscus* contains four or five species of handsome antelopes bearing horns with a short open spiral in both sexes. The coat of the topi *D. korrigum* has a peculiar and pleasing blue cast on the rich brown ground colour, and the faces of the blesbok and the bontebok which are races of a single species, *D. dorcas*, are ornamented with white patches; the little known Hunter's antelope *D. hunteri* of northern Kenya and Somaliland, has longer horns and a white facial spectacle. Several geographical races of the species in the genus were formerly regarded as separate species. The bontebok and blesbok, once plentiful on the plains of southern Africa, exist now only within fences, but are becoming semi-domesticated; they are increasing in numbers and will probably continue to do so, as they are a profitable source of supply for the venison market. The conservation and rational exploitation of these and other antelopes as converters of low class grazing into high class protein has been discussed in

Volume I. The two species and numerous subspecies of the hartebeests, *Alcelaphus*, could be similarly preserved and turned to good account. They are widely distributed through the open and the scrub covered plains of Africa south of the Sahara, and are still numerous in many areas. They are queer looking creatures, higher at the withers than the croup, with long, narrow, straight faces, and horns growing from a pedicel, often of exaggerated size, above. The horns are ringed and carried by both sexes; they spread outwards and then upwards, and resemble in shape the handle-bars of a bicycle. The muzzle is slightly flattened, with large crescentic nostrils.

Figure 26 Coke's hartebeest (*Alcelaphus*). The horns grow on a tall pedicel.

The muzzle is even broader and flatter, with crescentic lateral nostrils, in the two species of wildebeest or gnu, *Connochaetes*. *C. gnou*, the white-tailed gnu of the Cape is extinct as a wild animal but is preserved within fences in several parts of South Africa. The blue wildebeest, *C. taurinus*, however, remains common throughout much of eastern Africa; it is the subject of an interesting study on the social structure of its herds, and its general biology, which has been discussed in Volume I. Both species run to hair, not all over the body, but in long manes, beards, tufted tails and a tuft of bristles on the face, particularly in the white-tailed gnu. This and the shape of the horns, spreading and turned upwards in *C. taurinus*, and in the form of forwardly directed hooks in *C. gnou*, gives them a wild and even ferocious appearance. In *C. gnou* the appearance of the animals is equalled by their behaviour, for if they are disturbed they prance about, buck and rush round in a wild intimidating manner; *C. taurinus* is much less demonstrative (Plate 30).

The subfamily Antilopinae contains fourteen genera of antelopes and gazelles, all but three found only in Africa; all are small to medium in size. The single species of klipspringer, *Oreotragus*, with short upright spiky horns in the male, is a small animal that has remarkable agility in climbing among rocks. The hooves are small, and the animal walks on its points; it can make use of the smallest footholds and traverse rocks that appear to be completely inaccessible. The oribi, *Ourebia*, the only species of its genus, is a little larger; it stands only about eighteen inches high at the shoulder, and has short horns in the males. It is an inhabitant of long grass and dense herbage. The steenbok and two species of grysbok, genus *Raphicerus*, are similar in size, and bear short horns in the males. They inhabit arid and scrub areas in East and South Africa. The species of the genus *Neotragus* are considerably smaller – the two species of suni and the pygmy antelope, the last the smallest of the ruminants, no larger than a rabbit. The sunis inhabit eastern Africa, the pygmy antelope the west coast; they all live in dense bush or forest country – short spiky horns are borne only by the males. The pygmy antelope is also known as the royal antelope from the native legend that it is the king of the animals. About five species of dik-diks, genus *Madoqua*, are found on the east and west sides of Africa south of the Sahara. They are small antelopes, about the size of hares, with rather long soft coats, and a tuft of hair on the head which almost hides the short horns of the males. They, too, live among dense herbage. The beira, *Dorcatragus megalotis*, of north-eastern Africa is rather larger, and has large ears covered with white hair inside, and a white ring round the eye; the males have short horns. This species inhabits dry hills and plateaux, and resembles the klipspringer in its agility among rocks. The blackbuck, *Antilope cervicapra*, is an Asiatic species found on the plains and in open woodland in India. It shows marked sexual dimorphism – the underside and a patch round the eye are white in both sexes, the rest of the body is fawn coloured in the hornless females, and in young bucks, but in the adult buck the colour becomes almost black. The bucks carry long backwardly directed ringed horns, twisted in an extended spiral.

The impala, *Aepyceros melampus*, is a herding species of eastern and southern Africa; it is about the size of a sheep, and the males bear ringed horns spreading gracefully upwards in an open spiral. They live on plains and among open bush, often in large herds; they are able to make spectacular leaps in fleeing from danger. Schenkel [422] found that impala in East Africa do not move far from their restricted home range although they do not hold or mark territory. Herds consist of up to 100 females with their young herded by a dominant male, and of up to sixty males of all ages in separate bachelor herds. The composition of the herds is fairly stable, as the dominant male shepherds his females and is aggressive to other males; but bachelor herds are looser in structure, as there is no shepherding, and there is little aggressive behaviour between their members. Within a herd the activities such as standing, feeding, chewing cud, or movement in a particular direction, tend to be simultaneous – when one animal starts a new activity the others tend to follow it. The animals at the edge of the herd are more alert for signs of danger from predators than those within. There is a social hierarchy in the herds though it is seldom expressed in aggression. The courtship behaviour differs in many respects from that of other antelopes because the dominant males possess the female herds throughout the year and not seasonally.

The dibatag *Ammodorcas clarkei* and the gerenuk *Litocranius walleri* each the sole member of its genus, have exceedingly long slender legs and necks, correlated with their browsing habits. They not only reach up to branches with their long necks, but also gain extra height by standing on their hind legs. In their browsing habit they resemble the giraffe, and the conformation of the snout, lips and muzzle also bears a striking resemblance to that of the giraffe. They both inhabit semi-desert regions of northeast Africa, browsing on the often withered scrub and acacias; their ranges overlap, but the gerenuk extends southward into northern Kenya. In both species only the males bear horns; in the dibatag they are curved forwards like those of the reedbuck, and in the gerenuk they sweep back in rather less than one whole turn of an open spiral. In the beautifully marked springbuck, *Antidorcas marsupialis*, of southern Africa both sexes are horned; this species has a crest of very long white hairs along the hind part of the ridge of the back. If the animal is alarmed it erects the hairs of its crest or 'fan' and those of the rump patch to make a conspicuous signal to its companions, an effect similar to that given by the pronghorn of America. The extraordinary vertical leaps which it makes when displaying, and the spectacular migrations which it formerly made in vast numbers when driven by drought to seek new grazing, have been described in Volume I. It is much reduced in numbers today, and in the Republic most of the species exists only within fences, though wild herds are still to be found in southwest Africa, Botswana and southern Angola.

The gazelles, about a dozen species in the genus *Gazella* and two in the genus *Procapra*, are not solely African, and between them cover the drier parts of Asia from Arabia to Mongolia as well as northern and eastern Africa. They are graceful animals, much resembling the springbuck; in all species but one of *Gazella* both sexes carry short lyrate horns; in *Procapra* the females are hornless. They are in general animals of open plains, and

Figure 27 Gerenuk (*Litocranius*) browsing.

some species live in herds of considerable size. Many of the gazelles, including the beautifully patterned and graceful Thompson's and Grant's gazelles of East Africa, as well as many species of the Hippotraginae, live in arid waterless regions and are able to survive without drinking for long periods. Some of the adaptations for desert life have been discussed in Volume I; recently Taylor [470] has pointed out another way in which such animals obtain water besides eating the rhizomes of grasses which have some water content. Antelopes need water amounting to two to four per cent of their body weight a day, yet oryx and Grant's gazelle survive droughts in hot semi-deserts. They necessarily eat dry food, which contains only one per cent of free water, mainly a dry shrub *Disperma* sp. During a drought in the north of Kenya the animals were followed and not seen to drink for at least four days, but they were found to feed by night rather than by day. The plants, though dry by day, are very hygroscopic, and at night, when the temperature drops and the humidity rises they absorb so much water that the food contains about thirty per cent of free water, and the antelopes are therefore able to obtain their water requirement from it. Taylor points out that in contrast the local zebu cattle are penned at night, and therefore feed by day and have to drink or perish from thirst. Lamprey [286] found that impala and Grant's gazelles conserve water by refraining from micturition for periods of up to twelve hours. Impala and most game animals avoid excessive water loss during the day by taking shelter in the shade of trees, but Grant's gazelles stay in the open in the hottest weather. He found impala living in an area where the nearest free water was twenty-five miles away, and getting all their water requirements by drinking the dew, licking it from the grass early in the morning.

Among the African bovids Ansell found [14] that in Zambia Lichtenstein's hartebeest, the blue wildebeest, the tsesebe, and the impala all have definite breeding seasons, though the dates may vary in different parts of the country. All the others, as far as is known, have no definite breeding season, though some may have such a marked peak period as to constitute almost a season. In Rhodesia, however, Dasmann and Mossman [100] found that in addition to impala and wildebeest, the kudu, waterbuck, giraffe and perhaps the eland were seasonal in breeding, but differed in the time when the young are born. They were unable to correlate the birth seasons with the plant phenology.

The subfamily Caprinae contains the sheep, goats and their relations, some twenty-two species in thirteen genera. Members of the family are found in widely separated countries, but only two species live in Africa. Nearly all of them inhabit mountainous or cold regions. The chiru,

Pantholops hodgsoni, and saiga, *Saiga tatarica*, are Asiatic species of rather sheep-like appearance. Only the males carry horns, those of the chiru being long, almost straight and vertical, those of the saiga short and lyrate. The nose of the chiru is swollen, and that of the saiga inflated to form a short downwardly directed proboscis, peculiarities said to be adaptations to breathing cold dry air laden with dust. The chiru inhabits the high plateaux of Tibet and Ladak, but the saiga has a much greater though discontinuous distribution on the steppes from the lower Volga to Mongolia. The saiga had become a scarce animal, and Bannikov [19] relates how under complete protection its numbers have now increased to over a million on the Volga steppes alone, where the herds are cropped to prevent over-population degrading the habitat, and to contain the risk of spreading foot-and-mouth disease. He regards the Mongolian animals as a separate species or subspecies.

Figure 28 Saiga (*Saiga*), showing short proboscis.

The two species of goral, *Naemorhedus*, are goat-like animals with short curved horns in both sexes, as are the two species of serow, *Capricornis*. The gorals live in the mountains from Manchuria and China to northern India, and *C. sumatrensis* from northern India and southern China through Malaya to Sumatra, whereas *C. crispus* is confined to Formosa and Japan. *Oreamnos americanus*, the Rocky Mountain goat, lives in the mountains of western America from Alaska to Oregon. It is strikingly handsome animal with thick white coat and short curved black horns; it is as agile on rocky crags and precipices as the chamois, *Rupicapra*, of the mountain ranges of

397

Europe and Asia Minor, in which the short vertical horns are backwardly hooked at the tips. The takin, *Budorcas taxicolor*, is a heavily built goat-like animal of the mountain thickets of northern Burma and southern China; the stout horns spread outwards and then backwards, and the coat is thick and shaggy. The long-coated musk ox, *Ovibos moschatus*, is even shaggier and needs the protection of the dense woolly undercoat in the Arctic of Canada and Greenland. The bases of the horns meet on the forehead like those of the African buffalo, turn downwards and then hook upwards at the tip. Musk oxen are gregarious, and form a phalanx with lowered heads when defending themselves against predators; but the man with a gun can take advantage of this habit to 'account for' a whole troop. The three species of tahr, *Hemitragus*, inhabit the Himalayas, the Nilgiri Hills, and Arabia. They resemble beardless goats, and have short stout backwardly curving horns; the Himalayan species has a long shaggy coat. They are gregarious, and as agile as goats among the rocks of their habitats.

The goats, alternatively markhors, turs, or ibexes, five species of the genus *Capra*, are also mountain dwellers; between them their range covers Africa north of the Sahara, southern Europe, and stretches across Asia to Mongolia. The horns are large, spreading and backwardly curved, but are spirally twisted in some forms. The domestic goat is descended from the wild goat, *C. hircus aegagrus* of southeast Europe and southwest Asia, perhaps with some hybridisation with other species. Its hardiness and ability to live on poor quality grazing and browsing have enabled it to survive in a feral state in many parts of the world where it has wrought great destruction on the ecology of the lands and islands into which it has been introduced.

Three genera of sheep, *Pseudois, Ammotragus* and *Ovis*, complete this account of the living mammals. Horns are carried by both sexes of all species, but they are much larger in males than females. The bharal or blue sheep, *P. nayaur*, of the Himalayas, Tibet and China north to Mongolia, has outwardly spreading horns turned backwards towards the tips, as does the Barbary sheep or aoudad, *A. lervia*, of the north African mountains. In sheep of the genus *Ovis* the horns are directed outwardly in a spiral curve to produce the impressive heads of bighorn and argali. The classification of the wild sheep is difficult; some systematists see them as a few species with numerous geographical races, others split them into a great many separate species. For example the bighorn sheep of Siberia, *O. canadensis*, is now generally regarded as the same species as the bighorn of western North America and south to Mexico; and *O. ammon*, the argali or Marco Polo's sheep of the mountains of central Asia, is subdivided into

398

between fifteen and twenty subspecies. All these forms, together with those of the mouflons, *O. musimon* and *orientalis* of Sardinia and Corsica, and the mountains from Asia Minor to Kashmir respectively, and their subdivisions, are gregarious active mountain animals living in herds, varying from a few individuals to many, and are agile sure-footed climbers. Domestic sheep are probably derived from the mouflon, which has an undercoat of short wool, but there may have been hybridisation with other species. All the wild sheep have coats of hair; the fleece of domestic sheep has been developed by artificially selecting a form in which the hairy outer coat is replaced by the greatly overgrown undercoat of wool.

The great number of species among the bovids whose ranges overlap or even coincide could be expected to lead to an intense interspecific competition for food. The conventional reply to such a suggestion is that they do not compete because each occupies its own ecological niche and so exploits a different part of the environment from those of sympatric species. There is, however, often more than mere avoidance of each other's niches, and the presence of each species may be necessary for the others to flourish. A study of grazing ungulates in the Serengeti National Park has illustrated one facet of this relationship in one area of Africa; no doubt its conclusions will be found to apply much more widely. Gwynne and Bell [192] found that zebras lead the succession of grazing because they have incisor teeth in both jaws and consequently can nip off the tough herbage stems for which they have a high tolerance. When feeding by selecting coarse stems they increase the relative frequency of leaf, and their trampling opens up the herb layer. This makes the structure of the vegetation more suitable for the wildebeest which follow them. The wildebeest in turn graze the herbage in such a way as to make the use of dicotyledon plants by gazelles easier. They conclude that 'the relationship between species in such a grazing succession can thus be seen to be facilitative rather than competitive, with the result that, within limits, their numbers are positively rather than negatively interdependent'.

In a study of the natural pastures in the Rukwa valley of Tanzania during fourteen years Vesey-Fitzgerald [490] showed how the grazing pressure rejuvenates the pasture and produces a grazing mosaic. The seasonal growth of herbage is used by a sequence of animals, the heavier ones followed by the lighter ones, and the pastures are used in rotation during the year: 'alternate periods of optimum use and rest occur, and the harmful effects of over-grazing do not appear'. As the waters recede from the flood plains elephants wade among the emerging vegetation looking for particular food-species, and trample down the dense mat of plants thus forming paths later used by other animals. They are followed by buffalo which tread down large lawns, which shortly provide pasture for antelopes and lighter animals. At the end of the dry season the flood plains have been turned into short-grass pasture by the grazing succession; if this rotation does not occur the grasses grow eight to ten feet high and are lost to the animal ecology by fierce grass fires and the valuable pasture is wasted. Lamprey [285] also emphasises how the different species facilitate each other; they are very tolerant of each other's presence and often associate together. In particular many species are dependent upon the activities of elephants, which pull down trees and shrubs, thereby making their foliage and fruit accessible to smaller animals, and open glades and pathways through dense woodland, enabling impala and waterbuck to enter. During droughts elephants dig for water in the dried up river beds, and thus make water available 'for themselves and also for hundreds of buffalo, wildebeest, zebra and many other species'. These researches well illustrate how interference with, or destruction of, one species can completely upset the ecological relationships of the fauna over great areas of country.

References

1   Aellen, V. & Perret, J. 1958. Sur une nouvelle trouvaille de *Zenkerella insignis* Matschie, 1898 (Rodentia, Anomaluridae). *Säugetierk. Mitt.* **6**, 21
2   Aldous, C.M. 1951. The feeding habits of pocket gophers (*Thomomys talpoides moorei*) in the high mountain ranges of central Utah. *J. Mammal.* **32**, 85
3   Allen, G.M. 1939. *Bats*. Cambridge, Mass.
4   Allen, G.M. 1942. Extinct and vanishing mammals of the western hemisphere. American Committee for International Wild Life Protection. *Special publication No. 11*
5   Allen, J.A. 1917. The skeletal characters of *Scutisorex* Thomas. *Bull. Amer. Mus. Nat. Hist.* **37**, 769
6   Amoroso, E.C., Edholm, O.G. & Rewell, R.E. 1947. Venous valves in the giraffe, Okapi, camel and ostrich. *Proc. zool. Soc. Lond.* **117**, 435
7   Anderson, S. & Jones, J.K. Jr. 1967. Recent mammals of the world. *A synopsis of families*. New York
8   Angulo, J.J. 1945. Some anatomical characteristics of the hutia conga. *J. Mammal.* **26**, 424
9   Angulo, J.J. 1947. Teat location in the Cuban solenodon. *J. Mammal.* **28**, 298
10 Angulo, J.J. & Alvarez, M.T. 1948. The genital tract of the male conga hutia, *Capromys pilorides* (Say). *J. Mammal.* **29**, 277
11 Anon. 1959. Just 30,000 happy quills. *Life Magazine*. August 31, 53
12 Anon. 1963. Antwerp pangolin diet. *Int. Zoo Yrbk.* **4** (1962) 128
13 Anon. 1966. A relict marsupial. *Nature*, Lond. **212**, 225
14 Ansell, W.F.H. 1960. The breeding of some larger mammals in Northern Rhodesia. *Proc. zool. Soc. Lond.* **134**, 251
15 Attenborough, D. 1956. *Zoo Quest to Guiana*. London
16 Bailey, S.W. & Dunnet, G.M. 1960. The gaseous environment of the pouch young of the brush-tailed possum, *Trichosurus vulpecula* Kerr. *C.S.I.R.O. Wildl. Res.* **5**, 149
17 Bailey, V. 1920. Old and new horns of the prong-horned antelope. *J. Mammal.* **1**, 128
18 Banfield, A.W.F. 1954. The role of ice in the distribution of mammals. *J. Mammal.* **35**, 104
19 Bannikov, A.G. 1967. Biology of the saiga. Jerusalem
20 Barabash-Nikiforov, I. 1962. *The Sea Otter* (*Kalan*) (Moscow 1947) (English Translation) Jerusalem
21 Barnard, E.A. 1969. Biological function of pancreatic ribonuclease. *Nature*, Lond. **221**, 340
22 Barrett, O.W. 1935. Notes concerning manatees and dugongs. *J. Mammal.* **16**, 216
23 Bartholomew, G.A. & Cade, T.J. 1957. Temperature regulation, hibernation and aestivation in the little pocket mouse, *Perognathus longimembris*. *J. Mammal.* **38**, 60

REFERENCES

24 Bartholomew, G.A. Jr. & Cary, G.R. 1954. Locomotion in pocket mice. *J. Mammal.* **35**, 386

25 Bartholomew, G.A. Jr. & Caswell, H.H. Jr. 1951. Locomotion in kangaroo rats and its adaptive significance. *J. Mammal.* **32**, 155

26 Bashanov, B.S. & Belosludov, B.A. 1941. A remarkable family of rodents from Kasakhstan, U.S.S.R. *J. Mammal.* **22**, 311

27 Batchelder, C.F. 1948. Notes on the Canada porcupine. *J. Mammal.* **29**, 260

28 Beidleman, R.G. & Weber, W.A. 1958. Analysis of a pika hay pile. *J. Mammal.* **39**, 599

29 Bell, T. 1836–9. Article 'Insectivora' in Todd's *Cyclopaedia of Anatomy & Physiology* **2**, 994. London

30 Benedict, F.G. 1936. The physiology of the elephant. *Carnegie Inst. Wash. Pub.* 474

31 Benedict, F.G. & Lee, R.C. 1938. Further observations on the physiology of the elephant. *J. Mammal.* **19**, 175

32 Benedict, J.E. 1926. Notes on the feeding habits of *Noctilio. J. Mammal.* **7**, 58

33 Benirschke, K. 1969. Cytogenetics in the zoo. *New Scientist.* **41**, 132

34 Benson, S.B. 1946. Further notes on the Dall porpoise. *J. Mammal.* **27**, 368

35 Bere, R.M. 1959. Queen Elizabeth National Park; Uganda. The hippopotamus problem and experiment. *Oryx,* Lond. **5**, 116

36 Bertram, G.C.L. & C.K.R. 1963. The status of manatees in the Guianas. *Oryx,* Lond. **7**, 90

37 Bertram, G.C.L. & C.K.R. 1966. The dugong. *Nature,* Lond. **209**, 938

38 Bertram, G.C.L. & C.K.R. 1968. Bionomics of dugongs and manatees. *Nature,* Lond. **218**, 423

39 Bick, Y.A.E. & Jackson, W.D. 1967. DNA content of Monotremes. *Nature,* Lond. **215**, 192

40 Bierman, W.H. & Slijper, E.J. 1947–8. Remarks on the species of the genus *Lagenorhynchus* I & II. *Proc. Acad. Sci. Amst.* **50**, 1354, **51**, 127

41 Bishop, S.C. 1947. Curious behaviour of a hoary bat. *J. Mammal.* **28**, 293, 409

42 Blair, W.F. 1941. Observations on the life history of *Baiomys taylori subater. J. Mammal.* **22**, 378

43 Blancou, L. 1962. A propos des formes nains de l'éléphant d'Afrique. *Mammalia,* Paris. **26**, 343

44 Blandford, W.T. 1888–91. The fauna of British India. *Mammalia.* London

45 Bloedel, P. 1955. Hunting methods of fish-eating bats, particularly *Noctilio leporinus. J. Mammal.* **36**, 390

46 Bohlken, H. 1958. Vergleichende untersuchungen an wildrindern (Tribus *Bovini* Simpson 1945). *Zool. Jb. Phys.* **68**, 113

47 Bonner, W.N. 1968. The fur seal of South Georgia. *Sci. Repts. British Antarctic Survey* No. **56**

48 Borell, A.E. 1942. Feeding habit of the pallid bat. *J. Mammal.* **23**, 337

49 Bothma, J. du P. 1965, Random observations on the food habits of certain Carnivora (Mammalia in southern Africa.) *Fauna & Flora, Pretoria,* No. **16**, 16

50 Bott, E., Denton, D.A., Goding, J.R. & Sabina, J.R. 1954. Sodium deficiency and corticosteroid secretion in cattle. *Nature,* Lond. **202**, 461

51 Brazenor, C.W. 1962. Rediscovery of a rare Australian possum. *Proc. zool. Soc. Lond.* **139**, 529

52 Britton, S.W. & Atkinson, W.E. 1938. Poikilothermism in the sloth. *J. Mammal.* **19**, 94

53 Broom, R. 1896. On a small fossil marsupial with large grooved premolars. *Proc. Linn. Soc. N.S.W.* **10**, 563

54 Brown, C.E. 1936. Rearing wild animals in captivity, and gestation periods. *J. Mammal.* **17**, 10

55 Brown, D.H., Caldwell, D.K. & M.C. 1966. Observations on the behaviour of wild and captive false killer whales, with notes on associated behaviour of other genera of captive delphinids. Los Angeles County Museum: *Contributions in Science. No. 95*

56 Brown, J. Clevedon. 1964. Observations on the elephant shrews (Macroscelididae) of equatorial Africa. *Proc. zool. Soc. Lond.* **143**, 103

57 Buchanan, G.D. 1967. The presence of two conceptuses in the uterus of a nine-banded armadillo. *J. Reprod. Fert.* **13**, 329

58 Bucher, G.C. 1937. Notes on the life history and habits of *Capromys. Mem. Soc. Cubana. Hist. Nat.* **11**, 93

59 Burbank, R.C. & Young, J.Z. 1934. Temperature changes and winter sleep of bats. *J. Physiol.* **82**, 459

60 Burrell, H. 1927. *The Platypus.* Sydney

61 Burt, W.H. 1932. The fish-eating habits of *Pizonyx vivesi* (Menegaux). *J. Mammal.* **13**, 363

62 Buss, I.O. & Savidge, J.M. 1966. Change in population number and reproductive rate of elephants in Uganda. *J. Wildl. Mgmt.* **30**, 791

63 Buss, I.O. & Smith, N.S. 1966. Observations on reproduction and breeding behaviour of the African elephant. *J. Wildl. Mgmt.* **30**, 375

64 Cabrera, A. 1925. *Genera mammalium. Insectivora, Galeopithecia.* Madrid

65 Cabrera, A. & Yepes, J. 1940. *Mammiferos Sud-Americanos.* Buenos Aires

66 Cahn, A.R. 1940. Manatees and the Florida freeze. *J. Mammal.* **21**, 222

67 Calaby, J.H. 1960. Observations on the banded anteater *Myrmecobius f. fasciatus* Waterhouse (Marsupialia), with particular reference to its food habits. *Proc. zool. Soc. Lond.* **135**, 183

68 Campbell, B. 1938. A reconsideration of the shoulder musculature of the Cape golden mole. *J. Mammal.* **19**, 234

69 Carter, T.D. 1950. On the migration of the red bat, *Lasiurus borealis borealis. J. Mammal.* **31**, 349

70 Carvalho, C.T. de. 1960. Notes on the three-toed sloth, *Bradypus tridactylus. Mammalia,* Paris. **24**, 155

71 Caughley, G. 1964. Social organisation and daily activity of the red kangaroo and the grey kangaroo. *J. Mammal.* **45**, 429

72 Cave, A.J.E. 1950. On the liver and gall-bladder of the giraffe. *Proc. zool. Soc. Lond.* **120**, 381

73 Cave, A.J.E. & Aumonier, F.J. 1967. Observations on dugong histology. *J. roy. micro. Soc.* **87**, 113

74 Chalmers, G. 1963. Jackson's scaly-tail (*Anomalurus jacksoni*) in captivity. *Int. Zoo Yrbk.* **4**, 123

75 Clarke, J.R. 1953. The hippopotamus in Gambia, West Africa. *J. Mammal.* **34**, 299

76 Coe, M.J. 1962. Notes on the habits of the Mount Kenya hyrax (*Procavia johnstoni mackinderi* Thomas) *Proc. zool. Soc. Lond.* **138**, 639

77 Coe, M.J. 1967. 'Necking' behaviour in the giraffe. *J. Zool. Lond.* **151**, 313

78 Colbert, E.H. 1941. A study of *Orycteropus gaudryi* from the island of Samos. *Bull. Amer. Mus. nat. Hist.* **78**, 305

79 Conaway, C.H. 1952. Life history of the water shrew (*Sorex palustris navigator*). *Amer. Midl. Nat.* **48**, 219

80 Constantine, D.G. 1958. Bleaching of hair pigment in bats by the atmosphere in caves. *J. Mammal.* **39**, 513

81 Cook, D.B. 1943. History of a beaver colony. *J. Mammal.* **24**, 12

REFERENCES

82 Coolidge, H.J. Jr. 1940. The Indo-Chinese forest ox or kouprey. *Mem. Mus. comp. Zool. Harv.* **54**, 417

83 Cornwall, I.W. 1968. Prehistoric animals and their hunters. London

84 Couturier, M.A.J. 1954. *L'ours brun*. Grenoble.

85 Cowan, I.M. 1944. The Dall porpoise, *Phocoenoides dalli* (True) of the northern Pacific ocean. *J. Mammal.* **25**, 295

86 Cranbrook, The Earl of. 1959. The feeding habits of the water shrew, *Neomys fodiens bicolor* Shaw, in captivity and the effect of its attack upon its prey. *Proc. zool. Soc. Lond.* **133**, 245

87 Crandall, L.S. 1964. *Wild mammals in captivity*. Chicago & London

88 Crowcroft, P. 1954. The daily cycle of activity in British shrews. *Proc. zool. Soc. Lond.* **123**, 715

89 Crowcroft, P. 1955. Notes on the behaviour of shrews. *Behaviour* 8, 63

90 Crowcroft, P. 1957. *The life of the Shrew*. London

91 C.S.I.R.O. Division of Wildlife Research. *Annual Report 1965–66*. Canberra

92 Curtis, J.D. & Kozicky, E.L. 1944. Observations on the eastern porcupine. *J. Mammal.* **25**, 137

93 Dagg, A. Innis. 1960. Food preferences of the giraffe. *Proc. zool. Soc. Lond.* **135**, 640

94 Dagg, A. Innis. 1960. Gaits of the giraffe and okapi. *J. Mammal.* **41**, 282

95 Dagg, A. Innis. 1965. Sexual differences in giraffe skulls. *Mammalia.* **29**, 610

96 Dalquest, W.W. 1950. The genera of the chiropteran family Natalidae. *J. Mammal.* **31**, 436

97 Dalrymple, A. 1771. *An historical collection of Voyages in the South Pacific Ocean.* Vol. II. London

98 Dampier, W. 1717, 1709. *A new voyage round the world.* Vol. I. 6 ed. *A voyage to New Holland.* Vol. III. 2 ed. London

99 Daniel, M.J. 1960. Porcupine quills in viscera of fisher. *J. Mammal.* **41**, 133

100 Dasmann, R.F. & Mossman, A.S. 1962. Reproduction in some ungulates in Southern Rhodesia. *J. Mammal.* **43**, 533

101 Davies, J.L. & Guiler, E.R. 1957. A note on the pygmy right whale, *Caperea marginata* Gray. *Proc. zool. Soc. Lond.* **129**, 579

102 Davis, D.D. 1964. The giant panda; a morphological study of evolutionary mechanisms. *Fieldiana : Zool. Mem. 3*

103 Davis, J.A. Jr. 1965. A preliminary report of the reproductive behaviour of the small Malayan chevrotain, *Tragulus javanicus* at New York Zoo. *Int. Zoo Yrbk.* **5**, 42

104 Davis, W.H. & Lidicker, W.Z. 1956. Winter range of the red bat *Lasiurus borealis*. *J. Mammal.* **37**, 280

105 Dawbin, W.H. 1966. Porpoises and porpoise hunting in Malaita. *Australian Nat. Hist.* Sept. 1966, 207

106 Dennler de La Tour, G. 1954. The guanaco. *Oryx,* Lond. **2**, 273

107 Dennler de La Tour, G. 1954. The vicuña. *Oryx,* Lond. **2**, 347

108 Deraniyagala, P.E.P. 1951. *Elephus maximus,* the elephant of Ceylon. (Part II). *Spolia zeylanica* **26**, 161

109 Dexler, H. & Freund, L. 1906. Zur Biologie und Morphologie von *Halicore dugong*. *Arch. Natg. Berlin.* **72**, 77

110 Dickson, V. 1949. The Jerboa. *Zoo Life,* Lond. **4**, 104

111 Dimelow, E.J. 1963. The behaviour of the hedgehog (*Erinaceus europaeus* L.) in the routine of life in captivity. *Proc. zool. Soc. Lond.* **141**, 281

112 Dimelow, E.J. 1963. Observations on the feeding of the hedgehog (*Erinaceus europaeus* L.) *Proc. zool. Soc. Lond.* **141**, 291

113 Ditmars, R.L. & Greenhall, A.M. 1935. The vampire bat: a presentation of undescribed habits and a review of its history. *Zoologica,* N. York. **19**, 53

114 Dixon, J. 1931. Pika versus weasel. *J. Mammal.* **12**, 72
115 Dobson, G. E. 1882–92. *A monograph of the Insectivora.* London
116 Doctor, The. [Kingsley, G. H.] 1888. Changes of habits in animals. *The Field*, Lond. July 14 No. 1855, 53
117 Dorst, J. 1953. Considérations sur le genre '*Otomops*' et description d'une espèce nouvelle de Madagascar. *Mem. Inst. Sci. Madagascar.* A. **8**, 235
118 Dougall, H. W. & Sheldrick, D. L. W. 1964. The chemical composition of a day's diet of an elephant. *E. Afr. Wildl. J.* **2**, 51
119 Dudley, P. 1735. An essay upon the natural history of whales, with a particular account of the ambergris found in the sperma ceti whale. *Phil. Trans. Roy. Soc. Lond.* **33** (1724–5), 256
120 Dukelski, N. M. 1927. External characters in the structure of the feet and their value for the classification of the voles. *J. Mammal.* **8**, 133
121 Dunn, E. R. 1931. The disk-winged bat (*Thyroptera*) in Panama. *J. Mammal.* **12**, 429
122 Dunn, L. H. 1933. Observations on the carnivorous habits of the spear-nosed bat, *Phyllostomus hastatus panamensis* Allen in Panama. *J. Mammal.* **14**, 188
123 Dunnet, G. M. 1964. A field study of local populations of the brush-tailed possum *Trichosurus vulpecula* in eastern Australia. *Proc. zool. Soc. Lond.* **142**, 665
124 Durrell, G. 1954. *Three singles to Adventure.* London
125 Durrell, G. M. 1953. Giant water shrew. *Zoo Life*, Lond. **8**, 145
126 Dwyer, P. D. 1962. Studies on the two New Zealand bats. *Zool. Pub. Victoria Univ. Wellington.* No. 28
127 Eales, N. B. 1925. External characters, skin and temporal gland of a foetal African elephant. *Proc. zool. Soc. Lond.* **1925**, 445
128 Eaton, R. L. 1969. Cooperative hunting by cheetahs and jackals and a theory of domestication of the dog. *Mammalia*, Paris. **33**, 87
129 Edmond-Blanc, F. 1947. A contribution to the knowledge of the Cambodian wild ox or kouproh. *J. Mammal.* **28**, 245
130 Eisenberg, J. F. 1961. Observations on the nest building behaviour of armadillos. *Proc. zool. Soc. Lond.* **137**, 322
131 Eisenberg, J. & Gould, E. 1966. The behaviour of *Solenodon paradoxus* in captivity. *Zoologica*, N.Y. **51**, 49
132 Elsner, R. 1966. Diving bradycardia in the unrestrained hippopotamus. *Nature*, Lond. **212**, 408
133 Enders, R. K. 1930. Notes on some mammals from Barro Colorado Island, Canal Zone. *J. Mammal.* **11**, 280
134 Enders, R. K. 1940. Observations on sloths in captivity at higher altitudes in the tropics and in Pennsylvania. *J. Mammal.* **21**, 5
135 Engler, C. H. 1943. Carnivorous activities of big brown and pallid bats. *J. Mammal.* **24**, 96
136 Epling, G. P. 1956. Morphology of the scent gland of the javelina. *J. Mammal.* **37**, 246
137 Eschricht, D. F. 1862. On the species of the genus *Orca* inhabiting the northern seas. Translated from Danish in: *Recent Memoirs on the Cetacea.* Ray Soc. 1866
138 Eskelund, K. 1960. *Drums in Bahia.* London
139 Evans, G. H. 1910. *Elephants and their diseases.* Rangoon
140 Ewer, R. F. 1958. Adaptive features in the skulls of African Suidae. *Proc. zool. Soc. Lond.* **131**, 135
141 Ewer, R. F. 1969. The 'instinct to teach'. *Nature*, Lond. **222**, 698
142 Fay, F. H. 1960. Structure and function of the pharyngeal pouches of the walrus (*Odobenus rosmarus* L.) *Mammalia*, Paris. **24**, 361

REFERENCES

143 Fearnhead, R.W., Shute, C.C.D. & Bellairs, A. d'A. 1955. The temporo-mandibular joint of shrews. *Proc. zool. Soc. Lond.* **125**, 795

144 Feniuk, B.K. & Kazantzeva, J.M. 1937. The ecology of *Dipus sagitta. J. Mammal.* **18**, 409

145 Fitch, H.S. 1948. Habits and economic relationships of the Tulare kangaroo rat. *J. Mammal.* **29**, 5

146 Fleay, D.H. 1935. Breeding of *Dasyurus viverrinus* and general observations on the species. *J. Mammal.* **16**, 10

147 Fleay, D. 1944. *We breed the platypus.* Melbourne

148 Flower, W.H. & Lydekker, R. 1891. *An introduction to the study of mammals living and extinct.* London

149 Ford, C.E., Hamerton, J.L. & Sharman, G.B. 1957. Chromosome polymorphism in the common shrew. *Nature,* Lond. **180**, 392

150 Fosbrooke, H.A. 1968. Elephants in the Serengeti National Park: an early record. *E. Afr. Wildl. J.* **6**, 150

151 Foster, J.B. 1966. The giraffe of Nairobi National Park: home range, sex ratios, the herd, and food. *E. Afr. Wildl. J.* **4**, 139

152 Foster, J.B. 1967. The white rhino reserve. *E. Afr. Wildl. J.* **5**, 168

153 Fraser, F.C. 1948. *In* Norman, J.R. & Fraser, F.C. *Giant fishes whales and dolphins.* 279. London

154 Fraser, F.C. 1956. A new Sarawak dolphin. *Sarawak Mus. J.* **7**, 478

155 Frechkop, S. 1949. La locomotion et la structure des tatous et des pangolins. *Ann. Soc. roy. Belgique.* **80**, 5

156 Frechkop, S. 1949. Explication biologique, fournie par les tatous, d'un des caractères distinctifs des Xénarthres et d'un caractère adaptatif analogue chez les pangolins. *Bull, Inst. roy. Sci. nat. Belgique* **25**, No. 28

157 Frechkop, S. 1953. Notes sur les mammifères. XXXIX De quelques particularités adaptives du squelette des Paresseux. *Bull. Inst. roy. Sci. nat. Belgique.* **29**, No. 35

158 Frechkop, S. 1955. Sous-ordre des ruminants ou sélénodontes. *In Grassé, P.-P. ed. Traite de Zoologie.* **17** Fasc. 1. Paris.

159 Frechkop, S. 1965. La specifité du cheval de Prjewalsky. *Bull. Inst. roy. Sci. nat. Belgique.* **41** (29) 1

160 Garrod, A.H. 1877. Notes on the manatee (*Manatus americanus*) recently living in the Society's gardens. *Trans. zool. Soc. Lond.* **10**, 137

161 Gee, E.P. 1950. Wild elephants in Assam. *Oryx,* **1**, 16

162 Gee, E.P. 1959. Report on a survey of the rhinoceros area of Nepal. *Oryx.* **5**, 52

163 Geist, V. 1968. Horn-like structures as rank symbols, guards and weapons. *Nature,* Lond. **220**, 813

164 Genelly, R.E. 1965. Ecology of the common mole-rat (*Cryptomys hottentotus*) in Rhodesia. *J. Mammal.* **46**, 647

165 Gilmore, R.M. 1961. *The Story of the Gray Whale.* 2nd ed. Revised. San Diego.

166 Glover, J. 1963. The elephant problem at Tsavo. *E. Afr. Wildl. J.* **1**, 30

167 Glover, T.D. & Sale, J.B. 1968. The reproductive system of the male rock hyrax (*Procavia* and *Heterohyrax*) *J. Zool. Lond.* **156**, 351

168 Goddard, J. 1966. Mating and courtship of the black rhinoceros (*Diceros bicornis* L.) *E. Afr. Wildl. J.* **4**, 69

169 Goddard, J. 1967. Home range, behaviour and recruitment rates of two black rhinoceros populations. *E. Afr. Wildl. J.* **5**, 133

170 Goddard, J. 1968. Food preferences of two black rhinoceros populations. *E. Afr. Wildl. J.* **6**, 1

171 Godfrey, G. & Crowcroft, P. 1960. *The Life of the mole.* London

172 Goode, G.B. and others. 1887. *The Fisheries and Fishery industries of the United States. Sect V*, I I, 299. Washington

173 Goodwin, G.G. 1928. Observations on *Noctilio*. *J. Mammal.* **9**, 104

174 Goodwin, G.G. 1946. Mammals of Costa Rica. *Bull. Amer. Mus. nat. Hist.* **87**, Article 5

175 Goodwin, G.G. & Greenhall, A.M. 1961. A review of the bats of Trinidad and Tobago. *Bull. Amer. Mus. nat. Hist.* **122**, Article 3

176 Gosse, P.H. 1847. Brief notes on the habits of *Noctilio mastivus*. *Ann. Mag. nat. Hist.* **20**, 424

177 Gould, E. 1964. Evidence for echolocation in shrews. *J. exp. Zool.* **156**, 19

178 Gould, E. 1965. Evidence for echolocation in the Tenrecidae of Madagascar. *Proc. Amer. Phil. Soc.* **109**, 352

179 Gould, E. & Eisenberg, J.F. 1966. Notes on the biology of the Tenrecidae. *J. Mammal.* **47**, 660

180 Gowers, Sir William. 1948. African elephants and ancient authors. *African Affairs 1948*, 173.

181 Gowers, Sir William & Scullard, H.H. 1950. Hannibal's elephants again. *Numismatic Chron. Ser. 6.* **10**, 271

182 Graham, A. 1966. East African Wildlife Society cheetah survey: extracts from the report by Wildlife Services. *E. Afr. Wildl. J.* **4**, 50

183 Grassé, P.-P. (ed.) 1955. *Traité de zoologie. Vol. 17 Mammifères.* Paris

184 Griffiths, M. & Barker, R. 1966. The plants eaten by sheep and by kangaroos grazing together in a paddock in south-western Queensland. *C.S.I.R.O. Wildl. Res.* **11**, 145

185 Griffiths, M. & Simpson, K.G. 1966. A seasonal feeding habit of spiny anteaters. *C.S.I.R.O. Wildl. Res.* **11**, 137

186 Grimble, A. 1952. *A pattern of islands.* London.

187 Gudger, E.W. 1945. Fishermen bats of the Caribbean region. *J. Mammal.* **26**, 1

188 Guiler, E.R. 1961. Breeding season of the Thylacine. *J. Mammal.* **42**, 396

189 Guiraud, M. 1948. Contribution à l'étude du *Phacochoerus aethiopoicus* (Pallas). *Mammalia*, Paris. **12**, 54

190 Gunter, G. 1941. Occurrence of the manatee in the United States, with records from Texas. *J. Mammal.* **22**, 60

191 Guth, C., Heim de Balsac, H. & Lamotte, M. 1959, 1960. Recherches sur la morphologie de *Micropotamogale lamottei* et l'évolution des *Potamogalinae*. *Mammalia*, Paris. **23**, 423, **24**, 190

192 Gwynne, M.D. & Bell, R.H.V. 1968. Selection of vegetation components by grazing ungulates in the Serengeti National Park. *Nature*, Lond. **220**, 390

193 Haga, R. 1960. Observations on the ecology of the Japanese pika. *J. Mammal.* **41**, 201

194 Hall, F.G. 1937. Adaptations of mammals to high altitudes. *J. Mammal.* **18**, 468

195 Haltenorth, T. 1953. *Die Wildkatzen der Alten Welt.* Leipzig

196 Hamilton, W.J. Jr. 1931. Habits of the star-nosed mole, *Condylura cristata*. *J. Mammal.* **12**, 345

197 Hamilton, W.J. Jr. 1942. The buccal pouch of *Peromyscus*. *J. Mammal.* **23**, 449

198 Hamilton, W.J. Jr. 1946. Habits of the swamp rice rat, *Oryzomys palustris palustris* (Harlan). *Amer. Midl. Nat.* **36**, 730

199 Hamilton, W.J. Jr. 1946. The black persimmon as a summer food of the Texas armadillo. *J. Mammal.* **27**, 175

200 Hamlett, G.W.D. 1939. Identity of *Dasypus septemcinctus* Linnaeus with notes on some related species. *J. Mammal.* **20**, 328

201 Hancock, D. 1965. Killer whales kill and eat a minke whale. *J. Mammal.* **46**, 341

REFERENCES

202 Handley, C.O. Jr. 1966. A synopsis of the genus *Kogia* (Pygmy sperm whales). *In Norris, K. S. ed. Whales, Dolphins and Porpoises.* Univ. California Press

203 Hanney, P. 1962. Observations upon the food of the barn owl (*Tyto alba*) in southern Nyasaland, with a method of ascertaining population dynamics of rodent prey. *Ann. Mag. nat. Hist.* Ser. 13, **6**, 305

204 Hanström, B. 1956. Ytterligare björkmusfynd i Bergslagen. *Fauna och flora.* **51**, 10

205 Harper, F. 1945. *Extinct and vanishing mammals of the Old World.* New York

206 Harris, G.P. 1808. Description of two new species of *Didelphis* from Van Dieman's land. *Trans. Linn. Soc. Lond.* **9**, 174

207 Harris, W.C. 1840–3. *Portraits of the game and wild animals of Southern Africa.* London

208 Harris, W.C. 1844. *The wild sports of southern Africa.* (4th ed.) London

209 Harrison, D.L. 1956. Mammals from Kurdistan, Iraq, with description of a new bat. *J. Mammal.* **37**, 257

210 Harrison, D.L. 1956. Gerbils from Iraq, with description of a new gerbil. *J. Mammal* **37**, 417

211 Harrison, D.L. & Davies, D.V. 1949. A note on some epithelial structures in microchiroptera. *Proc. zool. Soc. Lond.* **119**, 351

212 Harrison, J.L. 1959. Defaecation in the flying lemur *Cynocephalus variegatus.* *Proc. zool. Soc. Lond.* **133**, 179

213 Harrison, J.L. & Traub, R. 1950. Rodents and insectivores from Selangor, Malaya. *J. Mammal.* **31**, 337

214 Harrison, R.J. & King, J.E. 1965. *Marine mammals.* London

215 Harvey, E.B. & Rosenberg, L.E. 1960. An apocrine gland complex of the pika. *J. Mammal.* **41**, 213

216 Hawkesworth, J. 1773. An account of the voyages undertaken by the order of his present Majesty . . . . . performed by Commodore Byron, Captain Wallis, Captain Carteret and Captain Cook. London

217 Hawkins, A.E. & Jewell, P.A. 1962. Food consumption and energy requirements of captive British shrews and the mole. *Proc. zool. Soc. Lond.* **138**, 137

218 Hawkins, A.E., Jewell, P.A. & Tomlinson, G. 1960. The metabolism of some British shrews. *Proc. zool. Soc. Lond.* **135**, 99

219 Hediger, H. 1950. La capture des éléphants au Parc National de la Garamba. *Bull. Inst. Royal. Col. Belge.* **21**, 218

220 Hediger, H. 1953. Ein symbioseartiges Verhältnis zwischen Flusspferd und Fisch. *Säugetierk. Mitt.* **1**, 75

221 Hediger, H. 1965. Man as a social partner of animals and vice-versa. *Symp. zool. Soc. Lond.* **14**, 291

222 Hediger, H. and others. 1950 [Okapi] *Acta tropica* **7**, Nr. 2

223 Heinsohn, G.E. 1966. Ecology and reproduction of the Tasmanian bandicoots (*Perameles gunni* and *Isoodon obesulus*) Univ. of California Pubs. in Zool. **80**

224 Hershkovitz, P. 1944. A systematic review of the neotropical water rats of the genus *Nectomys* (Cricetinae). *Misc. Publ. Mus. Zool. Univ. Mich.* No. **58**

225 Hershkovitz, P. 1955. On the cheek pouches of the tropical American paca, *Agouti paca* (Linnaeus, 1766). *Säugetierk. Mitt.* **3**, 67

226 Herter, K. 1963. *Igel.* Wittenberg Lutherstaat. (English translation: *Hedgehogs*, London 1965)

227 Hester, F.J., Hunter, J.R. & Whitney, R.R. 1963. Jumping and spinning behaviour in the spinner porpoise. *J. Mammal.* **44**, 586

228 Hill, J.P. 1895. Preliminary note on the occurrence of a placental connection in *Perameles obesula*, and on the foetal membranes of certain macropods. *Proc. Linn. Soc. N.S.W.* **20**, 578

229 Hill, W.C.O. 1945. Notes on the dissection of two dugongs. *J. Mammal.* **26**, 153

230 Hinton, M.A.C. 1936. Some interesting points in the anatomy of the freshwater dolphin *Lipotes* and its allies. *Proc. Linn. Soc. Lond.* **148**, 183

231 Hinton, M.A.C. & Pycraft, W.P. 1922. Preliminary note on the affinities of the genus *Lipotes*. *Ann. Mag. Nat. Hist. Ser.* 9, **10**, 232

232 Hirasaka, K. 1954. Basking habit of the Japanese bear. *J. Mammal.* **35**, 128

233 Hodge, W.H. 1947. The not so terrible mouse. *Nat. Hist. N.Y.* **56**, 310

234 Honigmann, H. 1936. Studies on the nutrition of mammals. Part 1. *Proc. zool. Soc. Lond.* 1936. 517

235 Hooper, E.T. 1968. Anatomy of the middle-ear walls and cavities in nine species of microtine rodents. *Occ. Pap. Mus. Zool. Univ. Mich.* No. 657

236 Hopwood, A.T. 1936. Zebrine and caballine horses and their former distribution in Europe. *Proc. zool. Soc. Lond.* **1936**, 897

237 Horner, B.E. 1947. Paternal care of young mice of the genus *Peromyscus*. *J. Mammal.* **28**, 31

238 Horst, C.J. van der. 1947. see van der Horst, C.J.

239 Howell, A.B. 1920. Contribution to the life-history of the Californian mastiff bat. *J. Mammal.* **1**, 111

240 Howell, A.B. 1920. Some Californian experiences with bat roosts. *J. Mammal.* **1**, 169

241 Howell, A.B. 1940. Cheek pouches of the paca. *J. Mammal.* **21**, 361

242 Howell, A.B. & Little, L. 1924. Additional notes on Californian bats, with observations upon the young of *Eumops*. *J. Mammal.* **5**, 261

243 Hubback, T. 1939. The Asiatic two-horned rhinoceros. *J. Mammal.* **20**, 1

244 Hubbard, W.D. 1929. Further notes on the mammals of Northern Rhodesia and Portuguese East Africa. *J. Mammal.* **10**, 294

245 Huey, L.M. 1925. Food of the Californian leaf-nosed bat. *J. Mammal.* **6**, 196

246 Huey, L.M. 1936. Desert pallid bats caught in mouse traps. *J. Mammal.* **17**, 285

247 Hunsaker, D. II & Hahn, T.C. 1965. Vocalization of the South American tapir, *Tapirus terrestris*. *Anim. Behav.* **13**, 69

248 Hurrell, E. 1962. Dormice. *Animals of Britain 10.* Sunday Times Publications. London

249 Hutchinson, G.E. & Ripley, S.D. 1954. Gene dispersal and the ethology of the Rhinocerotidae. *Evolution* **8**, 178

250 Ingles, G. 1953. Observations on Barro Colorado Island mammals. *J. Mammal.* **34**, 266

251 Ingles, L.G. 1960. Tree climbing by mountain beavers. *J. Mammal.* **41**, 120

252 Innis, A.C. 1958. The behaviour of the giraffe, *Giraffa camelopardalis*, in the eastern Transvaal. *Proc. zool. Soc. Lond.* **131**, 245

253 Jaczewski, Z. 1958. Reproduction of the European bison. *Bison bonasus* (L) in reserves. *Acta theriologica* **1**, 333

254 Jarman, P.J. 1966. The status of the dugong (*Dugong dugon* Müller); Kenya, 1961. *E. Afr. Wildl. J.* **4**, 82

255 Jepsen, G.L. 1966. Early Eocene bat from Wyoming. *Science* **154**, 1333

256 Jerdon, T.C. 1867. *The Mammals of India*. Roorkee

257 Jeuniaux, C. 1961. Chitinase: an addition to the list of hydrolases in the digestive tract of vertebrates. *Nature, Lond.* **192**, 135

258 Jones, F.Wood. 1923–5. *The Mammals of South Australia*. Adelaide

259 Jones, F. Wood. 1941. The external characters of a neonatal *Pedetes*. *Proc. zool. Soc. Lond. B.* **110**, 199

260 Jones, F. Wood. 1953. Some readaptations of the mammalian pes in response to arboreal habits (*Cyclopes*). *Proc. zool. Soc. Lond.* **123**, 33

REFERENCES

261 Jonsgård, Å. & Nordli, O. 1952. Concerning a catch of white-sided dolphins (*Lagenorhynchus acutus*) on the west coast of Norway, winter 1952. *Norsk. Hvalfangst. Tid. 1952*, 229

262 Kalmbach, E.R. 1943. The armadillo in relation to agriculture and game. *Game, Fish & Oyster Commission*, Austin, Texas, 1

263 Kellog, R. 1940. Whales, giants of the sea. *Nat. Geogr. Mag.* Jan. 1940, 35

264 Kermack, N.A. 1963. The cranial structure of the Triconodonts. *Phil. Trans. Roy. Soc. Lond. B.* **246**, 83

265 Kiley-Worthington, M. 1965. The waterbuck (*Kobus defassa* Ruppel 1935 and *K. ellipsiprimnus* Ogilby 1833) in East Africa: a study of the sexual behaviour. *Mammalia*, Paris. **29**, 177

266 Kilham, L. 1958. Territorial behaviour in pikas. *J. Mammal.* **39**, 307

267 King, J.E. 1964. *Seals of the world*. London

268 King, J.E. 1966. Relationships of the Hooded and Elephant Seals (genera *Cystophora* and *Mirounga*) *J. Zool. Lond.* **148**, 385

269 King, J.M. 1965. A field guide to the reproduction of the Grant's zebra and Grévy's zebra. *E. Afr. Wildl. J.* **3**, 99

270 Klingel, H. 1965. Notes on the biology of the plains zebra *Equus quagga boehmi* Matschie. *E. Afr. Wildl. J.* **3**, 86

271 Kock, D. & Schomber, H.W. 1961. Beitrag zur Kenntnis der Lebens- und Verhaltensweise des Gundis, *Ctenodactylus gundi* (Rothmann, 1776) *Säugetierk. Mitt.* **9**, 165

272 Koford, C.B. 1957. The vicuña and the puna. *Ecol. Monogr.* **27**, 153

273 Krefft, J.L.G. 1871. *The mammals of Australia*. Sydney

274 Krieg, H. 1939. Begegnungen mit Ameisenbären und Faultieren in freier Wildbahn. *Z. Tierpsychol.* **2**, 291

275 Krog, J., Reite, O.B. & Fjellheim, P. 1969. Vasomotor responses in the growing antlers of the reindeer, *Rangifer tarandus*. *Nature*, Lond. **223**, 99

276 Krott, P. 1959. Der Vielfrass (*Gulo gulo* L. 1758) *Monogr. der Wildsäugetiere* **13**

277 Krumbiegel, I. 1939. Die giraffe. *Monogr. d. wildsäugetiere* **8** Leipzig

278 Krumholz, L.A. 1943. Notes on manatees in Florida waters. *J. Mammal.* **24**, 272

279 Krumrey, W.A. & Buss, I.O. 1968. Age estimation, growth and relationships between body dimensions of the female African elephant. *J. Mammal.* **49**, 22

280 Krutzsch, P.H. 1954. North American Jumping Mice (Genus *Zapus*) *Univ. Kansas Pubs., Mus. Nat. Hist.* **7**, 349

281 Krutzsch, P.H. 1955. Observations on the Californian mastiff bat. *J. Mammal.* **36**, 407

282 Kruuk, H. 1966. A new view of the hyaena. *New Scientist.* **30**, 849

283 Kruuk, H. & Turner, M. 1967. Comparative notes on predation by lion, leopard, cheetah and wild dog in the Serengeti area, East Africa. *Mammalia*, Paris, **31**, 1

284 Lamont, J. 1876. *Yachting in Arctic Seas*. London

285 Lamprey, H.F. 1963. Ecological separation of the large mammal species in the Tarangire Game Reserve, Tanganyika. *E. Afr. Wildl. J.* **1**, 63

286 Lamprey, H.F., Glover, P.E., Turner, M.M. & Bell, R.H.V. 1967. Invasion of the Serengeti National Park by elephants. *E. Afr. Wildl. J.* **5**, 151

287 Lankester, E.R. 1910. *Monograph of the Okapi*. London.

288 Laurent, P. 1938. Observations sur le comportement des petits mammifères sauvages en captivité. *Mammalia*, Paris. **2**, 12

289 Lawrence, W.E. & Rewell, R.E. 1948. The cerebral blood supply of the Giraffidae. *Proc. zool. Soc. Lond.* **118**, 202

290 Laws, R.M. 1966. Age criteria for the African elephant, *Loxodonta a. africana*. *E. Afr. Wildl. J.* **4**, 1

291 Laws, R.M. 1967. Eye lens weight and age in African elephants. *E. Afr. Wildl. J.* **5**, 46

292 Laws, R.M. 1968. Dentition and ageing of the hippopotamus. *E. Afr. Wildl. J.* **6**, 19

293 Laws, R.M. & Clough, G. 1966. Observations on reproduction in the hippopotamus *Hippopotamus amphibius* Linn. *In* Rowlands, I.W. ed. Comparative biology of reproduction in mammals. *Symp. zool. Soc. Lond.* **15**, 117

294 Laws, R.M., Parker, I.S.C. & Archer, A.L. 1967. Estimating live weights of elephants from hind leg weights. *E. Afr. Wildl. J.* **5**, 106

295 Layne, J.N. 1958. Observations on freshwater dolphins in the upper Amazon. *J. Mammal.* **39**, 1

296 Leopold, A.S. & Darling, F.F. 1953. *Wildlife in Alaska.* New York

297 Le Souef, A.S. & Burrell, H. 1926. *The wild animals of Australasia.* London & Sydney

298 Lim Boo Liat, 1967. Notes on the food habits of *Ptilocercus lowii* Gray (Pentail tree-shrew) and *Echinosorex gymnurus* (Raffles) (Moonrat) in Malaya with remarks on 'ecological labelling' by parasite patterns. *J. Zool. Lond.* **152**, 375

299 Lindsey, A.A. 1937. The Weddell seal in the Bay of Whales, Antarctica. *J. Mammal.* **18**, 127

300 Liu, Ch'eng-chao. 1937. Notes on the food of Chinese Hedgehogs. *J. Mammal.* **18**, 355

301 Llanos, A.C. & Crespo, J.A. 1952. Ecologia de la viscacha (*Lagostomus maximus maximus* Blainer) en el nordeste de la provincia de Entre Rios. *Rev. Invest. Agricolas.* **6**, 289

302 Loukashkin, A.S. 1937. The mammals of North Manchuria. *J. Mammal.* **18**, 327

303 Loukashkin, A.S. 1940. On the pikas of North Manchuria. *J. Mammal.* **21**, 402

304 Loukashkin, A.S. 1944. The giant rat-headed hamster, *Cricetulus triton nestor* Thomas, of Manchuria. *J. Mammal.* **25**, 170

305 Loveridge, A. 1923. Notes on East African mammals collected 1920–3. *Proc. zool. Soc. Lond.* **1923** 685,

306 Lyall-Watson, M. 1963. A critical re-examination of food 'washing' behaviour in the raccoon (*Procyon lotor* Linn.) *Proc. zool. Soc. Lond.* **141**, 371

307 Lydekker, R. 1915. *Catalogue of the ungulate mammals in the British Museum (Natural History).* Vol. 4. London

308 Lyne, A.G., Pilton, P.E. & Sharman, G.B. 1959. Oestrous cycle, gestation period and parturition in the marsupial *Trichosurus vulpecula. Nature,* Lond. **183**, 622

309 Malzy, P. 1965. Un mammifère aquatique de Madagascar: le Limnogale. *Mammalia,* Paris. **29**, 400

310 Marshall, A.J. 1947. The breeding cycle of an equatorial bat (*Pteropus giganteus* of Ceylon). *Proc. Linn. Soc. Lond.* **159**, 103

311 Marshall, W.H. 1941. *Thomomys* as burrowers in the snow. *J. Mammal.* **22**, 196

312 Martin, H.T. 1892. *Castorologia, or the history and traditions of the Canadian beaver.* London & Montreal

313 Martin, K. 1943. The Colorado pika. *J. Mammal.* **24**, 394

314 Matthews, L. Harrison. 1939. Reproduction in the spotted hyaena, *Crocuta crocuta* Erxleben, *Phil. Trans. Roy. Soc. Lond.* B **230**, 1

315 Matthews, L. Harrison. 1939. The bionomics of the spotted hyaena, *Crocuta crocuta* Erxleben, *Proc. zool. Soc. Lond.* A **109**, 43

316 Matthews, L. Harrison. 1939. The subspecies and variation of the spotted hyaena, *Crocuta crocuta* Erxl. *Proc. zool. Soc. Lond.* B **109**, 237

317 Matthews, L. Harrison. 1941. Notes on the genitalia and reproduction of some African bats. *Proc. zool. Soc. Lond.* **111** B, 289

411

REFERENCES

318 McCann, C. 1940. The short-nosed fruit bat (*Cynopterus sphinx*) as an agent of seed dispersal in the wild date (*Phoenix sylvestris* L.) *J. Bombay Nat. Hist. Soc.* **42**, 184

319 Mearns, E.A. 1907. Mammals of the Mexican boundary of the United States. *Bull. U.S. nat. Mus.* **56**, Pt. 1

320 Merriam, C.H. 1918. Review of the grizzly and big brown bears of North America (genus *Ursus*) with description of a new genus *Vetularctos*. *North Amer. Fauna No.* **41**. Washington

321 Miller, F.W. 1930. Notes on some mammals of southern Matto Grosso, Brazil. *J. Mammal.* **11**, 10

322 Miller, G.S. Jr. 1899. Preliminary list of New York mammals. *Bull. New York State Mus.* **6**, 273

323 Miller, G.S. Jr. 1907. The families and genera of bats. *Bull. U.S. nat. Mus.* No. **57**

324 Miller, G.S. Jr. 1918. A new river-dolphin from China. *Smithsonian Misc. Coll.* **68**, No. 9

325 Minoprio, J.D.L. 1945. Sobre el *Chlamyphorus truncatus* Harlow. *Acta zool. Lilloana* **3**, 5

326 Mitchell, H.A. 1964. Investigations of the cave atmosphere of a Mexican bat colony. *J. Mammal.* **45**, 568

327 Mohr, E. 1936. Biologische Beobachtungen an *Solenodon paradoxus* Brandt. *Zool. Ang.* **113**, 176, **116**, 65, **117**, 233

328 Mohr, E. 1961. Schuppentiere. *Die Neue Brehm-Bücherei*. Wittenberg Lutherstadt.

329 Mohr, E. 1968. Spielbereitschaft beim wisent. *Z. Säugertierk.* **33**, 116

330 Montagu, I. 1957. Colour-film shots of the wild camel. *Proc. zool. Soc. Lond.* **129**, 592

331 Moore, A.W. 1933. Food habits of Townsend and coast moles. *J. Mammal.* **14**, 36

332 Moore, A.W. 1942. Shrews as a check on Douglas fir regeneration. *J. Mammal.* **23**, 41

333 Moore, J.C. 1951. The range of the Florida manatee. *Quart. J. Fla. Acad. Sci.* **14**, 1

334 Moore, J.C. 1951. The status of the manatee in the Everglades National Park, with notes on its natural history. *J. Mammal.* **32**, 22

335 Moore, W.G. 1948. Bat caves and bat bombs. *Turtox News*, Chicago. **26**, 262

336 Morejohn, G.V. 1968. A killer whale-gray whale encounter. *J. Mammal.* **49**, 327

337 Morris, D. 1962. The behaviour of the green acouchi (*Myoprocta pratti*) with special reference to scatter hoarding. *Proc. zool. Soc. Lond.* **139**, 70

338 Morrison, P.R. 1945. Acquired homiothermism in the pregnant sloth. *J. Mammal.* **26**, 272

339 Morrison, P. & Elsner, R. 1962. Influence of altitude on heart and breathing rates in some Peruvian rodents. *J. appl. Physiol.* **17**, 467

340 Morrison-Scott, T.C.S. 1947. A revision of our knowledge of African elephant's teeth, with notes on forest and 'pygmy' elephants. *Proc. zool. Soc. Lond.* **117**, 505

341 Murie, J. 1872. On the form and structure of the manatee (*Manatus americanus*). *Trans. zool. Soc. Lond.* **8**, 127

342 Mutere, F.A. 1965. Delayed implantation in an equatorial fruit bat. *Nature, Lond.* **207**, 780

343 Myers, G.T. & Vaughan, T.A. 1964. Food habits of the plains pocket gopher in Eastern Colorado. *J. Mammal.* **45**, 588

344 Napier-Bax, P. & Sheldrick, D.L.W. 1963. Some preliminary observations on the food of elephants in the Tsavo Royal National Park (East) of Kenya. *E. Afr. Wildl. J.* **1**, 40

345 Nevo, E. & Amir, E. 1964. Geographic variation in reproduction and hibernation patterns of the forest dormouse. *J. Mammal.* **45**, 69

346 Newman, C.C. & Baker, R.H. 1942. Armadillo eats young rabbits. *J. Mammal.* **23**, 450

347 Newman, H.H. 1913. The Natural history of the nine-banded armadillo of Texas. *Amer. Nat.* **67**, 513

348 Nishiwaki, M. 1966. A discussion of rarities among the smaller cetaceans caught in Japanese waters. *In* Norris, K.S. ed. *Whales, Dolphins and porpoises.* Berkeley and Los Angeles

349 Nishiwaki, M. & Norris, K.S. 1966. A new genus, *Peponocephala*, for the odontocete cetacean species *Electra electra. Sci. Rep. Whales Res. Inst.* **20**, 95

350 Norris, K.S. 1964. Some problems of echolocation in cetaceans. *In* Tavolga, W.N. ed. *Marine bio-acoustics.* 317. New York

351 Novick, A. 1962. Orientation in neotropical bats. 1. Natalidae and Emballonuridae. *J. Mammal.* **43**, 449

352 O'Donoghue, P.N. 1963. Reproduction in the female hyrax (*Dendrohyrax arborea ruwenzorii*) *Proc. zool. Soc. Lond.* **141**, 207

353 Olds, J.M. 1950. Notes on the hood seal (*Cystophora cristata*) *J. Mammal.* **31**, 450

354 Omura, H., Fujino, K. & Kimura, S. 1955. Beaked whales *Berardius bairdi* of Japan, with notes on *Ziphius cavirostris. Sci. Repts. Whales Res. Inst.* **10**, 89

355 Orr, R.T. 1954. Natural history of the pallid bat *Antrozous pallidus* (Le Conte) *Proc. Calif. Acad. Sci.* **28**, 165

356 Osborn, D.J. 1953. Age classes, reproduction and sex ratios of Wyoming beaver. *J. Mammal.* **34**, 27

357 Osborn, H.F. 1936–42. *Proboscidea. A monograph of the discovery, evolution, migration and extinction of the mastodonts and elephants of the world.* 2 vols. New York

358 Osgood, W.H. 1921. A monographic study of the American marsupial *Caenolestes. Publ. Field. Mus. Nat. Hist. zool. ser.* **14**, 1

359 Osgood, W.H. 1943. The mammals of Chile. *Field Mus. Nat. Hist. zool. ser.* **30**, 1

360 Osgood, W.H. 1943. Clinton Hart Merriam 1855–1942. *J. Mammal.* **24**, 421

361 Osgood, W.H. 1947. Cricetine rodents allied to *Phyllotis. J. Mammal.* **28**, 165

362 Owen, R. 1846. *A history of British fossil mammals and birds.* London

363 Owen, R. 1868. On the anatomy of vertebrates. Vol. 3 *Mammals.* London

364 Paradiso, J.L. 1967. A review of the wrinkle-faced bats (*Centurio senex* Gray), with description of a new subspecies. *Mammalia,* Paris. **31**, 595

365 Parker, G.H. 1922. The breathing of the Florida manatee *Trichechus latrostris. J. Mammal.* **3**, 127

366 Parrington, F.R. 1961. The evolution of the mammalian femur. *Proc. zool. Soc. Lond.* **137**, 185

367 Parsons, B.T. & Sheldrick, D.L.W. 1964. Some observations on biting flies (*Diptera, Muscidae,* sub-fam. *Stomoxydinae*) associated with the black rhinoceros (*Diceros bicornis* (L)). *E. Afr. Wildl. J.* **2**, 78

368 Pearson, O.P. 1942. On the cause and nature of a poisonous action produced by the bite of a shrew *Blarina brevicauda. J. Mammal.* **23**, 159

369 Pearson, O.P. 1947. The rate of metabolism of some small mammals. *Ecology.* **28**, 127

370 Pearson, O.P. 1948. Life history of mountain viscachas in Peru. *J. Mammal.* **29**, 345

371 Pearson, O.P. 1949. Reproduction of a South American rodent, the mountain viscacha. *Amer. J. Anat.* **84**, 143

372 Pearson, O.P. 1951. Mammals in the highlands of southern Peru. *Bull. Mus. Comp. Zool.* **106**, 117

373 Perry, J.S. 1952. The growth and reproduction of elephants in Uganda. *Uganda J.* **16**, 51

REFERENCES

374 Perry, J.A. 1953. The reproduction of the African elephant, *Loxodonta africana*. *Phil. Trans. Roy. Soc. Lond.* B **237**, 93

375 Petter, F. 1961. Repartition geographique et écologie des rongeurs désertiques (du Sahara occidental à l'Iran oriental) *Mammalia*, Paris, **25** (special no.)

376 Peyre, A. 1956. Ecologie et biogeographie du desman (*Galemys pyrenaicus* G.) dans les Pyrénées françaises. *Mammalia*, Paris. **20**, 403

377 Peyre, A. 1962. Recherches sur l'intersexualité specifique chez *Galemys pyrenaicus* G. (Mammifère. Insectivore). *Arch. Biol.* Paris. **73**, 1

378 Pocock, R.I. 1916. On the external characters of the mongooses (Mungotidae). *Proc. zool. Soc. Lond.* **1916**, 742

379 Poglayen-Neuwall, I. 1967. On the marking behaviour of the kinkajou (*Potos flavus* Schreber) *Zoologica.* **51**, 137

380 Prakash, I. 1959. Foods of the Indian false vampire. *J. Mammal.* **40**, 545

381 Pucek, Z. 1963. Season changes in the braincase of some representatives of the genus *Sorex* from the palearctic. *J. Mammal.* **44**, 523

382 Quay, W.B. and Reeder, W.G. 1954. The hemorrhagic and hemopoietic nodules in the alar and interfemoral membranes of *Pizonyx vivesi* (Chiroptera). *J. Morph.* **94**, 439

383 Quick, H.F. 1953. Occurrence of porcupine quills in carnivorous mammals. *J. Mammal.* **34**, 256

384 Quilliam, T.A. ed. 1966. The mole: its adaptation to an underground environment. (Proceeding of a Ciba Foundation Guest Meeting). *J. Zool., Lond.* **149**, 31

385 Quiring, D.P. & Harlan, C.F. 1953. On the anatomy of the manatee. *J. Mammal.* **34**, 192

386 Rabb, G.B. 1959. Toxic salivary glands of the primitive insectivore *Solenodon*. *Acad. Sci. Chicago Nat. Hist. Misc.* No. 170

387 Rae, B.B. 1965. The food of the common porpoise (*Phocaena phocaena*) *J. Zool. Lond.* **146**, 114

388 Rahm, U. 1956. Notes on pangolins of the Ivory Coast. *J. Mammal.* **37**, 531

389 Rahm, U. 1960. Note sur les spécimens actuellement connus de *Micropotamogale* (*Mesopotamogale*) *ruwenzorii* et leur répartition. *Mammalia*, Paris. **24**, 511

390 Rahm, U. 1962. L'élevage et la reproduction en captivité de l'*Atherurus africanus* (Rongeurs, Hystricidae). *Mammalia*, Paris. **26**, 1

391 Rand, A.L. 1935. On the habits of some Madagascar mammals. *J. Mammal.* **16**, 89

392 Ratcliffe, F.N. 1931. The flying fox (*Pteropus*) in Australia. *Bull. Commonwealth Council for Sci. and Indust. Res.* **53**

393 Ratcliffe, F.N. 1932. Notes on the fruit bats of Australia. *J. Anim. Ecol.* **1**, 32

394 Ray, C. 1967. Social behaviour and acoustics of the Weddell seal. *Antractic J. United States.* **2** (4), 105

395 Ray, C. 1968. *Marine parks for Tanzania.* Washington, The Conservation Foundation

396 Reed, C.A. 1944. Behaviour of a shrew-mole in captivity. *J. Mammal.* **25**, 196

397 Reed, C.A. 1958. Observations on the burrowing rodent *Spalax* in Iraq. *J. Mammal.* **39**, 386

398 Reeder, W.G. & Norris, K.S. 1954. Distribution, type locality and habits of the fish-eating bat, *Pizonyx vivesi*. *J. Mammal.* **35**, 81

399 Rice, D.W. 1968. Stomach contents and feeding behaviour of killer whales in the eastern north Pacific. *Norsk Hvalfangst-Tid.* 1968 (2). 35

400 Reinhardt, J. 1862. *Pseudorea crassidens*, a cetacean hitherto unknown in the Danish fauna. Trans. in *Recent Memoirs on the Cetacea*. London: Ray Society 1866

401 Rewell, R.E. 1949. Hypertrophy of sebaceous glands on the snout as a secondary

male sexual character in the capybara *Hydrochoerus hydrochaeris. Proc. zool. Soc. Lond.* **119**, 817

402   Ride, W.D.L. 1959. Mastication and taxonomy in the macropodine skull. *In* Function and Taxonomic Importance. *Publ. Syst. Ass.* **No. 3**, 33

403   Ride, W.D.L. 1965. Locomotion in the Australian marsupial *Antechinomys. Nature,* Lond. **205**, 199

404   Ripley, S.D. 1952. Territorial and sexual behaviour in the great Indian rhinoceros, a speculation. *Ecology.* **33**, 570

405   Ritchie, A.T.A. 1963. The black rhinoceros (*Diceros bicornis* L.) *E. Afr. Wildl. J.* **1**, 54

406   Roberts, A. 1951. *The Mammals of South Africa.* Cape Town

407   Rodahl, K. 1949. Toxicity of polar bear liver. *Nature,* Lond. **164**, 530

408   Romer, J.D. 1955. Cetaceans recorded from within or near Hong Kong territorial waters. *Mem. Hong Kong Biol. Circle* **No. 3**, 1

409   Sale, J.B. 1965. The feeding behaviour of rock hyraces (genera *Procavia* and *Heterohyrax*) in Kenya. *E. Afr. Wildl. J.* **3**, 1

410   Sale, J.B. 1965. Hyrax feeding on a poisonous plant. *E. Afr. Wildl. J.* **3**, 127

411   Sale, J.B. 1966. The habitat of the rock hyrax. *J. E. Afr. Nat. Hist. Soc.* **25**, 205

412   Sale, J.B. 1966. Daily food consumption and mode of ingestion in the hyrax. *J. E. Afr. Nat. Hist. Soc.* **25**, 215

413   Sanborn, C.C. 1930. Distribution and habits of the three-banded armadillo, (*Tolypeutes*). *J. Mammal.* **11**, 61

414   Sanborn, C.C. & Watkins, A.R. 1950. Notes on the Malay tapir and other game animals in Siam. *J. Mammal.* **31**, 430

415   Sanderson, I.T. 1940. The mammals of the north Cameroons forest area. *Trans. zool. Soc. Lond.* **24**, 623

416   Scammon, C.M. 1874. *The Marine Mammals of the north-western coast of North America.* San Francisco & New York

417   Schaller, G.B. 1967. *The deer and the tiger.* Chicago & London

418   Schaller, G.B. 1968. Hunting behaviour of the cheetah in the Serengeti National Park, Tanzania. *E. Afr. Wildl. J.* **6**, 95

419   Scheffer, V.B. 1947. The mystery of the mima mounds. *Sci. Monthly* **65**, 283

420   Scheffer, V.B. 1950. The striped dolphin, *Lagenorhynchus obliquidens* Gill 1865, on the coast of North America. *Amer. Midl. Nat.* **44**, 750

421   Scheffer, V.B. 1958. *Seals, sea lions and walruses.* Stanford & London

422   Schenkel, R. 1966. On sociology and behaviour in impala (*Aepyceros melampus* Lichtenstein) *E. Afr. Wildl. J.* **4**, 99

423   Schevill, W.E. & Watkins, W.A. 1965. Underwater calls of *Trichechus* (Manatee). *Nature,* Lond. **205**, 373

424   Schevill, W.E., Watkins, W.A. & Ray, C. 1966. Analysis of underwater *Odobenus* calls with remarks on the development and function of the pharyngeal pouches. *Zoologica.* N. York. **51**, 103

425   Schmidt-Nielsen, B., & K., Houpt, T.R. & Jarnum, S.A. 1956. Water balance of the camel. *Amer. J. Physiol.* **185**, 185

426   Schmidt-Nielsen, K., & B., Jarnum, S.A. & Houpt, T.R. 1957. Body temperature of the camel and its relation to water economy. *Amer. J. Physiol.* **188**, 103

427   Schneider, R. 1956. Untersuchungen über den feinbau der schäfendrüse beim Afrikanischen und Indischen elefanten, *Loxodonta africana* Cuvier und *Elephas maximus* Linnaeus. *Acta anat.* **28**, 303

428   Schusterman, R.J. 1966. Underwater click vocalisation by a Californian sea lion: effects of visibility. *Psychological Rec.* **16**, 129

REFERENCES

429 Schwarz, E. 1927. Paul Matschie. *J. Mammal.* **8**, 292
430 Sclater, W.L. 1900. *The Mammals of South Africa.* London
431 Scoresby, W. Jr. 1820. *An account of the Arctic Regions with a history and description of the Northern whale fishery.* Edinburgh
432 Shackleford, J.M. 1963. The salivary glands and salivary bladder of the nine-banded armadillo. *Anat. Rec.* **145**, 513
433 Shadle, A.R. 1950. Feeding, care, and handling of captive porcupines (Erethizon). *J. Mammal.* **31**, 411
434 Shadle, A.R. 1955. Removal of foreign quills by porcupines. *J. Mammal.* **36**, 463
435 Shadle, A.R., Nouth, A.M., Gese, E.C. & Austin, T.S. 1943. Comparison of tree cutting of six beaver colonies in Allegany State park, New York. *J. Mammal.* **24**, 32
436 Shadle, A.R., Smelzer, M. & Matz, M. 1946. The sex reactions of porcupines (*Erethizon d. dorsatum*) before and after copulation. *J. Mammal.* **27**, 116
437 Shapiro, J. 1949. Ecological and life history notes on the porcupine in the Adirondacks. *J. Mammal.* **30**, 247
438 Sharp, W.M. 1943. A pocket gopher (*Geomys lutescens*) travels backward. *J. Mammal* **24**, 99
439 Sheldon, W.G. 1937. Notes on the giant panda. *J. Mammal.* **18**, 13
440 Short, R.V. 1966. Oestrous behaviour, ovulation and the formation of the corpus luteum in the African elephant, *Loxodonta africana. East Afr. Wildl. J.* **4**, 56
441 Short, R.V. & Buss, I.O. 1965. Biochemical and histological observations on the corpora lutea of the African elephant *Loxodonta africana. J. Reprod. Fert.* **9**, 61
442 Shortridge, G.C. 1934. *The Mammals of South West Africa.* London
443 Silver, J. 1928. Pilot black-snake feeding on the big brown bat. *J. Mammal.* **9**, 149
444 Simpson, G.G. 1941. Vernacular names of South American mammals. *J. Mammal.* **22**, 1
445 Simpson, G.G. 1945. The principles of classification and a classification of mammals. *Bull. Amer. Mus. Nat. Hist.* **85**, 1
446 Sivertsen, E. 1941. On the biology of the harp seal *Phoca groenlandica* Erxl. *Hvalrådets Skrifter* No. **26**
447 Skinner, M.P. 1922. The prong-horn. *J. Mammal.* **3**, 82
448 Skinner, M.P. 1925. Winter movements of the cony. *J. Mammal.* **6**, 202
449 Slijper, E.J. 1962. *Whales.* London
450 Smith, N.S. & Ledger, H.P. 1965. A method of predicting live weight from dissected leg weight. *J. Wildl. Mgmt.* **29**, 504
451 Smith, R.E. 1967. Natural history of the prairie dog in Kansas. *Univ. Kansas Mus. nat. Hist. Misc. Pub. No.* **49**
452 Snyder, L.L. 1924. Some details on the life history and behaviour of *Napaeozapus insignis abictorum* (Preble) *J. Mammal.* **5**, 233
453 Soper, J.D. 1944. The mammals of southern Baffin Island, Northwest Territories, Canada. *J. Mammal.* **25**, 221
454 Spinage, C.A. 1968. Horns and other bony structures of the skull of the giraffe, and their functional significance. *E. Afr. Wildl. J.* **6**, 53
455 Starck, D. 1958. Beitrag zur Kenntnis der Armtaschen und anderer Hautdrüsenorgane von *Saccopteryx bilineata* Temminck 1838 (*Chiroptera, Emballonridae*) *Morph. Jahb.* **99**, 1
456 Steen, I. & G.B. 1965. The thermo-regulatory importance of the beaver's tail. *Comp. Biochem. Physiol.* **15**, 267
457 Sterndale, R.A. 1884. *Natural History of the Mammalia of India and Ceylon.* Calcutta
458 Stevenson-Hamilton, J. 1947. *Wild life in South Africa.* London

459   Stodart, E. 1966. Management and behaviour of breeding groups of the marsupial *Perameles nasuta* Geoffroy in captivity. *Aust. J. Zool.* **14**, 611

460   Stonehouse, B. 1968. Thermoregulatory function of growing antlers. *Nature*, Lond. **218**, 870

461   Struthers, P.H. 1928. Breeding habits of the Canadian porcupine (*Erethizon dorsatum*). *J. Mammal.* **9**, 300

462   Studier, E.H., Beck, L.R. & Lindeborg, R.G. 1967. Tolerance and initial metabolic response to ammonia intoxication in selected bats and rodents. *J. Mammal.* **48**, 564

463   Sturgess, I. 1948. The early embryology and placentation of *Procavia capensis*. *Acta zoologica.* **29**, 393

464   Svihla, A. & R.D. 1933. Notes on the jumping mouse *Zapus trinotatus trinotatui* Rhoads. *J. Mammal.* **14**, 131

465   Taber, F.W. 1945. Contribution on the life history and ecology of the nine-banded armadillo. *J. Mammal.* **26**, 211

466   Talbot, L.M. 1960. A look at threatened species. *Oryx* **5**, 152

467   Talmadge, R.V. & Buchanan, G.D. 1954. The armadillo (*Dasypus novemcinctus*) – a review of its natural history, ecology, anatomy and reproductive physiology. *Rice Inst. Pamphlet.* **41**, No. 2

468   Tate, G.H.H. 1931. Random observations on habits of South American mammals. *J. Mammal.* **12**, 248

469   Tate, G.H.H. 1948. Notes on the Haitian hutia, *Plagiodontia*, and extinct related genera. *J. Mammal.* **29**, 176

470   Taylor, C.R. 1968. Hygroscopic food: a source of water for desert antelopes? *Nature*, Lond. **219**, 181

471   Terres, J.K. 1956. Migration records of the red bat *Lasiurus borealis*. *J. Mammal.* **37**, 442

472   Tevis, L. Jr. 1950. Summer behaviour of a family of beavers in New York State. *J. Mammal.* **31**, 40

473   Thomas, O. 1904. On the osteology and systematic position of the rare Malagasy bat *Myzopoda aurita*. *Proc. zool. Soc. Lon.* **1904** pt 2, 2

474   Thurston, J.P., Noirot-Timothée, C. & Arman, P. 1968. Fermentative digestion in the stomach of *Hippopotamus amphibius* (Artiodactyla: Suiformes) and associated ciliate protozoa. *Nature*, Lond. **218**, 882

475   Todd, R.B. 1835–59. *The Cyclopaedia of Anatomy and Physiology*. London

476   Trapido, H. 1946. Observations on the vampire bat with special reference to longevity in captivity. *J. Mammal.* **27**, 217

477   Trapido, H. 1949. Gestation period, young, and maximum weight of the isthmian capybara, *Hydrochoerus isthmius* Goldman. *J. Mammal.* **30**, 433

478   Troughton, E. 1951. *The Furred Animals of Australia.* 4th ed. Sydney & London

479   Turner, M.I.M. & Watson, R.M. 1965. An introductory study on the ecology of hyrax (*Dendrohyrax brucei* and *Procavia johnstoni*) in the Serengeti National Park. *E. Afr. Wildl. J.* **3**, 49

480   Twigg, G.I. 1965. Studies on *Holochilus sciureus berbicensis*, a cricetine rodent from the coast region of British Guiana. *Proc. zool. Soc. Lond.* **145**, 263

481   van der Horst, C.J. 1947. A human-like embryo of *Elephantulus*. *Proc. zool. Soc. Lond.* **117**, 334

482   van der Horst, C.J. 1948. Some early embryological stages of the golden mole *Eremitalpa granti* (Broom). *Spec. Pub. roy. Soc. S. Afr. R. Broom Commemorative Vol.*, 225

483   van der Horst, C.J. 1949. An early stage of placentation in the aardvark, *Orycteropus*. *Proc. zool. Soc. Lond.* **119**, 1

CC

REFERENCES

484 Van Deusen, H.M. 1963. First New Guinea record of *Thylacinus*. *J. Mammal*. **44**, 279

485 Van Deusen, H.M. 1968. Carnivorous habits of *Hypsignathus monstrosus*. *J. Mammal* **49**, 335

486 Van Deusen, H.M. & George, G.G. 1969. Results of the Archbold Expeditions. No. 90. Notes on the Echidnas (Mammalia, Tachyglossidae) of New Guinea. *Amer. Mus. Novit*. No. 2383

487 Van Tyne, J. 1929. Notes on the habits of *Cyclopes dorsalis*. *J. Mammal*. **10**, 314

488 Verheyen, R. 1954. Monographie éthologique de l'hippopotame (*Hippopotamus amphibius* Linné) *Inst. Parc. Nat. Congo Belge. Exploration du Parc National Albert*. Bruxelles

489 Verschuren, J. 1958. Écologie et biologie des grands mammifères (primates, carnivores, ongulés). *Exploration du Parc National de la Garamba. Fasc*. 9. Bruxelles

490 Vesey-Fitzgerald, D.F. 1965. The utilization of natural pastures by wild animals in the Rukwa Valley, Tanganyika. *E. Afr. Wildl. J*. **3**, 38

491 Vincent, F. 1964. Quelques observations sur le comportement de deux Pangolins africains (Pholidota). *Mammalia*, Paris. **28**, 659

492 Vinogradov, B.S. & Argyropulo, A.I. 1931. Zur biologie der turkestanischen springmäuse (Dipodidae). *Z. Säugertierk*. Berlin. **6**, 164

493 Waldo, C.M. & Wislocki, G.B. 1951. Observations on the shedding of the antlers of Virginia deer (*Odocoileus virginianus borealis*). *Amer. J. Anat*. **88**, 351

494 Waldo, C.M., Wislocki, G.B. & Fawcett, D.W. 1949. Observations on the blood supply of growing antlers. *Amer. J. Anat*. **84**, 27

495 Walker, E.P. 1964. *Mammals of the World*. Baltimore

496 Wallace, A.R. 1869. *The Malay Archipelago*. London

497 Waterhouse, G.R. 1846. *A Natural History of the Mammalia. Vol. I. The Marsupiata, or pouched animals*. London

498 Waterhouse, G.R. 1848. *Natural History of the Mammalia. Vol. II. Rodentia or Gnawing Mammalia*. London.

499 Watkins, W.A. & Schevill, W.E. 1968. Underwater playback of their own sounds to *Leptonychotes* (Weddell seals). *J. Mammal*. **49**, 287

500 Watson, G.E. 3rd. 1961. Behavioural and ecological notes on *Spalax leucodon*. *J. Mammal*. **42**, 359

501 Weber, M. 1927–8. *Die Säugetiere*. Jena

502 Weddell, J. 1825. *A voyage towards the South Pole, performed in the years 1822–4*. London

503 Weitz, B. 1953. Serological relationships of hyrax and elephant. *Nature*, Lond. **171**, 261

504 Wharton, C.H. 1950. Notes on the life history of the flying lemur. *J. Mammal*. **31** 269

505 White, J. 1790. *Journal of a voyage to New South Wales, with sixty five plates of* ...... *Natural Productions*. London

506 Wilke, F., Taniwaki, T. & Kuroda, N. 1953. *Phocoenoides* and *Lagenorhynchus* in Japan, with notes on hunting. *J. Mammal*. **34**, 488

507 Wille, A. 1954. Muscular adaptation of the nectar-eating bats (subfamily Glossophaginae). *Trans. Kansas Acad. Sci*. 57, 315

508 Wilson, V.J. 1965. Observations on the greater kudu *Tragelaphus strepsiceros* from a tsetse control hunting scheme in Northern Rhodesia. *E. Afr. Wildl. J*. **3**, 27

509 Wilson, V.J. & Clarke, J.E. 1962. Observations on the common duiker *Sylvicapra grimmia* Linn., based on material collected from a tsetse control game elimination scheme. *Proc. zool. Soc. Lond*. **138**, 487

510 Wilson, V.J. & Roth, H.H. 1967. The effects of tsetse control operations on common duiker in eastern Zambia. *E. Afr. Wildl. J.* **5**, 53

511 Winter, W.H. 1964. Elephant behaviour. *E. Afr. Wildl. J.* **2**, 163

512 Wirtz, W.O. II. 1968. Reproduction, growth and development, and juvenile mortality in the Hawaiian monk seal. *J. Mammal.* **49**, 229

513 Wislocki, G.B. 1928. The placentation of hyrax (*Procavia capensis*) *J. Mammal.* **9**, 117

514 Wislocki, G.B. 1928. Nematode parasites in the ovaries of the anteater (*Tamandua tetradactyla*). *J. Mammal.* **9**, 318

515 Wislocki, G.B. 1930. On an unusual placental form in the Hyracoidea: its bearing on the theory of the phylogeny of the placenta. *Contr. Embryol. Carn. Instn.* **21**, 83

516 Wislocki, G.B. 1943. Studies on the growth of deer antlers II. *Essays in biology in honour of Herbert M. Evans.* 631. Univ. Calif. Press

517 Wislocki, G.B. 1949. Seasonal changes in the testes, epididymides and seminal vesicles of deer investigated by histochemical methods. *Endocrinology.* **44**, 167

518 Wislocki, G.B. 1954. Antlers in female deer, with a report of three cases in *Odocoileus*. *J. Mammal.* **35**, 486

519 Wislocki, G.B. & Enders, R.K. 1935. Body temperatures of sloths, anteaters and armadillos. *J. Mammal.* **16**, 328

520 Wislocki, G.B. & van der Westhuysen, O.P. 1940. The placentation of *Procavia capensis* with a discussion of the placental affinities of the Hyracoidea. *Contr. Embryol. Carn. Instn.* **28**, 65

521 Wislocki, G.B., Weatherford, H.L. & Singer, M. 1947. Osteogenesis of antlers investigated by histological and histochemical methods. *Anat. Rec.* **99**, 265

522 Witte, G.-F. de & Frechkop, S. 1955. Sur un espèce encore inconnue de mammifère Africain, *Potamogale ruwenzorii*, sp. n. *Bull. Inst. roy. Sci. nat. Belgique.* **31**, No. 84

523 Wolf, A. & O'Shea, T. 1969. Two bighorn sheep-coyote encounters. *J. Mammal.* **49**, 770

524 Wright, P.L. & Rausch, R. 1955. Reproduction in the wolverine, *Gulo gulo*. *J. Mammal.* **36**, 346

525 Wyman, J. 1863. Description of a 'White Fish', or 'White Whale' (*Beluga borealis*, Less.) *J. Boston Nat. Hist. Soc.* **7**, 603

526 Yalden, D.W. 1966. The anatomy of mole locomotion. *J. Zool. Lond.* **149**, 55

527 Yamada, M. 1954. Some remarks on the pygmy sperm whale, *Kogia*. *Sci. Repts. Whales Res. Inst.* **9**, 37

528 Yocom, C.F. 1946. Notes on the Dall porpoise off California. *J. Mammal.* **27**, 364

529 Zeuner, F.E. 1963. *A history of domesticated animals.* London

530 Zimmerman, K. 1964. Zur Säugetier-Fauna Chinas. Ergebnisse der Chinesisch-Deutscher Sammelreise durch Nord- und Nordost-China 1956 No. 15. *Mitt. zool. Mus., Berl.* **40**, 87

Index

beaver (*Castor*), 155–9; *see also* mountain beaver

beira (*Dorcatragus*), 393

beisa, *see* gemsbuck

beluga (*Delphinapterus leucas*), 235–6, 243–4, 247

Berardius, 233

Bettongia, 43

bharal (*Pseudois nayaur*), 398

Bibos, 388

bilbie or rabbit bandicoot (*Thylacomys*), 31

binturong (*Arctitis binturong*), 265, 273, 274

bison (American: buffalo) (*Bison*), 388

Bison bison, 388; – *bonasus*, 388

blackbuck (*Antelope cervicapra*), 393–4

blackfish (Orcinae), 244–5; many-toothed (Haiwaiian porpoise), 248

Blarina, 48, 59, 60, 62; – *brevicauda*, 61

Blarinella, 62

blesbok (*Damaliscus dorcas*), 391

blue buck (*Hippotragus*), 391

blue bull, *see* nilgai

blue doe, blue flyer; *see* kangaroo, red

bobcat (*Felis rufa*), 283

bongo (*Taurotragus*), 389

bontebok (*Damaliscus dorcas*), 391

Boselaphus, 388–9

Bos banteng, 388; – *gaurus*, 388; – *grunniens*, 388; – *primigenius*, 388; – *taurus*, 388

bouto (*Inia geoffrensis*), 239–40

Bovidae, 387–400

Bovinae, 388–9

Brachyphylla, 94; – *cavernarum*, 94

Brachytarsomys albicauda, 170

Brachyuromys, 170

Bradypodidae, 120–4

Bradypus, 120–4; – *cucilliger*, 121; – *griseus*, 121

brocket (*Mazama*), 380

Budorcas taxicolor, 398

bufeo (*Sotalia fluviatilis*), 241

buffalo (Bovinae), 388, 400

Buphagus, 340

Burmeistria, 130

Burramyinae, 37–8

Burramys parva, 37–8

bushbuck (*Tragelaphus*), 389

bush-cow (*Syncerus*), 388

Cabassous, 128–9

cacomistle or ring-tail cat (*Bassaricus*), 266

Caenolestes, 32; – *fuliginosus*, 32–3

Caenolestidae, 32–3

Callorhinus ursinus, 291

Callosciurus, 146

Calomyscus, 169

Caloprymnus, 43

Caluromys, 23

Caluromysiops, 23

camel (*Camelus*): Arabian (one-humped) and Bactrian (two-humped), 365–7

Camelidae, 364–71

Camelus bactrianus, 365–7; – *dromedarius*, 365–7

cane rat (*Thryonomys*), 196, 218–19

Canidae, 255–60

Canis, 256; – *latrans*, 256; – *lupus*, 256

Cannomys badius, 176, 177

Canoidea, 255

Caperea marginata, 228

Capra, 398; – *hircus aegagrus*, 398

Capreolus capreolus, 380

Capricornis crispus, 397; – *sumatrensis*, 397

Caprinae, 396–400

Caprolagus, 142

Capromyidae, 213–15

Capromys, 214; – *melanurus*, 214; – *nana*, 214; – *pilorides*, 214; – *prehensilis*, 214

capybara (*Hydrochoerus*), 204–6

caracal (*Felis caracal*), 283

Cardiocraniinae, 194

Cardiocranius, 194

Cardioderma, 85; – *cor*, 86

caribou, *see* reindeer

Carnivora: fissiped, 255–87; pinniped, 288–312

Carollia, 92–3; – *perspicillata*, 92

Carolliinae, 92–3

423